Control Systems:
Theory and Applications

RIVER PUBLISHERS SERIES IN AUTOMATION, CONTROL AND ROBOTICS

Series Editors:

ISHWAR K. SETHI
Oakland University
USA

TAREK SOBH
University of Bridgeport
USA

QUAN MIN ZHU
University of the West of England
UK

Indexing: All books published in this series are submitted to the Web of Science Book Citation Index (BkCI), to CrossRef and to Google Scholar.

The "River Publishers Series in Automation, Control and Robotics" is a series of comprehensive academic and professional books which focus on the theory and applications of automation, control and robotics. The series focuses on topics ranging from the theory and use of control systems, automation engineering, robotics and intelligent machines.

Books published in the series include research monographs, edited volumes, handbooks and textbooks. The books provide professionals, researchers, educators, and advanced students in the field with an invaluable insight into the latest research and developments.

Topics covered in the series include, but are by no means restricted to the following:

- Robots and Intelligent Machines
- Robotics
- Control Systems
- Control Theory
- Automation Engineering

For a list of other books in this series, visit www.riverpublishers.com

Control Systems:
Theory and Applications

Editors

Vsevolod M. Kuntsevich

Space Research Institute of NAS and NSA of Ukraine
Ukraine

Vyacheslav F. Gubarev

Space Research Institute of NAS and NSA of Ukraine
Ukraine

Yuriy P. Kondratenko

Petro Mohyla Black Sea National University
Ukraine

Dmitriy V. Lebedev

International Research and
Training Center for Information Technologies and Systems
Ukraine

Vitaliy P. Lysenko

National University of Life and Environmental Sciences
Ukraine

River Publishers

Routledge
Taylor & Francis Group

LONDON AND NEW YORK

Published 2018 by River Publishers
River Publishers
Alsbjergvej 10, 9260 Gistrup, Denmark
www.riverpublishers.com

Distributed exclusively by Routledge
4 Park Square, Milton Park, Abingdon, Oxon OX14 4RN
605 Third Avenue, New York, NY 10017, USA

First issued in paperback 2023

Control Systems: Theory and Applications / by Vsevolod M. Kuntsevich, Vyacheslav F. Gubarev, Yuriy P. Kondratenko, Dmitriy V. Lebedev, Vitaliy P. Lysenko.

Routledge is an imprint of the Taylor & Francis Group, an informa business

Publisher's Note
The publisher has gone to great lengths to ensure the quality of this reprint but points out that some imperfections in the original copies may be apparent.

ISBN 13: 978-87-7022-922-7 (pbk)
ISBN 13: 978-87-7022-024-8 (hbk)
ISBN 13: 978-1-00-333770-6 (ebk)

While every effort is made to provide dependable information, the publisher, authors, and editors cannot be held responsible for any errors or omissions.

Contents

Preface   **xiii**

List of Contributors   **xix**

List of Figures   **xxiii**

List of Tables   **xxix**

List of Abbreviations   **xxxi**

PART I: Theory of Automatic Control

1 Estimation of Impact of Bounded Perturbations on Nonlinear Discrete Systems   **3**

Vsevolod M. Kuntsevich

1.1   Introduction   3

1.2   Interval Invariant Sets of Nonlinear Control Systems with Linear Constraints   4

1.3   Invariant Sets of Systems with Two-Sided Nonlinear Constraints   9

1.4   Invariant Sets of Systems with Norm-Based Estimates of Nonlinear Functions   12

1.5   Conclusion   14

      References   14

2 Control of Moving Objects in Condition of Conflict   **17**

Arkadii A. Chikrii

2.1   Introduction   17

2.2   Formulation of the Problem: Scheme of the Method   18

2.3   Game Problems for Fractional Systems   25

2.4 Fractional Conflict-Controlled Processes with Integral Block
       of Control   31
2.5 Specific Case of Simple Matrix   33
2.6 Conclusion   40
      References   40

3 Identification and Control Automation of Cognitive Maps in Impulse Process Mode   **43**

Vyacheslav Gubarev, Victor Romanenko and Yurii Miliavskyi

3.1 Introduction   43
3.2 Cognitive Maps Identification with Full Information   46
      3.2.1 Parametric Identification in Deterministic Case . . .   46
      3.2.2 Identification in Case with Noise   48
              3.2.2.1 Combinatorial method of solution   48
              3.2.2.2 Identification problem solution with least
                         squares method   51
              3.2.2.3 Regularized solution of the identification
                         problem   52
3.3 Control Automation in Cognitive Maps in Impulse
      Process Mode   53
3.4 Experimental Research   55
      3.4.1 Research of the Algorithms of CM Incidence Matrix
                Identification   55
      3.4.2 Automating Control of Impulse Processes in CM of
                Human Resources Management in IT Company . .   59
3.5 Conclusion   62
      References   63

4 Decentralized Guaranteed Cost Inventory Control of Supply Networks with Uncertain Delays   **65**

Leonid M. Lyubchyk and Yuri I. Dorofieiev

4.1 Introduction   66
4.2 Problem Statement   69
4.3 Descriptor Transformation of Supply Network Model   75
4.4 Construction of Delay-Dependent Lyapunov-Krasovskii
      Functional   77
4.5 Guaranteed Cost Inventory Control Based on Invariant
      Ellipsoids Method   80
4.6 Synthesis of Guaranteed Cost Inventory Control   83
4.7 Stability Analysis of Controlled Supply Network   86

4.8 Numerical Example 89
4.9 Conclusion 93
     References 93

**5 Application of a Special Method of Nondimensionization in the
Solution of Nonlinear Dynamics Problems** **97**

*Maksym V. Maksymov, Olexander I. Brunetkin and
Oksana B. Maksymova*

5.1 Introduction 98
5.2 Models and Modeling 98
     5.2.1 Method of Nondimensionization of Mathematical
            Models 98
            5.2.1.1 Formulation of the problem 98
            5.2.1.2 Purpose and objectives of the study . . . 100
            5.2.1.3 Scheme for ensuring self-similarity by
                     criteria for models 100
            5.2.1.4 Concept of the procedure of ensuring
                     self-similarity by criteria 103
            5.2.1.5 Physical modeling 110
            5.2.1.6 Examples of reduction of models to a
                     nondimensionized form in problems of
                     technical systems dynamics 112
     5.2.2 Examples of Using the Results of Reducing Models
            to a Nondimensionized Form 120
            5.2.2.1 Compiling data when displaying solution
                     results Non-stationary heat transfer 120
            5.2.2.2 Approximate solution of nonlinear
                     ordinary differential equations.
                     Solution method 131
5.3 Conclusion 139
     References 140

PART II: Control Systems Applications

**6 Energy Efficiency of Smart Control Based on
Situational Models** **145**

Volodymyr M. Dubovoi and Mariya S. Yukhymchuk

6.1 Introduction 145
6.2 Situation-Event Graph Model 149

6.3   Energy Loss Model   151
    6.3.1   Formalizing of the Criterion of Energy Efficiency of
          Smart Control on the Basis of Situational Models .   151
    6.3.2   Development of the Method for Simulation of
          Systems with Smart Control Based on Situational
          Models in Conditions of Combined Uncertainty of
          the Initial Information   154
    6.3.3   Development of an Algorithm for Estimating
          the Energy Efficiency of Situational Model-Based
          Smart Control   160
    6.3.4   Experimental Investigations of Developed Model . .   161
6.4   Analysis of Simulation Results   163
6.5   Conclusion   164
    References   164

**7 Ellipsoidal Pose Estimation of an Uncooperative Spacecraft
from Video Image Data**   **169**
Vyacheslav F. Gubarev, Nikolay N. Salnikov
and Serhii V. Melnychuk
7.1   Introduction   169
7.2   Relative Pose Determination   171
7.3   Position and Attitude Determination by Learning   173
7.4   Informative Features of Image   175
7.5   Estimation Problem Statement   178
7.6   Method of Nonlinear Ellipsoidal Filtration   182
7.7   Estimation of Target Attitude and Angular Velocity   185
7.8   Numerical Simulation   188
7.9   Conclusion   191
    References   192

**8 Fuzzy Controllers for Increasing Efficiency of the Floating
Dock's Operations: Design and Optimization**   **197**
Yuriy P. Kondratenko, Oleksiy V. Kozlov and Andriy M. Topalov
8.1   Introduction   198
8.2   Design Features of the Automatic Control Systems of the
    Floating Docks Main Operations   200
8.3   Combined Design Approach of the FCs for the ACSs of the
    Floating Docks Main Operations   205

8.4 Design of the FC for the Draft ACS of the Floating Dock for Low-Tonnage Vessels 212
8.5 Conclusion 225
References 226

9 Efficiency Control for Multi-assortment Production Processes Taking into Account Uncertainties and Risks   **233**
Anatoliy Ladaniuk, Viacheslav Ivashchuk and Jaroslav Smityukh
9.1 Introduction 234
9.2 Decision-Making Under Uncertainty 236
9.3 The Estimation of Stability and Accuracy of Presented Solutions 238
9.4 Conclusion 241
References 241

10 On the Coordinate Determination of Space Images by Orbital Data   **245**
Dmitriy V. Lebedev
10.1 Introduction 245
10.2 Systems of Coordinates. Problem Statement of In-flight Calibration 247
10.3 Coordinate Binding of Space Images 249
    10.3.1 Estimation of the Point Landmark Coordinates by Stereoshooting Results 250
    10.3.2 Geo-referencing of Frame Elements 251
10.4 Factors Affecting the Accuracy of the Images Coordinate Reference. Assessment of the Accuracy of the Binding . . . 253
10.5 Algorithms of In-flight Geometric Calibration 255
    10.5.1 Calibration Algorithms Based on Photogrammetric Equations 255
        10.5.1.1 Algorithm 1 256
        10.5.1.2 Algorithm 2 257
    10.5.2 Calibration for a Priori Unknown Point Landmarks . 257
    10.5.3 Calibration Using a Virtual Reference System . . . 259
    10.5.4 Reduction of the Flight Calibration Algorithm . . . 261

10.6   Modeling of Algorithms of Calibration and Topographic
      Binding   262
      10.6.1   Calibration Algorithms 1 and 2   264
      10.6.2   Calibration of Algorithms B and R   265
10.7   Conclusion   267
     References   268

11 Algorithms of Robotic Electrotechnical Complex Control in Agricultural Production

    271

Vitaliy Lysenko, Igor Bolbot, Yuriy Romasevych,
Viatcheslav Loveykin and Valeriy Voytiuk

11.1   Introduction   271
11.2   An Overview of Recent Findings   272
11.3   The Purpose of the Research   273
11.4   Statement of Basic Materials   273
      11.4.1   Development of Upper-Level Control Algorithms .   273
      11.4.2   Development of Control Algorithms at the Lower
             Level   283
11.5   Conclusion   288
     References   288

12 Information Support of Some Automated Systems of Remote Monitoring of Planted Areas State

    291

Vitaliy Lysenko, Serhii Shvorov, Oleksii Opryshko,
Natalia Pasichnyk and Dmytro Komarchuk

12.1   Introduction   292
12.2   Formulation of the Problem. Scheme of the Method   293
      12.2.1   Investigated Objects   293
      12.2.2   Hardware for Spectral Monitoring   294
12.3   Obtained Results and Discussion   297
      12.3.1   Continuous Sowing Crops – Wheat   297
      12.3.2   Growing Crops   300
      12.3.3   Determination of Optimal Movement Routes for
             Harvesters to Harvest Energy Crops and for
             Necessary Transport Equipment to Transport Them   302
12.4   Conclusion   309
     References   310

13 Synthesis of an Optimal Combined Multivariable Stabilization System for Adsorption Process Control **315**

Sergey Osadchy, Valentina Zubenko and Marja Yakoreva

13.1 Introduction 315
13.2 Optimal Stabilization System Synthesis Problem
 Formulation 316
13.3 Justification of Choice and Description of the Algorithm for
 Optimal Systems Synthesis 319
13.4 Optimal Combined Stabilization System's Synthesis 321
13.5 Conclusion 323
 References 324

Index **325**

About the Editors **327**

Preface

Control theory and its implementation in practice are very important areas of the modern scientific activity. Every year, beginning from 1994, the Ukrainian Association on Automatic Control (being National Member organization of IFAC – International Federation on Automatic Control) holds an annual International conference on control problems and applications. The recent XXIV International conference "Automatics'2017", organized by the National University of Life and Environmental Sciences of Ukraine in September 2017, was held in Kiev with participation of the scientists from different countries (Ukraine, Poland, Norway, Belarus, Russia, and others) and representatives of such firms and companies as Bunge Ukraine, Siemens, S-Engineering, and others. More than 100 reports were presented at the conference and the best 13 (5 theoretical and 8 dealing with applied problems) were selected for publication in this book by the National Committee of Ukrainian Association on Automatic Control.

The subject of this book is to present a systematized research, description, and analysis of the modern theoretical results and applications of control theory and various types of control systems into different fields of human activity. For the first time, papers with specific research investigations on actual problems of automation in agriculture were included in the conference program. Two such papers can be found in this book. By the way, the above-mentioned firms and companies have demonstrated for conference participants a lot of successful examples of the control systems realization in the agricultural area.

In terms of structure, the 13 chapters of this book have been grouped into two sections: (a) Theory of Automatic Control and (b) Control Systems Applications.

The first part, "Theory of Automatic Control", includes five contributions:

The chapter "Estimation of Impact of Bounded Perturbations on Nonlinear Discrete Systems", by Vsevolod Kuntsevich, deals with nonlinear discrete dynamic systems (consisting of linear and nonlinear parts) subjected to bounded perturbations. This chapter presents the original theoretical methods for calculation of the interval sets and the radii of these sets.

Arkadii Chikrii, in "Control of Moving Objects in Condition of Conflict", considers the quasilinear conflict-controlled processes with a cylindrical terminal set. The mathematical method of resolving functions which allows obtaining sufficient conditions for the solvability of the pursuit problem at a certain guaranteed time in the class of strategies that use information on the behavior of the opponent in the past as well as in the class of stroboscopic strategies is presented in this chapter in detail.

In "Identification and Control Automation of Cognitive Maps in Impulse Process Mode", Vyacheslav Gubarev, Victor Romanenko, and Yurii Miliavskyi discuss the original results concerning identification and control problems in cognitive maps (CM). Identification problem is solved in several statements depending on whether the CM node coordinates are measured precisely or with noise. Control problem is solved for the case when two types of controls are available. A closed-loop control system is simulated for the CM of IT company HR (human resources) management.

The chapter "Decentralized Guaranteed Cost Inventory Control of Supply Networks with Uncertain Delays", by Leonid Lyubchyk and Yuri Dorofieiev, devoted to the problem of guaranteed cost inventory control of supply networks with uncertain transport delays in the presence of disturbances modeling the changing demand. The solution of the decentralized guaranteed cost inventory control problem under conditions of uncertain demand is obtained on the basis of invariant ellipsoids method. The results of computer simulation of supply network with a decentralized guaranteed cost control are presented.

Application of the special method of nondimensionization in the solution of nonlinear dynamics problem, considered in chapter "Application of a special method of nondimensionization in the solution of nonlinear dynamics problems", by Maxim Maksymov, Olexander Brunetkin and Oksana Maksymova, allows reducing the number of criteria compared to its number predicted according to the dimension theory. This makes it possible to define the view of normalizing quantities equating all similarity criteria in a nondimensionized model to one. The potential possibilities of this approach are demonstrated by some examples connected with approximation problems.

The second part, "Control Systems Applications", includes eight contributions:

Volodymyr Dubovoi and Mariya Yukhymchuk in the chapter "Energy Efficiency of Smart Control Based on Situational Models" estimate the energy losses during logical, fuzzy, and statistical controls using a proposed mathematic model, which allows estimating advantages of fuzzy and statistical situational control, considering the near and far horizon of events.

In " Ellipsoidal Pose Estimation of an Uncooperative Spacecraft from Video Image Data", V. Gubarev, N. Salnikov, and S. Melnychuk consider the ellipsoidal pose estimation of uncooperative spacecraft from video image data during autonomous approach and docking used for this computer vision system and dynamic filtering. Pose determination is based on informative features of images with the use of machine learning that reduces computational complexity of the proposed algorithm.

The chapter "Fuzzy Controllers for Increasing Efficiency of the Floating Dock's Operations: Design and Optimization", by Yuriy Kondratenko, Oleksiy Kozlov, and Andriy Topalov, presents a combined step-by-step approach to design of fuzzy controllers for the automatic control systems of the floating docks main operations. The proposed approach allows performing the synthesis of the structure and parameters of the Mamdani type fuzzy controllers and to design automatic control systems of the main control parameters of the floating docks operations using expert knowledge in conjunction with certain optimization procedures on the basis of mathematical programming methods.

Anatoliy Ladaniuk, Viacheslav Ivaschuk, and Jaroslav Smityukh in the chapter "Efficiency control for Multi-assortment Production Processes taking into Account Uncertainties and Risks" describe the method of efficiency control for monitoring and automation of multi-assortment processes applied to control of food production technological complexes. The functional structure and main components of the control system for technological complexes that are generalized for the production of multi-assortment products are presented. The approach is based on the uses of logical-dynamic models, which is adapted for batch process.

The chapter "On the Coordinate Determination of Space Images by Orbital Data", by Dmitriy Lebedev, is devoted to coordinate determination of the objects located on the space images only by considered orbital data. Special attention is given to describing in-flight geometric calibration algorithms. The characteristics of coordinate binding accuracy of satellite images, which is stated by proposed algorithms usage of in-flight geometric calibration, are investigated. Computer modeling confirms the efficiency of the proposed algorithms.

In "Algorithms of robotic electrotechnical complex control in agricultural production," V. Lysenko, I. Bolbot, Yu. Romasevych, V. Loveykin, and V. Voytiuk present the mobile robotic electrotechnical complex which can move greenhouse area by means of technological guideways, thereby providing monitoring of the main parameters of the greenhouses atmosphere as well as phytomonitoring including the quality of products while identifying its zones. The strategies of controlling electrotechnical complexes which provide the growth technology are formed on the basis of information coming from the robotic electrotechnical complexes, maximizing the product profit at the current moment. Such a complex should be characterized by individual features of intelligence and operate according to determined management algorithms, to monitor the phytosanitary and atmospheric conditions, to move around the area, the traversed path, avoiding obstacles, etc.

V. Lysenko, S. Shvorov, N. Pasichnik, O. Opryshko, and D. Komarchuk in the chapter "Information support of some automated systems of remote monitoring of planted areas state" consider an automated system for biogas reactors with plants raw materials as a material for anaerobic fermentation. It is proposed to use biomass of planted areas, the further cultivation of which is economically inexpedient for cold spring climate regions. The problem deals with the metrological support of sensory equipment using cameras with standard and infra-red lenses. The spectral channels laboratory evaluation was carried on the example of the Go PRO HERO 4 camera. The problem to choose the best routes for harvesting field plants on the basis of electronic maps is also considered.

The chapter "Synthesis of an Optimal Combined Multivariable Stabilization System for Adsorption Process Control", by Serhiy Osadchy, Valentina Zubenko and Marja Yakoreva, deals with an increasing efficiency of short-cycle adsorption of impurities from hydrogen which is obtained by the conversion method. The proposed approach is based on the approximation of the complete thermodynamic work to the minimum required one due to the introduction of combined control system. The task is reduced to design optimal combined multivariable stochastic stabilization system which consists of two control circuits.

The chapters selected for this book provide an overview of some problems in the area of control systems design, modeling, engineering and implementation and the approaches and techniques that relevant research groups within this area are employing to try to solve them.

We would like to express our deep appreciation to all authors for their contributions as well as to reviewers for their timely and interesting comments and suggestions. We certainly look forward to working with all contributors again in nearby future.

July 20, 2018

Editors
Vsevolod M. Kuntsevich
Vyacheslav F. Gubarev
Yuriy P. Kondratenko
Dmitriy V. Lebedev
Vitaliy P. Lysenko

List of Contributors

Bolbot, Igor, *National University of Life and Environmental Sciences of Ukraine, Kyiv, Ukraine; E-mail: igor-bolbot@ukr.net*

Brunetkin, Olexander I., *Odesa National Polytechnic University, Odesa, Ukraine; E-mail: a.i.brunetkin@gmail.com*

Chikrii, Arkadii A., *Glushkov Institute of Cybernetics NAS of Ukraine, Kyiv, Ukraine; E-mail: g.chikrii@gmail.com*

Dorofieiev, Yuri I., *National Technical University "Kharkiv Polytechnic Institute", Kharkiv, Ukraine; E-mail: dorofeev@kpi.kharkiv.edu*

Dubovoi, Volodymyr M., *Department "Computer Control Systems" in Vinnytsia National Technical University, 95 Khmelnytske Shose, Vinnytsia, 21021, Ukraine; E-mail: v.m.dubovoy@gmail.com*

Gubarev, Vyacheslav F., *Space Research Institute, National Academy of Sciences of Ukraine and State Space Agency of Ukraine, Glushkov, Ukraine; E-mail: v.f.gubarev@gmail.com*

Ivashchuk, Viacheslav, *National University of Food Technologies, Kyiv, Ukraine; E-mail: ivaschuk@nuft.edu.ua*

Komarchuk, Dmytro, *National University of Life and Environmental Sciences of Ukraine, Kyiv, Ukraine; E-mail: dmitruyk@gmail.com*

Kondratenko, Yuriy P., *Petro Mohyla Black Sea National University, Mykolaiv, Ukraine; E-mail: yuriy.kondratenko@chmnu.edu.ua*

Kozlov, Oleksiy V., *Admiral Makarov National University of Shipbuilding, Mykolaiv, Ukraine; E-mail: oleksiy.kozlov@nuos.edu.ua*

Kuntsevich, Vsevolod M., *Space Research Institute of NAS and NSA of Ukraine, Kyiv, Ukraine; E-mail: vsevolod.kuntsevich@gmail.com*

Ladaniuk, Anatoliy, *National University of Food Technologies, Kyiv, Ukraine; E-mail: ladanyuk@ukr.net*

Lebedev, Dmitriy V., *International Research and Training Center for Information Technologies and Systems, Ukraine; E-mail: ldv1491@gmail.com*

Loveykin, Viatcheslav, *National University of Life and Environmental Sciences of Ukraine, Kyiv, Ukraine; E-mail: lovvs@ukr.net*

Lysenko, Vitaliy P., *National University of Life and Environmental Sciences of Ukraine, Kyiv, Ukraine; E-mail: lysenko@nubip.edu.ua*

Lyubchyk, Leonid M., *National Technical University "Kharkiv Polytechnic Institute", Kharkiv, Ukraine; E-mail: lyubchyk@kpi.kharkiv.edu*

Maksymov, Maksym V., *Odesa National Polytechnic University, Odesa, Ukraine; E-mail: prof.maksimov@gmail.com*

Maksymova, Oksana B., *Odesa National Academy of Food Technologies, Odesa, Ukraine; E-mail: m.oxana.b@gmail.com*

Melnychuk, Serhii V., *Space Research Institute, National Academy of Sciences of Ukraine and State Space Agency of Ukraine, Glushkov, Ukraine; E-mail: melnychuk89s@mail.com*

Miliavskyi, Yurii, *"Institute for Applied System Analysis" of National Technical University of Ukraine "Igor Sikorsky Kyiv Polytechnic Institute", 37a Peremohy av., Kyiv, 03056, Ukraine; E-mail: yuriy.milyavsky@gmail.com*

Pasichnyk, Natalia, *National University of Life and Environmental Sciences of Ukraine, Kyiv, Ukraine; E-mail: n.pasichnyk@nubip.edu.ua*

Opryshko, Oleksii, *National University of Life and Environmental Sciences of Ukraine, Kyiv, Ukraine; E-mail: ozon.kiev@gmail.com*

Osadchy, Sergey, *Central Ukrainian National Technical University, Kropivnitsky, Ukraine; E-mail: srg2005@ukr.net*

Romanenko, Victor, *"Institute for Applied System Analysis" of National Technical University of Ukraine "Igor Sikorsky Kyiv Polytechnic Institute", 37a Peremohy av., Kyiv, 03056, Ukraine; E-mail: ipsa@kpi.ua*

Romasevych, Yuriy, *National University of Life and Environmental Sciences of Ukraine, Kyiv, Ukraine; E-mail: romasevichyuriy@ukr.net*

Salnikov, Nikolay N., *Space Research Institute, National Academy of Sciences of Ukraine and State Space Agency of Ukraine, Glushkov, Ukraine; E-mail: salnikov.nikolai@gmail.com*

Shvorov, Serhii, *National University of Life and Environmental Sciences of Ukraine, Kyiv, Ukraine; E-mail: sosdok@i.ua*

Smityukh, Jaroslav, *National University of Food Technologies, Kyiv, Ukraine; E-mail: smityuh1@gmail.com*

Topalov, Andriy M., *Admiral Makarov National University of Shipbuilding, Mykolaiv, Ukraine; E-mail: topalov_ua@ukr.net*

Voytiuk, Valeriy, *National University of Life and Environmental Sciences of Ukraine, Kyiv, Ukraine; E-mail: vdv-tsim@ukr.net*

Yakoreva, Marja, *Central Ukrainian National Technical University, Kropivnitsky, Ukraine; E-mail: gvf-52@gmail.com*

Yukhymchuk, Mariya S., *Department "Computer Control Systems" in Vinnytsia National Technical University, 95 Khmelnytske Shose, Vinnytsia, 21021, Ukraine; E-mail: umcmasha@gmail.com*

Zubenko, Valentina, *Central Ukrainian National Technical University, Kropivnitsky, Ukraine; E-mail: zub_valya@ukr.net*

List of Figures

Figure 1.1	An illustration of multi-valued mapping (1.5).	5
Figure 3.1	CM of a commercial bank.	55
Figure 3.2	Identification of CM weights.	57
Figure 3.3	Identification of CM weights with using regularization.	58
Figure 3.4	CM of HR management in IT company.	59
Figure 3.5	CM nodes coordinates in a closed-loop control system.	62
Figure 4.1	Graphical representation of the supply network model.	72
Figure 4.2	The structure of the supply network: a – existing; b – proposed.	89
Figure 4.3	Changes in stock levels of product x_{11}.	91
Figure 4.4	Changes in stock levels of product x_{12}.	92
Figure 4.5	Changes in stock levels of product x_2.	92
Figure 5.1	Temperature profile for asymmetric plate heating: a – examples of the results of numerical calculation of the relative temperature Θ as a function of the relative coordinate X for different values of Fo; b – accepted scheme for analytical calculations. . .	123
Figure 5.2	Calculation scheme of motion of a fluid with a free surface.	126
Figure 5.3	Scheme of the tank.	130
Figure 5.4	Influence of filling depth on frequency of natural oscillations of liquid between coaxial cylinders. . .	130
Figure 5.5	The nondimensionized coordinates x of the position of the pendulum as a function of the nondimensionized time t obtained on the basis of various methods for solving the nonlinear nondimensionized equation (5.21).	136

Figure 5.6   The nondimensionized coordinates x of the position of the pendulum as a function of the time t, obtained on the basis of various methods for solving nonlinear nondimentionized equation (5.29): $a = 0.4; b = 20.$ 139

Figure 6.1   The growth of the number of objects connected with Internet. 146

Figure 6.2   Classification of the models of fuzzy situational-event networks. 148

Figure 6.3   Complex control system. 150

Figure 6.4   Situation-event graph. 151

Figure 6.5   The case of incorrect identification of situation. . . 155

Figure 6.6   Probabilities of errors and correct solution. 157

Figure 6.7   Algorithm for estimating the energy efficiency of intelligent control based on situational models. . . . 160

Figure 6.8   Structure of a multi-room heating control system. . 162

Figure 6.9   Changing the time of the normalized value of the average energy loss in the control of the logical automaton, the fuzzy controller and the statistical SDM with the dimensions of the situational graph 4, 8, and 16. 163

Figure 6.10   Changing the normalized value of average energy losses with increasing system dimension and control of a logical automaton, fuzzy controller, and statistical SDM with horizons 1, 30, and 120 events. 163

Figure 7.1   Camera, Target and associated reference frames. . . 172

Figure 7.2   Variance of the brightness histogram to the changing lighting direction. 176

Figure 7.3   Texture informative features. 177

Figure 7.4   Outer contour invariance to ambient light variation. 177

Figure 7.5   Change of body components of angular velocity in time. 189

Figure 7.6   Change of quaternion components in time (red stands for q_0, blue – for q_1, green – q_2, turquoise – q_3) 189

Figure 7.7   Change of norm of angular velocity estimation error in time 190

Figure 7.8   Change of norm of quaternion estimation error in time 190

Figure 8.1   Two towers floating dock with the vessel: a –
submerged; b – after surfacing; 1 – pontoon; 2 –
dock supports; 3 – right tower; 4 – left tower; 5 –
vessel. 201

Figure 8.2   The layout of the ballast tanks. 203

Figure 8.3   Functional structure of the generalized system of
automatic control of the main parameters of the
floating dock. 203

Figure 8.4   Structure of the FC. 204

Figure 8.5   FC linguistic terms and their parameters at
$m_i = w = 3$. 216

Figure 8.6   FC linguistic terms and their optimized parameters
at $m_i = w = 3$. 218

Figure 8.7   ACS transients at the floating dock submerging:
yellow line – RM; blue line – with optimally
tuned conventional PD-controller; red line – with
non-optimized FC at $m_i = w = 3$; green line – with
optimized FC at $m_i = w = 3$. 219

Figure 8.8   FC linguistic terms and their parameters at
$m_i = w = 5$. 220

Figure 8.9   FC linguistic terms and their optimized parameters
at $m_i = w = 5$. 222

Figure 8.10   ACS transients at the floating dock submerging:
yellow line – RM; green line – with optimally
tuned conventional PD-controller; red line – with
non-optimized FC at $m_i = w = 5$; blue line – with
optimized FC at $m_i = w = 5$. 222

Figure 8.11   Diagram of the evaluation functional of the
influence of RB rules of the FC on the control
process of the floating dock draft. 223

Figure 8.12   ACS transients at the floating dock submerging:
yellow line – RM; blue line – with optimized FC
at $m_i = w = 5$ with full RB; red line – with optimized
FC at $m_i = w = 5$ with reduced RB. 224

Figure 9.1   The Control Process of Water Boiling by uses of
Linear Model. 235

Figure 9.2   The structure of decision-making for control of the
MAP. 238

Figure 9.3   The structure of decision-making for control of the
                  MAP.   241
Figure 10.1   Connection between the bases **I, G, O, T, K** and **E**.   248
Figure 10.2   Forming of the stereopair.   250
Figure 10.3   Shooting formation.   258
Figure 10.4   Position of a virtual point.   260
Figure 11.1   The movement algorithm of the electrotechnical
                  complex of phytomonitoring in a greenhouse with
                  the use of 6D-SLAM.   275
Figure 11.2   The algorithm for measuring the parameters of
                  phytosanitary and atmospheric conditions.   276
Figure 11.3   Flow diagram of the control algorithm of the mobile
                  robot with the aim of monitoring phytosanitary and
                  atmospheric conditions in closed-ground structures.   278
Figure 11.4   Flow diagram of content management.   280
Figure 11.5   A flow diagram for output of data to the screen and
                  storing information in a file.   280
Figure 11.6   Interface of the control system of the electrotechnical
                  complex of the strategic level.   281
Figure 11.7   The appearance of a mobile robot.   282
Figure 11.8   Graphs of the function obtained by simulating the
                  mode of working out the set point of the PI regulator
                  with: (a) classical; (b) modified structure. . . .   286
Figure 12.1   Results of objects spectral indicators measurements
                  with FC200 and GoPro HERO4 cameras use. . . .   296
Figure 12.2   Dependence of intensity of components of color
                  from wavelength for GoPro HERO4 camera with
                  standard lens (channels R, G, B) and IR lens MR
                  (channels iR1, iR2, iR3).   297
Figure 12.3   The photos of wheat crops on the experimental
                  stationary field to be received from the standard
                  lenses and IR lenses (2017.05.19).   298
Figure 12.4   Dependence of the components color intensity
                  on nitrogen amount of dry matter for the plants
                  above-ground part (x_all) and three upper
                  leaves (x_3).   299
Figure 12.5   The photos of maize on the experimental stationary
                  field to be received from the standard lenses and IR
                  lenses (2017.05.24).   300

Figure 12.6   Dependence of the IR2_PR integral index on the amount of nitrogen in the maize ground part. 302

Figure 12.7   Photo of the complicated area landscape with UAV. 303

Figure 12.8   Results of simulation of the routes of harvesting and transport vehicles for the first experimental field at the enterprise TDV ≪Terezone≫. 307

Figure 12.9   Actual fuel consumption of fuel for a harvest campaign for each field using and with and without the application of the proposed method to be implemented in the SPPR. 308

Figure 12.10 Total fuel consumption of fuel for a harvest campaign for each field using and with and without the application of the proposed method to be implemented in the SPPR. 309

Figure 13.1   The functional structure of the system for adsorption process control. 317

Figure 13.2   The structural scheme of the combined stabilization system. 317

Figure 13.3   The structural diagram of the stabilization system. . 320

List of Tables

Table 5.1   Geometric characteristics and shape factors 121

Table 5.2   The value of coefficients for calculating the natural frequencies of fluid oscillations in tanks 128

Table 5.3   Coefficients of stretching and error of analytical solutions for different initial angles of deviation (angular amplitudes) in the absence of energy dissipation 135

Table 5.4   Coefficients of stretching and errors of analytic solutions for different initial angles of deviation in the presence of the environment resistance 138

Table 6.1   (not complete) of the logical identification of the situation 157

Table 6.2   (incomplete) of fuzzy identification of the situation . 158

Table 8.1   Rule base of the FC at $m_i = w = 3$ 217

Table 8.2   Quality indicators comparative analysis of the ACS of the floating dock draft 219

Table 8.3   Rule base of the FC at $m_i = w = 5$ 221

Table 8.4   Quality indicators comparative analysis of the ACS of the floating dock draft 223

Table 8.5   Quality indicators comparative analysis of the ACS of the floating dock draft with optimized FC with full and reduced RB 224

Table 10.1   Results of Algorithm 2 simulation 265

Table 10.2   Results of B- and R-algorithms simulation 266

Table 11.1   Parameters of the robotic system 284

Table 11.2   Automatic control quality indicators for set point processing 285

Table 11.3   Integrated parameters of the random disturbance function 286

Table 11.4   Quality indicators of automatic regulation for the disturbances neutralization mode 287

Table 12.1   Perspective spectral indices for winter wheat   299

Table 12.2   Spectral indices and dimensions of maize plants . . .   301

Table 12.3   Data of modeling the routes of harvesting and
transport equipment for the first experimental field   .   308

List of Abbreviations

CB	Coordinate binding
CCS	Chaser's coordinate system
CGS	Centimeter, gram, second system
Chaser	Active spacecraft during docking
CM	Cognitive map
CS	Coordinate system
CVS	Computer vision system
DMP	Decision maker person
EC	Energy crops
ERS	Earth remote sensing
FC	Fuzzy (smart) control
FGC	Flight geometric calibration
HR	Human resources
ICS	Inertial coordinate system
IT	Information technologies
L	Length
LA	Logical automaton
LF	Lyapunov function
LIDAR	Light Identification, Detection and Ranging
LKF	Lyapunov-Krasovskii functional
LMI	Linear matrix inequality
LSM	Least squares method
M	Mass
MAP	Multi-assortment production
MM	Mathematical models
MPC	Model predictive control
NDVI	Normalized Difference Vegetation Index
NIR	Near-infrared
RMSE	Root-mean-square error
SC	Spacecraft
SCA	Short-cycle adsorption

SDM	Statistical decision-making subsystem
SDP	Semidefinite programming
SI	International system
SN	Supply network
SP	Sub-satellite polygon
SS	Star sensor
T	Time
Target	Passive spacecraft during docking
TC	Technological complex
TCS	Target's coordinate system
TP	Technological process
UAV	Unmanned aerial vehicle
WGS-84	World Geodetic System 1984

PART I

Theory of Automatic Control

1

Estimation of Impact of Bounded Perturbations on Nonlinear Discrete Systems

Vsevolod M. Kuntsevich

Space Research Institute of NAS and NSA of Ukraine, Kyiv, Ukraine
Corresponding Author: vsevolod.kuntsevich@gmail.com

We consider a class of nonlinear discrete dynamical systems consisting of linear and nonlinear parts, subject to bounded perturbations. For nonlinear functions only, the estimates are given either in the form of multivalued mappings or by the norm bounds.

The chapter presents the methods for calculation of the interval invariant sets and the radii of these sets, which are analogues of the variance in case of probabilistic nature of perturbations.

1.1 Introduction

One often needs to evaluate the extent to which the designed control system compensates uncontrolled set-valued perturbations affecting a plant. Analysis of discrete-time dynamical systems is important in itself, the more so it has been recently stimulated by the use (almost exclusively) of digital technologies for control of dynamic systems of various nature and for various purposes. It is also worth to mention that measurements and control inputs are applied at discrete instants of time. This is why the only discrete mathematical models of dynamic systems are considered below.

The most comprehensive analysis of the influence of bounded perturbations on linear continuous and discrete dynamic systems is presented in the monograph [1], where one can find a fairly complete bibliography (see also the review [2]).

The situation is quite different with analysis of the influence of bounded perturbations on nonlinear systems. There are a number of publications on solution of particular problems (see, for instance, [3]), but nearly all of them consider nonlinear systems represented in the quasilinear form, which applies certain limits on the class of systems under consideration.

In applications, the exact form of nonlinear functions is unknown quite often, and only some of the estimates are known a priori. The chapter concerns this case and it continues, to a certain extent, the research begun in [4].

1.2 Interval Invariant Sets of Nonlinear Control Systems with Linear Constraints

Recall the definition of an invariant set. An invariant set of dynamic plant subject to bounded (set-valued) perturbations is such a set that if the motion of system starts at any point of this set, then the entire trajectory remains in this set for all feasible values of perturbations.

The objective is calculation of an invariant set for the system

$$X_{n+1} \quad AX_n + f(X_n)B + Z_n, \tag{1.1}$$

where $X_n \in \mathbf{R}^m$, A is an $m \times m$ matrix, $B^T \quad (1, 0, \ldots, 0)$, and $Z_n \in \mathbf{R}^m$ is a perturbation vector at discrete time n estimated a priori as

$$Z_n \in \mathbf{Z} \quad \mathbf{z}_1 \times \mathbf{z}_2 \times \ldots \times \mathbf{z}_m, \tag{1.2}$$

where $\mathbf{z}_i \quad \{ _i : | _i| \le \sigma_i \}; i \quad \overline{1; m}$, $f(X_n)$ is a nonlinear function of scalar argument, $f(X_n) \quad f[\sigma(X_n)]$, $\sigma(\cdot) \quad C^T X_n$, $\|C\| \ / \ 0$, and $f(0) \quad 0$.

The function $f[\sigma(X_n)]$ satisfies the linear constraints,

$$0 < \underline{k} \cdot \sigma(\cdot) \le f[\cdot] \le \overline{k} \cdot \sigma(\cdot) \quad \text{with } \sigma(\cdot) \quad C^T X_n \ge 0, \tag{1.3}$$

$$\overline{k} \cdot \sigma(\cdot) \ge f[\cdot] \ge \underline{k} \cdot \sigma(\cdot) \quad \text{with } \sigma(\cdot) \quad C^T X_n \le 0. \tag{1.4}$$

The two-sided constraints (1.3) and (1.4) determine the estimate of function $f(\cdot)$ as multivalued mapping in the form of its first-order cone (see Figure 1.1),

$$f[\sigma(X_n)] \in \varphi(X_n) \quad \varphi^+(X_n) \bigcup \varphi^-(X_n); \quad i \quad \overline{1; m}; \tag{1.5}$$

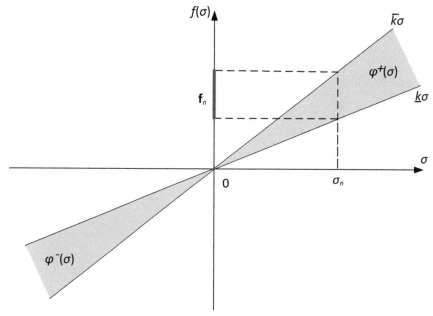

Figure 1.1 An illustration of multi-valued mapping (1.5).

where

$$\varphi^+(X_n) \qquad \{\sigma(\cdot) : \underline{k}\sigma(\cdot) \le f(\cdot) \le \overline{k}\sigma(\cdot)\} \quad \text{with } \sigma(\cdot) \ge 0;$$

$$\varphi^-(X_n) \qquad \{\sigma(\cdot) : \overline{k}\sigma(\cdot) \ge f(\cdot) \ge \underline{k}\sigma(\cdot)\} \quad \text{with } \sigma(\cdot) \le 0;$$

with the axis given by the equation $f_1[\sigma_1(X)] \qquad \overset{\circ}{k}\sigma_1(X)$, where $\overset{\circ}{k}$ $0, 5(\overline{k} + \underline{k})$.

Note that the class of autonomous systems,

$$X_{n+1} \qquad AX_n + f[\sigma(X_n)]B,$$

with a nonlinear vector function subject to constraints (1.3) and (1.4), has been studied since the pioneer work by A. I. Lurie and V. N. Postnikov [5], also in the publications by M. A. Aizerman and F. R. Gantmacher [6], R. Kalman, V. A. Yakubovich, V. Popov, Ya. Z. Tsypkin, E. Jury, and others when solving the problem known to the community as a problem of absolute stability.

The multivalued mapping $\varphi(X)$ with linear bounds maps the values σ_n on the intervals $\mathbf{f}_n$ (see Figure 1.1) with the sizes (radii) dependent linearly on σ_n. Hence, the systems (1.1), (1.2), and (1.5) describe the family of linear systems,

$$X_{n+1} \quad (A + kBC^T)X_n + Z_n, \underline{k} \leq k \leq \bar{k}. \tag{1.6}$$

Assume vector X_n is estimated with the interval set,

$$X_n \in \bar{\mathbf{X}}_n \quad \bar{\mathbf{x}}_{1,n} \times \bar{\mathbf{x}}_{2,n} \times \cdots \times \bar{\mathbf{x}}_{m,n};$$
$$\bar{\mathbf{x}}_{i,n} \quad \{x_i : \underline{x}_{i,n} \leq x_i \leq \bar{x}_{i,n}\}, \quad i \quad \overline{1;m}. \tag{1.7}$$

As it follows from (1.1), (1.2), (1.6), and (1.7), the system is described by the linear difference inclusion,

$$X_{n+1} \in \mathbf{X}_{n+1} \quad \bigcup_{\substack{X_n \in \bar{\mathbf{X}}_n \\ \underline{k} \leq k \leq \bar{k}}} (A + kBC^T)X_n + \mathbf{Z}. \tag{1.8}$$

As seen from (1.8), the set

$$\mathbf{X}'_{n+1} \quad \bigcup_{\substack{X_n \in \bar{\mathbf{X}}_n \\ \underline{k} \leq k \leq \bar{k}}} (A + kBC^T)X_n$$

is not an interval. Calculation and further use of an exact approximation of set $\mathbf{X}'_n$ in general requires significant computations. Hence, it is reasonable to approximate these sets with the minimum-volume outbound interval sets as done in [7],

$$\bar{\mathbf{X}}'_{n+1} \quad \bar{\mathbf{x}}'_{1,n+1} \times \bar{\mathbf{x}}'_{2,n+1} \times \cdots \times \bar{\mathbf{x}}'_{m,n+1}, \quad i \quad \overline{1;m}, \tag{1.9}$$

where

$$\bar{\mathbf{x}}'_{i,n+1} \quad \{x_i : \underline{x}'_{i,n+1} \leq x_i \leq \bar{x}'_{i,n+1}\}, i \quad \overline{1;m}, \tag{1.10}$$

$$\bar{x}'_{1,n+1} \quad \min_{\substack{X_n \in \bar{\mathbf{X}}_n \\ \underline{k} \leq k \leq \bar{k}}} \{\gamma_1(\cdot) \quad (A_1^T + kC^T)X_n\},$$

$$\bar{x}'_{1,n+1} \quad \max_{\substack{X_n \in \bar{\mathbf{X}}_n \\ \underline{k} \leq k \leq \bar{k}}} \{\gamma_1(\cdot) \quad (A_1^T + kC^T)X_n\}, \tag{1.11}$$

$$\underline{x}'_{i,n+1} \quad \min_{\substack{X_n \in \bar{\mathbf{X}}_n \\ \underline{k} \leq k \leq \bar{k}}} \{\gamma_i(\cdot) \quad A_i^T X_n\},$$

$$\bar{x}'_{i,n+1} \quad \max_{\substack{X_n \in \bar{\mathbf{X}}_n \\ \bar{k} \leq k \leq \bar{k}}} \{\gamma_i(\cdot) \quad A_i^T X_n\}, \quad i \quad \overline{2;m}. \tag{1.12}$$

Since the functions $\gamma_i(\cdot)$, $i = \overline{1;m}$, are bilinear, the solutions of problems (1.11) and (1.12) are found at the vertices X_n^s of the set $\bar{\mathbf{X}}_n$. We represent the interval set $\bar{\mathbf{X}}_n$ (1.7) in the form

$$\bar{\mathbf{X}}_n = \operatorname*{conv}_{s=\overline{1;S}}\{X_n^s\},$$

where X_n^s is the s-th vertex of the polytope $\bar{\mathbf{X}}_n$ and re-write problems (1.11) and (1.12) as follows:

$$\underline{\bar{x}}'_{1,n+1} = \min_{\substack{s=\overline{1;S}\\ j=\overline{1;2}}}\{(A_1^T + k_j C^T)X_n^s\}$$

$$\overline{\bar{x}}'_{1,n+1} = \max_{\substack{s=\overline{1;S}\\ j=\overline{1;2}}}\{(A_1^T + k_j C^T)X_n^s\},$$

(1.13)

$$\underline{\bar{x}}'_{i,n+1} = \min_{s=\overline{1;S}}\{A_i^T X_n^s\}$$

$$\overline{\bar{x}}'_{i,n+1} = \max_{s=\overline{1;S}}\{A_i^T X_n^s\}, \qquad i = \overline{2;m}.$$

(1.14)

The dimensions of problems (1.13) and (1.14) are insignificant. Therefore, we can easily find the solutions by searching all vertices. The solutions, $\underline{x}'_{i,n+1}, \overline{\mathbf{x}}'_{i,n+1}, i = \overline{1;m}$, define the desired set

$$\bar{\mathbf{X}}'_{n+1} = \Gamma(\bar{\mathbf{X}}_n).$$

(1.15)

Further, we shall consider the equation of evolution of interval sets instead of equation (1.8),

$$\bar{\mathbf{X}}_{n+1} = \Gamma(\mathbf{X}_n) + \mathbf{Z}.$$

(1.16)

A bounded invariant set of system (1.16) exists if the family of autonomous systems (1.15) is asymptotically stable. With the aim of simplification of analysis of the robust stability of system (1.15), we shall use the sufficient conditions of the robust stability,

$$\max_{\underline{k}\leq k\leq \bar{k}} |A_1^T + k C^T| \leq q < 1,$$

(1.17)

$$\max |A_i^T| \leq q_i < 1, \qquad i = \overline{2;m}.$$

(1.18)

Remark 1. *The fulfilment of inequality* $\|A_m^T\| < 1$ *for the systems with the Frobenius matrices,*

$$A \quad \left\|\begin{array}{cc} 0 & \vdots \\ \vdots & \vdots \; I_{m-1} \\ 0 & \vdots \\ \hline & A_m^T \end{array}\right\|,$$

is the necessary and sufficient condition of the robust stability despite the determinant of A equals 1. Hence, inequalities (1.18) in case of the stable Frobenius matrix and with inequalities (1.17) reduce to the equalities,

$$\max \|A_i^T\| \quad 1, \quad i \quad \overline{2; m}.$$

Assume further that inequalities (1.17) and (1.18) are fulfilled. Next, we calculate the invariant set of systems (1.16) and (1.2). Below, by an invariant set of this family of systems, we shall mean such a set $\overset{*}{\mathbf{X}}$ with the variable parameter k that the radius $r(\overset{*}{\mathbf{X}})$, which is defined as

$$r(\overset{*}{\mathbf{X}}) \quad \max_{X \in \overset{*}{\mathbf{X}}} \|X\|_\infty,$$

reaches its maximum.

Assume $\overline{\mathbf{X}}_{n+1} \quad \overline{\mathbf{X}}_n \quad \overset{*}{\mathbf{X}}$ in (1.16) and obtain the equation

$$\overset{*}{\mathbf{X}} \quad \Gamma(\overset{*}{\mathbf{X}}) + \mathbf{Z}, \tag{1.19}$$

which defines the target interval invariant set of the considered family of systems.

An analytical solution of the nonlinear equation (1.19) is hardly possible. Therefore, we apply the following trick. Consider m-dimensional cube $\tilde{\mathbf{X}}$ of the radius m with the center at the origin and calculate the value of α in the equation

$$r[\Gamma(\tilde{\mathbf{X}})] \quad \alpha r(\tilde{\mathbf{X}}).$$

For the matrix $\tilde{\mathbf{A}}$ satisfying conditions (1.17) and (1.18), the radius value is bounded by the inequality $r[\Gamma(\overset{*}{\mathbf{X}})] \leq mq$, where q satisfies the inequalities $\alpha \leq q < 1$ with respect to (1.17) and (1.18). As $\overset{*}{\mathbf{X}}$ is an interval set and the radius of Minkowski sum of two interval sets equals the sum of their radii, i.e., $r[\Gamma(\overset{*}{\mathbf{X}}) + \mathbf{Z}] \quad r[\Gamma(\overset{*}{\mathbf{X}})] + r(\mathbf{Z})$, we obtain

$$r(\overset{*}{\mathbf{X}}) \quad (1-\alpha)^{-1} r(\mathbf{Z}), \quad \alpha < 1.$$

The radius $r(\overset{*}{\mathbf{X}})$ of invariant set is an analogue of the variance value in case of consideration of stochastic nature of perturbations.

1.3 Invariant Sets of Systems with Two-Sided Nonlinear Constraints

Consider the same family of systems (1.1) and (1.2),

$$X_{n+1} \quad AX_n + f(X_n)B + Z_n, Z_n \in \mathbf{Z},$$

but with the different properties of the nonlinear function $f(X_n)$. Assume that this function is given by the lower and upper bounds as follows,

$$\underline{k}\overset{*}{f}[\sigma(X_n)] \leq f[\sigma(X_n)] \leq \overline{k}\overset{*}{f}[\sigma(X_n)], \tag{1.20}$$

where k takes the values in the interval $0 < \underline{k} \leq k \leq \overline{k}$ and $\sigma(X_n)$ $C^T X_n$ as above. The function $\overset{*}{f}(\cdot)$ is monotone and it satisfies the conditions $\overset{*}{f}(0) \quad 0$ and $\overset{*}{f}[-\sigma(X_n)] \quad -\overset{*}{f}[\sigma(X_n)]$.

The two-sided constraints (1.20) determine the multivalued mapping $\eta(\sigma_n)$. As above, we analyze system (1.1) with the nonlinear function $f(X)$ satisfying the conditions (1.20) and with the multivalued mapping $\eta(\sigma_n)$ defined by the "curved" cone of the first order,

$$f[\sigma(X_n) \quad \sigma_n] \in \eta(X_n) \quad \{\sigma_n : 0 < \underline{k}\overset{*}{f}(\sigma_n) \leq f(\sigma_n) \leq \overline{k}\overset{*}{f}(\sigma_n)\}. \tag{1.21}$$

The multivalued mapping $\eta(X)$ with the nonlinear bounds transforms the values σ_n into the intervals $\mathbf{f}_n$ with the sizes (radii) dependent nonlinearly on σ_n. Hence, systems (1.1), (1.2), and (1.21) define *the family of nonlinear systems*

$$X_{n+1} \quad AX_n + k\overset{*}{f}(\sigma_n \quad C^T X_n)B + Z_n, \quad \underline{k} \leq k \leq \overline{k}. \tag{1.22}$$

As the vector X_n is estimated by (1.7), the dynamics of family of systems (1.22) and (1.2) is described by the nonlinear difference inclusion as follows:

$$X_{n+1} \in \mathbf{X}_{n+1} \quad \bigcup_{\substack{X_n \in \bar{\mathbf{X}}_n \\ \underline{k} \leq k \leq \overline{k}}} [AX_n + k\overset{*}{f}(\sigma_n \quad C^T X_n)B + \mathbf{Z}].$$

Since the set

$$\mathbf{X}_{n+1} \quad \bigcup_{\substack{X_n \in \bar{\mathbf{X}}_n \\ \underline{k} \le k \le \bar{k}}} [AX_n + k \overset{*}{f}(\sigma_n \quad C^T X_n)B]$$

is not interval, we approximate it with the upper-bound interval set (1.9) of the minimum volume as above. Here the values of $\underline{x}''_{1,n+1}$ and $\bar{x}''_{1,n+1}$ in contrary to (1.11) are calculated as follows,

$$\underline{x}''_{1,n+1} \quad \min_{X_n \in \mathbf{X}_n} \{\psi_1(\cdot) \quad A_1^T X_n + \eta(\sigma_n)\}, \tag{1.23}$$

$$\bar{x}''_{1,n+1} \quad \max_{X_n \in \mathbf{X}_n} \{\psi_1(\cdot) \quad A_1^T X_n + \eta(\sigma_n)\}, \tag{1.24}$$

and the values $\underline{x}''_{i,n+1}, \bar{x}''_{i,n+1}, i \quad \overline{2; m}$, satisfy the equalities $\underline{x}''_{i,n+1}$ $\underline{x}'_{i,n+1}, \bar{x}''_{i,n+1} \quad \bar{x}'_{i,n+1}, i \quad \overline{2; m}$ (see (1.12)).

The extremums of monotone function $\psi_1(\cdot)$ are found at the vertices X_n^s of the convex centrally symmetric set $\bar{\mathbf{X}}_n$ and at the bounds $k_1 \quad \underline{k}$ and $k_2 \quad \bar{k}$ of the feasible interval of values k. Hence, problems (1.23) and (1.24) reduce to the combinatorial ones,

$$\underline{x}_{1,n+1} \quad \min_{\substack{s=\overline{1;2^m} \\ l=\overline{1;2}}} \{A_1^T X_n^s + k_l \overset{*}{f}(\sigma_n^s)\},$$

$$\bar{x}_{1,n+1} \quad \max_{\substack{s=\overline{1;2^m} \\ l=\overline{1;2}}} \{A_1^T X_n^s + k_l \overset{*}{f}(\sigma_n^s)\},$$

where $\sigma_n^s \quad C^T X_n^s$.

These problems can be easily solved by direct search as the dimensions are low.

The solutions $\underline{x}_{i,n+1}, \bar{x}_{i,n+1}, i \quad \overline{2; m}$, and the values $\underline{x}'_{i,n+1}, \bar{x}'_{i,n+1}$ determine the interval set $\bar{\mathbf{X}}_{n+1}$. Further, we consider the evolution of interval sets

$$\bar{\mathbf{X}}_{n+1} \quad G(\bar{\mathbf{X}}_n) + \mathbf{Z}, \tag{1.25}$$

where $G(\bullet)$ denotes the nonlinear mapping (1.22).

System (1.25) has a bounded invariant set if the nonlinear autonomous system,

$$X_{n+1} \in \bar{\mathbf{X}}_{n+1} \quad G(\bar{\mathbf{X}}_n), \tag{1.26}$$

is asymptotically stable either in the large or in the target domain $\overset{\circ}{\mathbf{X}}$ which contains the origin. We do not consider here the excessive (in the applications) requirement of maintaining the stability in the large, but restrict the analysis to the given domain

$$X_n \in \overset{\circ}{\mathbf{X}} \quad \{X : \|X\|_\infty \le \rho\}.$$

With the aim of stability analysis of an autonomous system (1.26), we use the radius $r(\mathbf{X})$ of set $\mathbf{X}$ and the results published in [8], where the contraction mapping principle was extended to the class of nonlinear difference inclusions.

Theorem 1. *Consider the radius $r(\mathbf{X}_n)$ as Lyapunov function. If the first difference of this function calculated along the trajectory of system (1.26) is negative definite, i.e.,*

$$\Delta v_n \quad v_{n+1} - v_n \quad r[\mathbf{F}(\mathbf{X}_n)] - r(\mathbf{X}_n) < 0 \quad \forall n, \tag{1.27}$$

then the trivial solution to the difference inclusion (1.26) is asymptotically stable.

Corollary 1. *If the center of minimum-radius outbound sphere for the set $\mathbf{X}_n$ coincides the origin, then the value $r(\mathbf{X}_n)$ calculated from equation (1.26) with the use of norm $\|X\|_\infty$ equals the radius of outbound sphere and the fulfilment of inequality (1.27) provides the fulfilment of inclusion*

$$\mathbf{X}_{n+1} \subset \mathbf{X}_n. \tag{1.28}$$

It is straightforward that for the case of interval sets $\mathbf{X}_n$ и $\mathbf{X}_{n+1}$ with the centers at the origin, a strict inclusion (1.28) takes place if at least a single non-strict inclusion among these ones, $\underline{x}_{i,n+1} \ge \underline{x}_{i,n}, \bar{x}_{i,n+1} \le \bar{x}_{i,n}$, $i \quad \overline{1; m}$, becomes strict.

Assume the matrix A and multivalued mapping (1.21) provide the fulfilment of inclusion $\bar{\mathbf{X}}_{n+1} \subset \bar{\mathbf{X}}_n$ for the set $\bar{\mathbf{X}}_{n+1}$. This provides both sufficient condition of the stability of nonlinear difference inclusions (1.26) and the necessary condition of the existence of bounded invariant set of system (1.25).

Assume $\bar{\mathbf{X}}_{n+1} \quad \bar{\mathbf{X}}_n \quad \overset{*}{\mathbf{X}}$, substitute $\overset{*}{\mathbf{X}}$ into (1.25) and obtain the equation

$$\overset{*}{\mathbf{X}} \quad G(\overset{*}{\mathbf{X}}) + \mathbf{Z}, \tag{1.29}$$

with the solution $\overset{*}{\mathbf{X}}$ which is a desired invariant set.

Analytical solution of equation (1.29) is hardly possible, and hence we apply the iterative procedure described in [8]. The efficiency of this (and any other iterative) procedure essentially depends on the starting point $\overset{*}{\mathbf{X}}_0$. As it follows from (1.29), $r(\overset{*}{\mathbf{X}}) > r(\mathbf{Z})$; however, one can hardly estimate a prori the difference between $r(\overset{*}{\mathbf{X}})$ and $r(\mathbf{Z})$. Hence, it is reasonable to start with $\overset{*}{\mathbf{X}}_0$ $\mathbf{Z}$. Next, we calculate $G(\overset{*}{\mathbf{X}}_0)$, $r_i(\overset{*}{\mathbf{X}}_0)$, i $\overline{1;m}$, and Δr_i

$r_i(\overset{*}{\mathbf{X}}_0) - r_i[\mathbf{X}_{0i}$ $r(\ _i)]$; i $\overline{1;m}$. Further, we use bisection of intervals $\Delta r_i(\cdot)$ and calculate $r_i(\mathbf{X}^*_{i,p})$; i $\overline{1;m}$, where p is the iteration number, as follows,

$$r_{i,p+1}(\mathbf{X}^*_{i,p+1})\quad r_i(\mathbf{X}^*_{i,p}) + 0,5\Delta r_i,\quad i\quad \overline{1;m}.$$

The iterations stop as $|\Delta r_{i,p}| \leq \varepsilon_{i,i}$ $\overline{1;m}$, where ε is the given feasible error.

The described algorithm converges at the rate of geometric progression.

The calculated radius $r(\overset{*}{\mathbf{X}})$ of interval set $\overset{*}{\mathbf{X}}$ equals

$$r(\overset{*}{\mathbf{X}})\quad \sum_{i=1}^{m} r_i(\overset{*}{\mathbf{X}}).$$

1.4 Invariant Sets of Systems with Norm-Based Estimates of Nonlinear Functions

Consider the same systems (1.1) and (1.2) in case of estimation of $f(X_n)$ by its norm, i.e., the given bounds are the following,

$$\overset{*}{f}(X_n)\quad k\|X_n\|_\infty \operatorname{sign} \sigma(X_n). \tag{1.30}$$

Here $\sigma(X_n)$ $C^T X_n$, $\|C\|$ $/$ 0, and $\underline{k} \leq k \leq \bar{k}$ as above.

Assume that X_n is estimate a priori by (1.7) as above and represent (1.7) in the following form,

$$X_n \in \mathbf{X}_n\quad \underset{s=\overline{1;2^m}}{\operatorname{conv}} \{X_n^s\},$$

where X_n^s is the s-th vertex of polytope $\mathbf{X_n}$ same as above.

We shall calculate the invariant set of systems (1.1) and (1.30) in the class of interval sets $\bar{\mathbf{X}}_n$.

Introduce the set

$$\mathbf{X}'_{n+1} \quad \bigcup_{\substack{X_n \in \bar{\mathbf{X}}_n \\ \underline{k} \le k \le \bar{k}}} \{AX_n + k\|X_n\|_\infty \operatorname{sign} C^T X_n B\}. \tag{1.31}$$

Approximate set (1.31) with the outbound minimum-volume interval set $\bar{\mathbf{X}}'_{n+1}$ (see (1.9), (1.10)), where the bounds $\underline{x}_{i,n+1}$, $\bar{x}_{i,n+1}$, $i \quad \overline{2;m}$, of intervals $\bar{\mathbf{x}}_{i,n+1}$, $i \quad \overline{2;m}$, are calculated in accordance with (1.12), and the bounds $\underline{x}_{1,n+1}$, $\bar{x}_{1,n+1}$ of the interval $\bar{\mathbf{x}}_{1,n+1}$ are solutions to the following problems,

$$\underline{x}_{1,n+1} \quad \min_{\substack{X_n \in \bar{\mathbf{X}}_n \\ \underline{k} \le k \le \bar{k}}} \{\gamma(\cdot)\}, \bar{x}_{1,n+1} \quad \max_{\substack{X_n \in \bar{\mathbf{X}}_n \\ \underline{k} \le k \le \bar{k}}} \{\gamma(\cdot)\}, \tag{1.32}$$

$$\gamma(\cdot) \quad A_1^T X_n + k \left(\sum_{i=1}^m |x_{i,n}| \right) \operatorname{sign} C^T X_n. \tag{1.33}$$

The solutions of problems (1.32) are obviously found at the vertices X_n^s of set $\bar{\mathbf{X}}_n$ and at the bounds $k_1 \quad \underline{k}$, $k_2 \quad \bar{k}$, of the feasible interval of values k. Hence, problems (1.32) reduce to the search among the vertices of X_n^s and the bounds k_1 and k_2.

Since the function $\overset{*}{f}(X)$ satisfies the equality $\overset{*}{f}(-X) \quad - \overset{*}{f}(X)$, the complexity of search problems can be reduced. As $\underline{x}_{1,n+1} \quad -\bar{x}_{1,n+1}$, we have to calculate the objective function only for the points X_n^s, where $C^T X_n^s \ge 0$. Therefore, calculation of $\underline{x}_{i,n+1}$, $\bar{x}_{i,n+1}$, $i \quad \overline{1;m}$, is reduced to solution of the problem

$$\bar{x}_{i,n+1} \quad \max_{\substack{s=\overline{1;2^m} \\ l=\overline{1;2}}} \left\{ A_1^T X_n^s + k_l \left(\sum_{i=1}^m |x_{i,n}^s| \right) \operatorname{sign}(C^T X_n^s) \right\}.$$

The calculated optimal values $\underline{x}_{i,n+1}$, $\bar{x}_{i,n+1}$, and $i \quad \overline{1;m}$, define the interval set $\bar{\mathbf{X}}'_{n+1} \quad \Phi(\bar{\mathbf{X}}_n)$. Below, we consider the evolution of interval sets described by the equation

$$\bar{\mathbf{X}}_{n+1} \quad \Phi(\bar{\mathbf{X}}_n) + \mathbf{Z}. \tag{1.34}$$

System (1.35) has a bounded invariant set if the autonomous system,

$$\bar{\mathbf{X}}_{n+1} \quad \Phi(\bar{\mathbf{X}}_n), \tag{1.35}$$

is asymptotically stable either in the large or in the given domain $\overset{\circ}{\mathbf{X}}$.

Check the sufficient conditions of stability of system (1.35) derived from the contracting mapping principle generalized in [8]. Assume the matrix A and parameter $\bar{k}$ provide its fulfilment.

Remark 2. *The function $f(x)$ $|x|$ can be represented in the quasilinear form, $f(x)$ $a(x)x$, where $a(x)$ $|x|/x$, since $a(0)$ is undefined. As the function $f(x)$ $|x|$ is not differentiable at the origin, one cannot define $a(0)$.*

Assume $\mathbf{X}_{n+1}$ $\mathbf{X}_n$ $\overset{*}{\mathbf{X}}$, substitute into (1.34) and obtain the equation

$$\overset{*}{\mathbf{X}}   \Phi(\overset{*}{\mathbf{X}}) + \mathbf{Z}. \tag{1.36}$$

The solution to equation (1.36) cannot be calculated analytically. Therefore, we apply the same iterative procedure as above.

1.5 Conclusion

In applications, the exact values of nonlinear functions at a point are unknown. More realistic, we can obtain certain (set-valued) estimates a priori. The same concerns perturbations that affect dynamical systems. We considered above the interval sets as estimates of uncertain bounded perturbations.

The chapter provides the approach to calculation of the interval invariant sets in three cases. Namely, the unknown values of nonlinear functions are estimated a priori

(a) with the two-sided *linear* constraints,
(b) with the two-sided *nonlinear* constraints, and
(c) with the *norm* bounds.

Obviously, the obtained results can be generalized to the cases of several nonlinear functions with the estimates considered in the chapter.

References

[1] B. T. Polyak, M. V. Khlebnikov, P. S. Shcherbakov, 'Control of Linear Systems Subject to Exogenous Disturbances: The Linear Matrix Inequalities Technique', Moscow, 2014.

[2] M. V. Khlebnikov, B. T. Polyak, V. M. Kuntsevich, 'Optimization of linear systems subject to bounded exogenous disturbances: The invariant ellipsoid technique', *Automation and Remote Control*, 2011, vol. 72, no. 11, pp. 2227–2275, https://doi.org/10.1134/S0005117911110026.

[3] A. G. Mazko, 'Robust stability and stabilization of dynamic systems. Methods of matrix and cone inequalities', *Proceedings of the Institute of Mathematics of NAS of Ukraine. Mathematics and its Applications*, vol. 102, 2016.

[4] A. V. Kuntsevich, V. M. Kuntsevich, 'Invariant sets for families of linear and nonlinear discrete systems with bounded disturbances', *Automation and Remote Control*, 2012, vol. 73, no. 1, pp. 83–96, https://doi.org/10.1134/S0005117912010067.

[5] A. I. Lurie, V. N. Postnikov, 'On the stability theory of regularized systems', *J. Appl. Math. Mech.*, 1944, vol. 8, no. 3, pp. 245–248.

[6] M. A. Aizerman, F. R. Gantmacher, 'Absolute stability of regularized systems', Moscow: Acad. Nauk SSSR, 1963.

[7] V. M. Kuntsevich, A. B. Kurzhanski, 'Attainability domains for linear and some classes of nonlinear discrete systems and their control', *Journal of Automation and Information Sciences,* 2010, vol. 42, no. 1, pp. 1–18, https://doi.org/10.1615/JAutomatInfScien.v42.i1.10.

[8] A. V. Kuntsevich, V. M. Kuntsevich, 'Stability in the domain of nonlinear difference inclusions', *Cybernetics and Systems Analysis*, 2010, vol. 46, no. 5, pp. 691–698, https://doi.org/10.1007/s10559-010-9249-3.

2

Control of Moving Objects in Condition of Conflict

Arkadii A. Chikrii

Glushkov Institute of Cybernetics NAS of Ukraine, Kyiv, Ukraine
Corresponding Author: g.chikrii@gmail.com

This chapter is devoted to quasilinear conflict-controlled processes with a cylindrical terminal set. A specific feature is that, instead of a dynamical system, we start with representation of a solution in a form that allows one to include an additive term with the initial data and a control unit. This makes it possible to consider a broad spectrum of dynamic processes in a unified scheme. Our study is based on the method of resolving functions. We obtain sufficient conditions for the solvability of the pursuit problem at a certain quaranteed time in the class of strategies that use information on the behavior of the opponent in the past as well as in the class of stroboscopic strategies. We also find conditions under which information on the prehistory of the evader does not matter.

In more detail, we study game problems for systems with Riemann-Liouville fractional derivatives and regularized derivatives of Dshrbashyan-Nersesyan–Caputo. For illustration of the method, we consider the case of a simple matrix as a dynamic system.

2.1 Introduction

A number of fundamental methods [1–5] exist in the theory of differential games that allow to formulate conditions for solvability of the problems of approach and avoidance in one or another class of strategies. Various techniques are used depending on the kind of exchange of information between the players about the process state and also on how the player, standing on which side the game is analyzed, chooses his control. In this chapter, as a tool for investigation, the method of resolving functions [4–13] was chosen based on using the inverse functionals of Minkowski [10] and substantiating the classic rule of parallel pursuit. Under different forms of Pontryagin's condition, this method was successfully applied in the study of the game problems with groups of participants [4, 7–9], games with the terminal functional, with state constraints and with imperfect information as well as in study of the processes with more complicated than ordinary differential equation dynamics.

In this chapter, we use the basic ideas of the method of resolving function to obtain sufficient conditions for solvability of the game problems. Minor assumptions are made about the dynamic process to encompass a wide range of conflict-controlled processes as wide as possible.

An attractive feature of the method of resolving functions is that it fully justifies the classical rule of parallel approach and allows one to efficiently apply the modern technique of set-valued mappings and their selections [14–16] when validating game constructions and obtaining meaningful results on the basis of these constructions.

An important class of conflict-controlled processes that can be embedded in the general scheme of the method is the game problems for systems with fractional derivatives. The systems with fractional derivatives were studied earlier, i.e., in the works [17–21]. Special game problems with incomplete information were studied in [22, 23].

2.2 Formulation of the Problem: Scheme of the Method

Let us denote by R^n the real n- dimensional Euclidian space and by $R_+ = \{t : t \geq 0\}$ the positive semi-axis. Consider the process with evolution described by the equation

$$z(t) = g(t) + \int_0^t \Omega(t, \tau)\varphi(u(\tau), v(\tau))d\tau, t \geq 0 \qquad (2.1)$$

Function $g(t), g : R_+ \to R^n$, is Lebesque measurable and bounded for $t>0$, matrix function $\Omega(t, \tau), t \geq \tau \geq 0$, is measurable in τ and also summable in τ for any $t \in R_+$. The control block is given by function $\varphi(u, v), \varphi : U \times V \to R^n$, that is assumed to be jointly continuous in its variables on the direct product of non-empty compacts U and V, i.e., $U, V \in K(R^n)$.

Control actions of the players - $u(\tau), u : R_+ \to U$, and $v(\tau), v : R_+ \to V$, are measurable functions.

In addition to the process (2.1), a terminal set is given having a cylindrical form

$$M^* = M_0 + M, \tag{2.2}$$

where M_0 is a linear subspace from R^n and $M \in K(L)$, where L – an orthogonal complement to M_0 in R^n.

The goals of the first (u) and the second (v) players are opposite. The first one strives in the shortest time to drive a trajectory of the process (2.1) to the set (2.2), the second one strives to maximally postpone the instant of time when the process trajectory hits the set M^*.

Let us take the side of the first player and assume that his opponent choose as controls an arbitrary measurable functions with values from V. We also assume that the game (2.1)–(2.2) take place on the interval $[0, T]$ and that the first player chooses as controls measurable functions of the form:

$$u(t) = u(g(T), v_t(\cdot)), t \in [0, T], u(t) \in U, \tag{2.3}$$

where $v_t(\cdot) = \{v(s) : 0 \leq s \leq t\}$ is a pre-history of the second player's control up to the instant t. If, for example, $g(t) = A(t)z_0$, where $A(t)$ is a matrix function such that $A(0) = E$ (E is a unit matrix) and $z(0) = z_0$, then we may consider that $u(t) = u(z_0, v_t(\cdot))$, i.e., control of the first player appears as a special type quasistrategy [4, 5].

The goal of the chapter is, under the information condition (2.3), to develop sufficient conditions for solvability of the problem in favor of the first player in some garanteed time, as well as to estimate this time and to find the control of first player that allows for the realization of this result.

Now let us describe the method of solving this problem. Original assumptions about functions $g(t), \Omega(t, \tau), \varphi(u, v)$ and sets U, V, M^* allow us to realize already known for differential games construction [4, 6] that we briefly outline below.

Define by π the orthoprojector acting from R^n onto L.

Setting

$$\varphi(U, v) = \{\varphi(u, v) : u \in U\}$$

let us consider the following set-valued mappings

$$W(t, \tau, v) = \pi\Omega(t, \tau)\varphi(U, v),$$

$$W(t, \tau) = \bigcap_{v \in V} W(t, \tau, v),$$

defined on sets $\Delta \times V$ and Δ, respectively, where

$$\Delta = \{(t,\tau) : 0 \le \tau \le t < \infty\}.$$

Pontryagin's condition. *Set-valued mapping* $W(t,\tau)$ *takes nonempty values on set* Δ.

By virtue of continuity of the function $\varphi(u,v)$ and the condition $U \in K(R^n)$, the mapping $\varphi(U,v)$ is continuous in v in Hausdorff metric [16].

Taking into account the assumptions concerning matrix function $\Omega(t,\tau)$, one can infer that the set-valued mappings $W(t,\tau,v)$ and $W(t,\tau)$ are measurable in τ [14]. Recall that a set-valued mapping $F(t), F : [0,T] \to 2^{R^n}$ is called measurable if for any open set $Y, Y \subset R^n$, the set $\{t \in [0,T] : F(t) \bigcap Y \ne \varnothing\}$ is measurable.

Let us denote by $P(R^n)$ a set of all non-empty closed sets from space R^n. Then, obviously,

$$W(t,\tau,v) : \Delta \times V \to P(R^n),$$

$$W(t,\tau) : \Delta \to P(R^n).$$

In this case, the set-valued mappings $W(t,\tau,v)$ and $W(t,\tau)$ are usually referred to as normal in τ [14].

It follows from Pontryagin's condition and some results of the papers [14–16] that for any $t \ge 0$, there exists at least one τ-measurable selection $\gamma(t,\tau) \in W(t,\tau)$. By assumptions, concerning the parameters of process (1.1), such selection $\gamma(t,\tau)$ is a function which is summable in τ for any fixed $t \ge 0, \tau \in [0,t]$. Denote

$$\xi(t,g(t),\gamma(t,\cdot)) = \pi g(t) + \int_0^t \gamma(t,\tau)d\tau$$

Now let us define a function

$$\alpha(t,\tau,v) = \sup\left\{\alpha \ge 0 : [W(t,\tau,v) - \gamma(t,\tau)]\bigcap\right.$$

$$\left. \alpha[M - \xi(t,g(t),\gamma(t,\cdot))] \ne \varnothing\right\} \tag{2.4}$$

and call it the resolving function. This function will play a key role in the sequel.

By virtue of assumptions concerning the parameters of process (2.1) and some known results from [4], one can infer that function (2.4) is measurable in τ and upper semicontinuous in v.

In what follows our prime concern will be with the joint dependence of function $\alpha(t, \tau, v)$ in variables τ and v. Let us fix some t and set $\alpha(\tau, v) = \alpha(t, \tau, v)$. We will say that function $\alpha(\tau, v)$, $\alpha : [0, T] \times V \to R_+$, is super-prositionally measurable if for any measurable function $v(\tau)$, $v : [0, T] \to V$, the superposition $\alpha(\tau, v(\tau))$, $\alpha : [0, T] \to R_+$, is a τ-measurable function. Sufficiently general assumption ensuring function $\alpha(\tau, v)$ to be superpositionally measurable is that of its $L \times B$ measurability [15, 16], i.e., of measurability with respect to σ-algebra being a product of σ-algebras $L([0, T])$ and $B(R^n)$. This σ-algebra consists of subsets of the set $[0, T] \times R^n$ generated by sets of the form $X \times Y$, where X is Lebesque measurable subset of the interval $[0,T]$ and Y-Borel measurable subset of R^n.

Denote $W(t, \tau, v) - \gamma(T, \tau) = H(\tau, v)$, $M - \xi(T, g(T), \gamma(T, \cdot)) = M_1$ and introduce a set-valued mapping

$$\Xi(\tau, v) = \{\alpha \in R_+ : H(\tau, v) \bigcap \alpha M_1 \neq \varnothing\} \tag{2.5}$$

Then

$$\alpha(\tau, v) = \sup\{\alpha \in R_+ : \alpha \in \Xi(\tau, v)\}.$$

Let us study the properties of the set-valued mapping (2.5). The following general result is true which generalizes the known statement from [16].

Lemma 1. *Let for $X \in P(R^n)$ the set-valued mappings $F(\omega)$, $F : X \to P(R^k)$ and $H(\omega)$, $H : X \to P(R^n)$, be normal and let $M(\omega, x)$, $M : X \times R^k \to P(R^n)$, be a Caratheodory mapping (measurable in ω and continuous in x).*

Then the mapping

$$\Xi(\omega) = \left\{x \in F(\omega) : H(\omega) \bigcap M(\omega, x) \neq \varnothing\right\}$$

is normal.

Setting in the statement of Lemma 1 $\omega = (\tau, v)$, $x = \alpha$ and respectively $F(\omega) = R_+$ and $M(\omega, x) = \alpha M_1$ and understanding by measurability the $L \times B$ measurability, we infer that the mapping $\Xi(\tau, v)$ is $L \times B$ measurable, since the mapping $H(\tau, v)$ is $L \times B$ measurable by virtue of its Lebesque measurability in τ and continuity in v [15].

Now let us show that the function $\alpha(\tau, v)$ is $L \times B$ measurable. Indeed, since the formula is true:

$$\alpha(\tau, v) = \sup_{\alpha \in \Xi(\tau, v)} \alpha = C(\Xi(\tau, v); 1),$$

where $C(X, p)$ is a support function of set X in direction p [24]; it follows from the $L \times B$ measurability of the set-valued mapping $\Xi(\tau, v)$ [14].

Thus, the function $\alpha(\tau, v)$ is $L \times B$ measurable, bounded below by zero and semicontinuous in v.

Let us show that the function $\inf\limits_{v \in V} \alpha(\tau, v)$ is measurable. To do this, we will treat V as a constant set-valued mapping. It is a measurable mapping [14]. The approximation set in V can be formed, for instance, by functions $v_m(\tau) = v_m$, where $V_* = \{v_1, v_2, \ldots\}$ is a countable dense subset of set V. Then, by virtue of $L \times B$ measurability of the considered function, it is superpositionally measurable whence follows that functions $\alpha(\tau, v_m)$ are measurable in τ. Let us now show that

$$\inf\limits_{v \in V} \alpha(\tau, v) = \sup\limits_{v_m} \alpha(\tau, v_m).$$

For this purpose we set $\alpha(\tau) = \inf\limits_{v \in V} \alpha(\tau, v)$ and fix τ. By definition of the greatest lower bound, for any $\varepsilon > 0$ there exists an element $v_\varepsilon \in V$ such that

$$\alpha(\tau, v_\varepsilon) \leq \alpha(\tau) + \varepsilon.$$

On the other hand, from the upper semicontinuity in v of the function $\alpha(\tau, v)$, it follows that a neighborhood $O(v_\varepsilon)$ of element v_ε exists, such that for any $v \in O(v_\varepsilon)$

$$\alpha(\tau, v) \leq \alpha(\tau, v_\varepsilon) + \varepsilon.$$

In its turn, from here and from the definition of set V_*, it follows that an element $v_m \in V_* \bigcap O(v_\varepsilon)$ exists such that

$$\alpha(\tau, v_m) \leq \alpha(\tau, v_\varepsilon) + \varepsilon \leq \alpha(\tau) + 2\varepsilon.$$

Then,

$$\inf\limits_{v_m} \alpha(\tau, v_m) \leq \alpha(\tau).$$

What is more, since the inverse inequality is always true in view of the inclusion $V_* \subset V$ then

$$\alpha(\tau) = \inf\limits_{v \in V} \alpha(\tau, v) = \inf\limits_{v_m} \alpha(\tau, v_m)$$

and therefore function $\alpha(\tau)$ is measurable as the greatest lower bound of countable set of measurable functions [14].

The following statement is a consequence of formula (2.4). If for some t, the inclusion $\xi(t, g(t), \gamma(t, \cdot)) \in M$ is satisfied, then function $\alpha(t, \tau, v)$ turns into infinity for all $\tau \in [0, t]$, $v \in V$.

Let us introduce a mapping

$$T(g(\cdot), \gamma(\cdot, \cdot)) = \left\{ t \geq 0 : \int_0^t \inf_{v \in V} \alpha(t, \tau, v) d\tau \geq 1 \right\} \qquad (2.6)$$

If for some t, the integral in expression (2.6) turns into infinity, then the inequality in braces is readily satisfied. If, on the other hand, the inequality in (2.6) fails for any t, then we set $T(g(\cdot), \gamma(\cdot, \cdot)) = \varnothing$.

We can now formulate the main result of the chapter.

Theorem 1. *Let in the games (2.1) and (2.2), Pontryagin's condition hold, $M = \text{co} M$ and let for some bounded function $g(t), t > 0$, and some measurable in τ selection $\gamma(t, \tau), t \geq \tau \geq 0$, of the set-valued mapping $W(t, \tau)$ the following relations be true:*

$$T(g(\cdot), \gamma(\cdot, \cdot)) = \varnothing \quad and \quad T \in T(g(\cdot), \gamma(\cdot, \cdot)), T < +\infty.$$

Then a trajectory of process (2.1) can be brought from the initial state $g(T)$ to the terminal set in time T, using control of the form (2.3).

Proof. Consider the case $\xi(T, g(T), \gamma(T, \cdot)) \notin M$. Let $v_T(\cdot)$ be an arbitrary measurable function with values in V. Analogously to [4], we introduce a test function

$$h(t) = 1 - \int_0^t \alpha(T, \tau, v(\tau)) d\tau, t \in [0, T].$$

Since the function $\alpha(T, \tau, v)$ is $L \times B$ measurable, it is superpositionally measurable as well, i.e., function $\alpha(T, \tau, v(\tau))$ is measurable. On the other hand, by assumptions concerning the parameters of process (2.1)–(2.2), the latter is bounded for almost all $\tau < T$ and therefore integrable on any finite interval of time. From this, it follows that the function $h(t)$ is continuous, nonincreasing and $h(0) = 1$. Therefore, these exists an instant $t_* = t(v(\cdot)), t_* \in (0, T]$, such that $h(t_*) = 0$.

In the sequel, the segments $[0, t_*)$ and $[t_*, T]$ will be referred to as "active" and "passive", respectively. Let us describe how the first player chooses his control on each of them. For this purpose, consider a set-valued mapping

$$U(\tau, v) = \{ u \in U : \pi\Omega(T, \tau)\varphi(u, v) - \gamma(T, \tau)$$

$$\in \alpha(T, \tau, v)[M - \xi(T, g(T), \gamma(T, \cdot))] \}. \qquad (2.7)$$

Since the function $\alpha(T, \tau, v)$ is $L \times B$ measurable, $M \in K(R^n)$, and the vector $\xi(T, g(T), \gamma(T, \cdot))$ is bounded, then the mapping

$$\alpha(T, \tau, v)[M - \xi(T, g(T), \gamma(T, \cdot))]$$

is $L \times B$ measurable. In addition, it is obvious that the left side of inclusion in (2.7) is jointly $L \times B$ measurable function in τ and v and continuous in u. From here, in view of the known statement from [14], it follows that the mapping $U(\tau, v)$ is $L \times B$ measurable. Therefore, its selection

$$u(\tau, v) = lex \min U(\tau, v) \tag{2.8}$$

is $L \times B$ measurable function. On the active segment $[0, t_*)$ control of the first player is set equal to

$$u(\tau) = u(\tau, v(\tau)). \tag{2.9}$$

By virtue of the function $u(\tau, v)$ $L \times B$ measurability, it is superpositionally measurable that implies the measurability of function $u(\tau)$, (2.10).

Let us analyze the "passive" segment $[t_*, T]$. We set $\alpha(T, \tau, v) \equiv 0$ in expression (2.7) for $u(\tau)\tau \in [t_*, T], v \in V$, and choose control of the first player in accordance with which the above-outlined scheme using expressions (2.7–2.9).

In the case $\xi(T, g(T), \gamma(T, \cdot)) \in M$, control of the first player on the interval $[0, T]$ is chosen from the same relations as on the passive segment, i.e., by the schemes (2.7–2.9) with $\alpha(T, \tau, v) \equiv 0, \tau \in [0, T], v \in V$.

Let us show that if the control of the first player is chosen in the form (2.9), then in both cases, in view of relations (2.7) and (2.8), a trajectory of process (2.1) will be brought to set M at instant T for any control of the second player.

From expression (2.1), we have

$$\pi z(T) = \pi g(T) + \int_0^T \pi \Omega(T, \tau) \varphi(u(\tau), v(\tau)) d\tau. \tag{2.10}$$

Let us first analyze the case $\xi(T, g(T), \gamma(T, \cdot)) \notin M$. To do this, add and substract from the right side of equality (2.10) the term $\int_0^T \gamma(T, \tau) d\tau$. Using the above-outlined rule for control choice of the first player, we obtain from (2.10) the inclusion

$$\pi z(T) \in \xi(T, g(T), \gamma(T, \cdot)) \left[1 - \int_0^{t_*} \alpha(T, \tau, v(\tau)) d\tau \right]$$

$$+ \int_0^{t_*} \alpha(T, \tau, v(\tau)) M d\tau.$$

Since M is a convex compact , $\alpha(T, \tau, v(\tau))$- nonnegative function for $\tau \in [0, t_*)$, and

$$\int_0^{t_*} \alpha(T, \tau, v(\tau))d\tau = 1,$$

then $\int_0^{t_*} \alpha(T, \tau, v(\tau))M d\tau = M$ and, consequently, $\pi z(T) \in M$ and $z(T) \in M^*$.

Suppose that $\xi(T, g(T), \gamma(T, \cdot)) \in M$. Then, taking into account the rule for control of the first player, from equality (2.10), one can immediately deduce the inclusion $\pi z(T) \in M$.

2.3 Game Problems for Fractional Systems

In this section, we introduce in a standard way the classic notions of Riemann-Liouville fractional integral and fractional derivative. To them corresponds the equation with fractional derivative in which instead of standard Cauchy condition at the initial instant $t = 0$ the fractional integral of appropriate fractional order is given. The reason is that, generally speaking , the solution of such equation has singularity at $t = 0$ and then only generalized initial conditions have sense here. However, from the physical point of view, it is desirable to have a standard Cauchy problem for equations with fractional derivatives.

In [18], Dzhrbashyan and Nersesyan introduced an equation with fractional derivative, in which instead of Riemann-Liouville derivative its regularized value is used and a standard Cauchy condition stands for the initial condition. Later, the new notion of fractional derivative was called a Dzhrbashyan and Nersesyan regularized derivative.

Let us define the fractional Riemann-Liouville integral of order $\beta, \beta \in (0, 1)$, of a function $z(t), t > 0$, by the formula [17]

$$(I_{0+}^\beta z)(t) = \frac{1}{\Gamma(\beta)} \int_0^t \frac{z(s)}{(t - s)^{1-\beta}} ds,$$

where $\Gamma(\beta)$ is Euler γ-function. Then, the fractional Riemann-Liouville derivative of order β [17] has the form

$$(D_{0+}^\beta z)(t) = \frac{d}{dt}(I_{0+}^{1-\beta})(t),$$

and the regularized Dzhrbashyan–Nersesyan-Caputo fractional derivative of order β [18, 19]

$$(D_{0+}^{(\beta)}z)(t) = (D_{0+}^{\beta}z)(t) - \frac{t^{-\beta}}{\Gamma(1-\beta)}z(+0).$$

We will associate each of the fractional derivatives with appropriate game problem.

Thus, let in the first problem the evolution of a conflict-controlled process be described by the system of differential equations

$$D^{\beta}\hat{z} = A\hat{z} + \varphi(u,v), \hat{z} \in R^{n}, u \in U, v \in V, \tag{2.11}$$

under the initial condition

$$I^{1-\beta}\hat{z}|_{t=0} = \hat{z}_0, \tag{2.12}$$

and in the second problem by the system

$$D^{(\beta)}z = Az + \varphi(u,v), z \in R^{n}, u \in U, v \in V, \tag{2.13}$$

under the initial condition

$$z|_{t=0} = z_0. \tag{2.14}$$

In the notations of fractional derivatives in (2.11) and (2.13), some symbols are omitted for the simplicity of exposition.

In addition to the dynamics of processes (2.11) and (2.13), the terminal set of the form (2.2) is given. The goals of the players in each of the cases are the same as in the general problem statement. Note, that in the problems (2.11), (2.12) and (2.13), (2.14), the first player (u) chooses his control in the form of measurable functions $u(t) = u(\hat{z}_0, v_t(\cdot))$ and $u(t) = u(z_0, v_t(\cdot))$ respectively, with values in the domain U.

Let us proceed to the deduction of integral representations for functions $\hat{z}(t)$ and $z(t)$. For this purpose, any arbitrary positive number ρ and complex number μ, we define a generalized matrix function of Mittag-Leffler [21]

$$E_{\rho}(B;\mu) = \sum_{k=0}^{\infty} \frac{B^{k}}{\Gamma(k\rho^{-1}+\mu)},$$

where B is an arbitrary square matrix of order n with complex-valued elements. Matrix function $E_{\rho}(B;\mu)$ is an integer function of argument B.

Theorem 2. *The players' controls chosen, the solution $\hat{z}(t)$ of the problem (2.11), (2.12) is defined by the formula*

$$\hat{z}(t) = t^{\beta-1}E_{1/\beta}(At^\beta; \beta)\hat{z}_0$$

$$+ \int_0^t (t - \tau)^{\beta-1}E_{1/\beta}(A(t - \tau)^\beta; \beta)\varphi(u(\tau), v(\tau))d\tau,$$

$$\tag{2.15}$$

while the solution $z(t)$ of the problem (2.13), (2.14) by the formula

$$z(t) = E_{1/\beta}(At^\beta; 1)\hat{z}_0$$

$$+ \int_0^t (t - \tau)^{\beta-1}E_{1/\beta}(A(t - \tau)^\beta; \beta)\varphi(u(\tau), v(\tau))d\tau.$$

$$\tag{2.16}$$

Proof. Let us first note that the function $F(\tau) = \varphi(u(\tau), v(\tau)), \tau > 0$, is measurable and essentially bounded. This implies that the integrals in formulas (2.15), (2.16) converge absolutely. The proof consists of two parts. In the first one, we will prove that the first terms in formulas (2.15), (2.16) are solutions of the homogeneous equations, satisfying the initial conditions (2.12), (2.14), respectively. In the second part, we will show that the second term in formulas (2.15), (2.16)

$$z_2(t) = \int_0^t (t - \tau)^{\beta-1}E_{1/\beta}(A(t - \tau)^\beta; \beta)F(\tau)d\tau \tag{2.17}$$

is a solution of non-homogeneous equations (2.15), (2.16).

The fact that $z_2(t)$ satisfies the zero initial condition immediately follows from the boundedness of the functions $E_{1/\beta}(A(t-\tau)^\beta; \beta)$ and $F(\tau)$ and that $\beta > 0$.

Denoting

$$\hat{z}_1(t) = t^{\beta-1}E_{1/\beta}(At^\beta; \beta)\hat{z}_0,$$

we proceed to calculations:

$$(D^\beta \hat{z}_1)(t) \equiv D^\beta[t^{\beta-1}E_{1/\beta}(At^\beta; \beta)\hat{z}_0]$$

$$= \frac{1}{\Gamma(1 - \beta)} \frac{d}{dt} \left(\int_0^t (t - \tau)^{-\beta}\tau^{\beta-1} \sum_{k=0}^\infty \frac{A^k\tau^{\beta k}}{\Gamma(\beta(k + 1))}d\tau \right)$$

$$= \frac{1}{\Gamma(1-\beta)} \sum_{k=0}^{\infty} \frac{A^k \hat{z}_0}{\Gamma(\beta k + \beta)} \frac{d}{dt} \int_0^t (t-\tau)^{-\beta} \tau^{\beta(k+1)-1} d\tau$$

$$= \frac{1}{\Gamma(1-\beta)} \sum_{k=0}^{\infty} \frac{A^k \hat{z}_0 B(1-\beta, \beta k + \beta)}{\Gamma(\beta k + \beta)} \frac{d}{dt} t^{\beta k}$$

$$= \frac{1}{\Gamma(1-\beta)} \sum_{k=1}^{\infty} \frac{A^k \Gamma(1-\beta)\Gamma(\beta k + \beta)}{\Gamma(\beta k + \beta)\Gamma(\beta k + 1)} \beta k t^{\beta k - 1} \hat{z}_0$$

$$= \sum_{k=1}^{\infty} \frac{\beta k A^k t^{\beta k - 1}}{\Gamma(\beta k + 1)} \hat{z}_0 = \sum_{k=1}^{\infty} \frac{A^k t^{\beta k - 1}}{\Gamma(\beta k)} \hat{z}_0$$

$$\overset{k=k'+1}{=} A t^{\beta - 1} \sum_{k'=0}^{\infty} \frac{A^{k'} t^{\beta k'}}{\Gamma(\beta k' + \beta)} \hat{z}_0 = A \hat{z}_1(t).$$

Here $B(z, \omega) = \int_0^1 x^{z-1}(1-x)^{\omega-1} dx = \frac{\Gamma(z)\Gamma(\omega)}{\Gamma(z+\omega)}$ is Euler β-function.

Let us now show that $\hat{z}_1(t)$ satisfies the initial condition (2.12)

$$(I^{1-\beta}\hat{z}_1)(t) = \frac{1}{\Gamma(1-\beta)} \int_0^t \frac{\hat{z}_1(\tau)}{(t-\tau)^\beta} d\tau$$

$$= \frac{1}{\Gamma(1-\beta)} \int_0^t \frac{\tau^{\beta-1}}{(t-\tau)^\beta} \sum_{k=0}^{\infty} \frac{A^k \tau^{\beta k}}{\Gamma(\beta k + \beta)} \hat{z}_0 d\tau$$

$$= \frac{1}{\Gamma(1-\beta)} \sum_{k=0}^{\infty} \frac{A^k \hat{z}_0}{\Gamma(\beta k + \beta)} \int_0^t \tau^{\beta(k+1)-1}(t-\tau)^{-\beta} d\tau$$

$$= \frac{1}{\Gamma(1-\beta)} \sum_{k=0}^{\infty} \frac{A^k t^{\beta k}}{\Gamma(\beta k + \beta)} \frac{\Gamma(\beta(k+1))\Gamma(1-\beta)}{\Gamma(\beta k + 1)} \hat{z}_0$$

$$= \sum_{k=0}^{\infty} \frac{A^k t^{\beta k}}{\Gamma(\beta k + 1)} \hat{z}_0 \xrightarrow{t \to 0} \hat{z}_0.$$

Consider the function

$$z_1(t) = E_{1/\beta}(At^\beta; 1) z_0 \equiv E_{1/\beta}(At^\beta) z_0,$$

where $E_{1/\beta}(At^\beta)$ is the matrix function of Mittag-Leffler.

Then

$$(D^\beta \hat{z}_1)(t) = \frac{1}{\Gamma(1-\beta)} \left[\frac{d}{dt} \left(\int_0^t (t-\tau)^{-\beta} \sum_{k=0}^\infty \frac{A^k \tau^{\beta k}}{\Gamma(\beta k+1)} d\tau \right) - t^{-\beta} \right] z_0$$

$$= \frac{1}{\Gamma(1-\beta)} \left[\sum_{k=0}^\infty \frac{A^k}{\Gamma(\beta k+1)} \frac{d}{dt} \int_0^t (t-\tau)^{-\beta} \tau^{\beta k} d\tau - t^{-\beta} \right] z_0$$

$$= \frac{1}{\Gamma(1-\beta)} \left[\sum_{k=0}^\infty \frac{A^k}{\Gamma(\beta k+1)} \frac{d}{dt} B(1-\beta, \beta k+1) \right.$$

$$\left. \times t^{1-\beta+\beta k} - t^{-\beta} \right] z_0$$

$$= \frac{1}{\Gamma(1-\beta)} \left[\sum_{k=1}^\infty A^k t^{\beta k - \beta} \frac{\Gamma(1-\beta)\Gamma(\beta k+1)}{\Gamma(\beta k+1)\Gamma(2+\beta k-\beta)} \right.$$

$$\left. \times (1-\beta+\beta k) - t^{-\beta} \right] z_0$$

$$= \frac{1}{\Gamma(1-\beta)} \left[\Gamma(1-\beta) \sum_{k=0}^\infty \frac{A^k t^{\beta(k-1)}}{\Gamma(1+\beta(k-1))} - t^{-\beta} \right] z_0$$

$$= A \sum_{k'=1}^\infty \frac{A^{k-1'} t^{\beta(k-1)'}}{\Gamma(1+\beta(k-1))} z_0 = A z_1(t).$$

Moreover, $z_1(t)$ satisfies the initial condition (2.14) since

$$\lim_{t\to\infty} z_1(t) = \lim_{t\to 0} \sum_{k=0}^\infty \frac{A^k t^{\beta k}}{\Gamma(\beta k+1)} z_0 = z_0.$$

Let us analyze function $z_2(t)$, defined by formula (2.17), and show that it satisfies equations (2.11) and (2.13) under the zero initial condition.

We have

$$(D^\beta z_2)(t) = (D^{(\beta)} z_2)(t)$$

$$= \frac{1}{\Gamma(1-\beta)} \frac{d}{dt} \int_0^t (t-\tau)^{-\beta}$$

$$\times \left(\int_0^\tau (\tau-s)^{\beta-1} E_{1/\beta}(A(\tau-s)^\beta; \beta) F(s) ds \right) d\tau.$$

$$(2.18)$$

Separately we will study the function

$$\psi(t) = \int_0^t (t-\tau)^{-\beta} \left(\int_0^\tau (t-\tau)^{\beta-1} \sum_{k=0}^\infty \frac{A^k(\tau-s)^{\beta k}}{\Gamma(k\beta+\beta)} F(s) ds \right) d\tau$$

$$= \sum_{k=0}^\infty \frac{A^k}{\Gamma(k\beta+\beta)} \int_0^t (t-\tau)^{-\beta} \left(\int_0^\tau (\tau-s)^{\beta(k+1)-1} F(s) ds \right) d\tau.$$

$$(2.19)$$

For this purpose, we consider the following integrals

$$I_k = \int_0^t \int_0^\tau (t-\tau)^{-\beta} (\tau-s)^{\beta(k+1)-1} F(s) ds d\tau$$

$$= \iint_{(\Delta_t)} (t-\tau)^{-\beta} (\tau-s)^{\beta(k+1)-1} F(s) d\tau ds,$$

$$\Delta_t = \{(s,\tau) : 0 \le s \le \tau \le t\}.$$

The latter double integral converges absolutely that allows, by virtue of Fubini theorem, to change the order of integration using Dirichle formula.

Then

$$I_k = \int_0^t \left(\int_s^t (t-\tau)^{-\beta} (\tau-s)^{\beta(k+1)-1} d\tau \right) F(s) ds$$

$$= B(1-\beta, \beta k+\beta) \int_0^t (t-s)^{\beta k} F(s) ds$$

$$= \frac{\Gamma(1-\beta)\Gamma(k\beta+\beta)}{\Gamma(k\beta+1)} \int_0^t (t-s)^{\beta k} F(s) ds.$$

$$(2.20)$$

From equalities (2.19) and (2.20), it follows that

$$\psi(t) = \Gamma(1-\beta) \sum_{k=0}^\infty \frac{A^k}{\Gamma(k\beta+1)} \int_0^t (t-s)^{\beta k} F(s) ds.$$

Since the function F(t) is measurable and bounded, then $\Psi(t)$ has the derivative almost everywhere:

$$
\begin{aligned}
\frac{d\psi}{dt} &= \Gamma(1-\beta)\left\{F(t) + \sum_{k=1}^{\infty}\frac{A^k\beta k}{\Gamma(k\beta+1)}\int_0^t (t-s)^{k\beta-1}F(s)ds\right\} \\
&= \Gamma(1-\beta)\left\{F(t) + \int_0^t \sum_{k=1}^{\infty}\frac{A^k(t-s)^{k\beta-1}}{\Gamma(\beta k)}F(s)ds\right\} \\
&= \Gamma(1-\beta)\left\{F(t) + A\int_0^t (t-s)^{\beta-1}E_{1/\beta}(A(t-s)^{\beta};\beta)F(s)ds\right\}.
\end{aligned}
$$

$$(2.21)$$

Substituting (2.21) into (2.18), we obtain the equalities

$$
D^\beta z_2 = D^{(\beta)}z_2 = Az_2 + \varphi(u,v).
$$

2.4 Fractional Conflict-Controlled Processes with Integral Block of Control

Alongside with the conflict-controlled processes (2.11), (2.12) and (2.13), (2.14), we will treat the processes differing from the above-mentioned in that the block of controls appears in them in the integral form. To be specific, in the case of Riemann-Liouville derivative, we will study the process

$$
D^\beta\hat{y} = A\hat{y} + \int_0^t (t-\tau)^{\gamma-1}\varphi(u(\tau),v(\tau))d\tau,\ 0 < \gamma < 1,\ 0 < \beta < 1,
$$

$$(2.22)$$

under the initial condition

$$
I^{1-\beta}\hat{y}|_{t=0} = \hat{y}_0
$$

$$(2.23)$$

and in the case of regularized Dzhrbashyan-Nersesyan-Caputo derivative the process

$$
D^{(\beta)}y = Ay + \int_0^t (t-\tau)^{\gamma-1}\varphi(u(\tau),v(\tau))d\tau
$$

$$(2.24)$$

under the initial condition

$$
y|_{t=0} = y_0.
$$

$$(2.25)$$

Theorem 3. *Controls of the players chosen, the solution $\hat{y}$ to the problem* (2.22), (2.23) *is given by the formula*

$$\hat{y}(t) = t^{\beta-1}E_{1/\beta}(At^{\beta};\beta)\hat{y}_0$$

$$+ \int_0^t \Gamma(\gamma)(t-\tau)^{\gamma+\beta-1}E_{1/\beta}(A(t-\tau)^{\beta};\gamma+\beta)\varphi(u(\tau),v(\tau))d\tau$$

(2.26)

and the solution $y(t)$ to the problem (2.24), (2.25) *by the formula*

$$y(t) = E_{1/\beta}(At^{\beta};1)y_0$$

$$+ \int_0^t \Gamma(\gamma)(t-\tau)^{\gamma+\beta-1}E_{1/\beta}(A(t-\tau)^{\beta};\gamma+\beta)\varphi(u(\tau),v(\tau))d\tau.$$

(2.27)

Proof. Taking into account the reasoning, presented in the proof of the Theorem 3, it suffices to show that the function

$$y_2(t) = \int_0^t \Gamma(\gamma)(t-\tau)^{\gamma+\beta-1}E_{1/\beta}(A(t-\tau)^{\beta};\gamma+\beta)\varphi(u(\tau),v(\tau))d\tau$$

is a solution of equations (2.26), (2.27) under the zero initial condition.

After application of formulas (2.15), (2.16) to systems (2.22), (2.24) with the zero initial conditions, we have

$$\hat{y}(t) \equiv y(t) = \int_0^t (t-\tau)^{\beta-1}E_{1/\beta}(A(t-\tau)^{\beta};\beta)\int_0^\tau (\tau-s)^{\gamma-1}F(s)dsd\tau$$

$$= \int_0^t \int_s^t (t-\tau)^{\beta-1}E_{1/\beta}(A(t-\tau)^{\beta};\beta)(\tau-s)^{\gamma-1}d\tau F(s)ds.$$

Let us calculate the integral

$$I(t-s) = \int_0^t (t-\tau)^{\beta-1}E_{1/\beta}(A(t-\tau)^{\beta};\beta)(\tau-s)^{\gamma-1}d\tau$$

$$\overset{\tau-s=\hat{\tau}}{=} \int_0^{t-s} (t-s-\hat{\tau})^{\beta-1}E_{1/\beta}(A(t-s-\hat{\tau})^{\beta};\beta)\hat{\tau}^{\gamma-1}d\hat{\tau}$$

$$= \int_0^{t-s} \hat{\tau}^{\beta-1}E_{1/\beta}(A\hat{\tau}^{\beta};\beta)(t-s-\hat{\tau})^{\gamma-1}d\hat{\tau}.$$

From here, we eventually obtain

$$I(t - s) = \Gamma(\gamma) \int_0^t (t - s)^{\gamma + \beta - 1} E_{1/\beta}(A(t - s)^\beta; \gamma + \beta),$$

whence follows that

$$\hat{y}(t) = y(t) = \Gamma(\gamma) \int_0^t (t - s)^{\gamma + \beta - 1} E_{1/\beta}(A(t - s)^\beta; \gamma + \beta) F(s) ds.$$

Remark 1. *If* $\gamma + \beta \geq 1$, *then the solutions* (2.26), (2.27) *turn out to be absolutely continuous functions [17], having bounded derivatives almost everywhere.*

Remark 2. *Equations* (2.22), (2.24) *can be considered as incorporating integrals, which have arbitrary* τ-*summable kernels.*

Thus, for the game problems with the fractional derivatives of Riemann-Liouville and Dzhrbashyan-Nersesyan-Caputo of the types (2.11)–(2.12), (2.13)–(2.14), (2.22)–(2.23), (2.24)–(2.25) the solutions can be presented by formulas (2.15), (2.16), (2.26) and (2.27), which are specific cases of representation (2.1).

The above-outlined general method can be applied for solution to each of the mentioned game problems.

2.5 Specific Case of Simple Matrix

For illustration of the method, we now analyze various specific cases, in which solution can be obtained in an analytic form.

In a sequel, for the brevity of exposition and the unification of notions, we will distinguish the four above -outlined problems by assigning to their parameters the values of indices $i, j : i = 1, 2, j = 1, 2$. Then a trajectory $z_{11}(t)$ corresponds to the process with Riemann-Liouville derivative and conventional block of control (2.11), while $z_{12}(t)$ to that with the integral block of control (2.22). In the turn, a trajectory $z_{21}(t)$ corresponds to the process with the regularized Dzhrbashyan-Nersesyan-Caputo derivative and the block of control in conventional form, and $z_{22}(t)$ to that with the integral block of control (2.24).

Thus, we have the four processes

$$z_{ij}(t) = g_{ij}(t) + \int_0^t \Omega_{ij}(t, \tau)\varphi(u(\tau), v(\tau))d\tau, \quad i = 1, 2, j = 1, 2,$$

$$(2.28)$$

where

$$g_{11}(t) = G_{11}(t)\hat{z}_0, G_{11}(t) = t^{\beta-1}E_{1/\beta}(At^\beta; \beta),$$
$$\Omega_{11}(t, \tau) = (t - \tau)^{\beta-1}E_{1/\beta}(A(t - \tau)^\beta; \beta),$$
$$g_{12}(t) = G_{12}(t)\hat{y}_0, G_{12}(t) = t^{\beta-1}E_{1/\beta}(At^\beta; \beta),$$
$$\Omega_{12}(t, \tau) = \Gamma(\gamma)(t - \tau)^{\gamma+\beta-1}E_{1/\beta}(A(t - \tau)^\beta; \gamma + \beta),$$
$$g_{21}(t) = G_{21}(t)z_0, G_{21}(t) = E_{1/\beta}(At^\beta; 1),$$
$$\Omega_{21}(t, \tau) = (t - \tau)^{\beta-1}E_{1/\beta}(A(t - \tau)^\beta; \beta),$$
$$g_{22}(t) = G_{22}(t)y_0, G_{22}(t) = E_{1/\beta}(At^\beta; 1),$$
$$\Omega_{22}(t, \tau) = \Gamma(\gamma)(t - \tau)^{\gamma+\beta-1}E_{1/\beta}(A(t - \tau)^\beta; \gamma + \beta).$$

Let

$$A = \lambda E, \quad \varphi(u, v) = u - v, \quad M^* = \{0\}, \quad U = aS, a > 1, V = S,$$

where λ is an integer and S is the unit ball centered at the origin. Then, the orthoprojector π appears as the operator of an identical transformation, defined by the matrix. All the matrix functions $G_{ij}(t)$ and $\Omega_{ij}(t, \tau)$ have the forms

$$G_{ij}(t) = \hat{g}_{ij}(t)E, \Omega_{ij}(t, \tau) = \omega_{ij}(t, \tau)E, \quad i = 1, 2, \quad j = 1, 2,$$

where $\hat{g}_{ij}(t)$ and $\omega_{ij}(t, \tau)$ are scalar functions and

$$E_\rho(B; \mu) = E_\rho(\lambda; \mu)E,$$

where $E_\rho(\lambda; \mu)$ is the generalized scalar function of Mittag-Leffler [17].
Then

$$W_{ij}(t, \tau, v) = \omega_{ij}(t, \tau)(aS - v)$$
$$W_{ij}(t, \tau) = |\omega_{ij}(t, \tau)|(a - 1)S.$$

Consequently, Pontryagin condition holds if $a \geq 1$.
Set $\gamma_{ij}(t, \tau) \equiv 0$. Then

$$\xi_{ij}(t, g_{ij}(t), \gamma_{ij}(t, \tau)) = g_{ij}(t) = g_{ij}(\hat{t})z_{ij}^0, \quad z_{ij}^0 \neq 0,$$

and

$$\alpha_{ij}(t, \tau, v) = \sup\{\alpha \geq 0 : \alpha\hat{g}_{ij}(t)z_{ij}^0 \in \omega_{ij}(t, \tau)(aS - v)\}$$

is the greatest root of the square equation for α:

$$\|\omega_{ij}(t,\tau)v - \alpha\hat{g}_{ij}(t)z_{ij}^0\| = |\omega_{ij}(t,\tau)|a.$$

From here, it follows that

$$\alpha_{ij}(t,\tau,v) = \frac{(v_0, q) + \sqrt{(v_0, q)^2 + \|q\|^2(a_0^2 - \|v_0\|^2)}}{\|q\|^2},$$

where $v_0 = \omega_{ij}(t,\tau)v, q = \hat{g}_{ij}(t)z_{ij}^0, a_0 = |\omega_{ij}(t,\tau)|a$.

It should be noted that $\hat{g}_{ij}(t) \neq 0$ up to the instant of the game termination. This function vanishing points to the possibility for the game termination with the help of the first direct method. It is evident that

$$\min_{\|v\| \le 1} \alpha_{ij}(t,\tau,v) = \frac{|\omega_{ij}(t,\tau)|(a-1)}{\|\hat{g}_{ij}(t)z_{ij}^0\|},$$

where the minimum is furnished by the element

$$v_{ij}(t,\tau) = -sign\{\hat{g}_{ij}(t)\omega_{ij}(t,\tau)\}\frac{z_{ij}^0}{\|z_{ij}^0\|}.$$

Then the time of game termination appears as the least root of the equation

$$\int_0^t \frac{(a-1)|\omega_{ij}(t,\tau)|}{|\hat{g}_{ij}(t)|\|z_{ij}^0\|}d\tau = 1,$$

as functions $\omega_{ij}(t,\tau)$ are continuous in t.

Let us introduce functions

$$\Phi_{ij}(t) = \int_0^t \frac{|\omega_{ij}(t,\tau)|}{|\hat{g}_{ij}(t)|}d\tau.$$

Then the time of the game termination can be given by the formula

$$T_{ij}(z_{ij}^0, 0) = \min\{t \ge o : \Phi_{ij}(t) \ge \frac{\|z_{ij}^0\|}{a-1}\}, \tag{2.29}$$

where functions $\Phi_{ij}(t)$ have the forms

$$\Phi_{11}(t) = \int_0^t |\tau^{\beta-1}E_{1/\beta}(\lambda\tau^\beta; \beta)|d\tau/|t^{\beta-1}E_{1/\beta}(\lambda t^\beta; \beta)|,$$

$$\Phi_{12}(t) = \Gamma(\gamma) \int_0^t |\tau^{\gamma+\beta-1} E_{1/\beta}(\lambda\tau^\beta; \beta+\gamma)| d\tau / |t^{\beta-1} E_{1/\beta}(\lambda t^\beta; \beta)|,$$

$$\Phi_{21}(t) = \int_0^t |\tau^{\beta-1} E_{1/\beta}(\lambda\tau^\beta; \beta)| d\tau / E_{1/\beta}(\lambda t^\beta; 1),$$

$$\Phi_{22}(t) = \Gamma(\gamma) \int_0^t |\tau^{\gamma+\beta-1} E_{1/\beta}(\lambda\tau^\beta; \beta+\gamma)| d\tau / E_{1/\beta}(\lambda t^\beta; 1).$$

To determine whether (or not) the time of the game termination $T_{ij}(z_{ij}^0, 0)$ is finite, an asymptotic representation of the generalized scalar function of Mittag-Leffler plays a key role. We take interest in specification of formulas (2.23) and (2.24) from [25, p. 134] giving such representation $E_\rho(x; \mu)$ for real x; $\rho > \frac{1}{2}$ and arbitrary μ.

From these formulas, it follows that for positive x

$$E_\rho(x; \mu) = \rho x^{\rho(1-\mu)} e^{x^\rho} - \sum_{k=1}^\rho \frac{x^{-k}}{\Gamma(\mu - k\rho^{-1})} + O(|x|^{-1-\rho}), \qquad (2.30)$$

and for negative x,

$$E_\rho(x; \mu) = -\sum_{k=1}^\rho \frac{x^{-k}}{\Gamma(\mu - k\rho^{-1})} + O(|x|^{-1-\rho}). \qquad (2.31)$$

As seen from asymptotic representations (2.30) and (2.31), in our example, it is reasonable to analyze two cases: $\lambda > 0$ and $\lambda < 0$.

Let $\lambda > 0$. Then the generalized functions of Mittag-Leffler, appearing in the formulas for $\Phi_{ij}(t)$, are positive. From this and from formula (1.15) [25, p. 120], here having the form

$$\int_0^x E_\rho(\lambda x^{1/\beta}; \mu) \tau^{\mu-1} d\tau = x^\mu E_\rho(\lambda x^{1/\beta}; \mu+1), (\mu > 0), \lambda \in R,$$

we infer formulas for functions $\Phi_{ij}(t)$

$$\Phi_{11}(t) = t^\beta E_{1/\beta}(\lambda t^\beta; \beta+1)/t^{\beta-1} E_{1/\beta}(\lambda t^\beta; \beta),$$

$$\Phi_{12}(t) = \Gamma(\gamma) t^{\gamma+\beta} E_{1/\beta}(\lambda t^\beta; \beta+\gamma+1)/t^{\beta-1} E_{1/\beta}(\lambda t^\beta; \beta),$$

$$\Phi_{21}(t) = t^\beta E_{1/\beta}(\lambda t^\beta; \beta+1)/E_{1/\beta}(\lambda t^\beta; 1),$$

$$\Phi_{22}(t) = \Gamma(\gamma) t^{\gamma+\beta} E_{1/\beta}(\lambda t^\beta; \beta+\gamma+1)/E_{1/\beta}(\lambda t^\beta; 1). \qquad (2.32)$$

Set $\rho = \frac{1}{\beta}$, $x = \lambda t^{\beta}$ in formula (2.30). It should be noted that since $\beta \in (0,1)$, then $\rho \in (1, \infty)$ and therefore $\rho > \frac{1}{2}$. From this it follows, the asymptotic representation

$$E_{1/\beta}(\lambda t^{\beta}; \mu) = \frac{1}{\beta}(\lambda t^{\beta})^{\frac{1}{\beta}(1-\mu)}e^{(\lambda t^{\beta})^{1/\beta}} - \sum_{k=1}^{\rho}\frac{(\lambda t^{\beta})^{-k}}{\Gamma(\mu - k\beta)} + O((t^{\beta})^{-1-\rho})$$

$$= \frac{1}{\beta}\lambda^{\frac{1}{\beta}(1-\mu)}t^{1-\mu}e^{\lambda^{1/\beta}t} + \cdots . \tag{2.33}$$

Using this representation, the following relations can be deduced

$$t^{\beta}E_{1/\beta}(\lambda t^{\beta}; \beta + 1) = \frac{1}{\beta}\lambda^{\frac{1}{\beta}(1-(\beta+1))}t^{\beta}t^{-\beta}e^{\lambda^{1/\beta}} + \cdots$$

$$= \frac{1}{\beta}\lambda^{-1}e^{\lambda^{1/\beta}t} + \cdots$$

$$t^{\beta-1}E_{1/\beta}(\lambda t^{\beta}; \beta) = \frac{1}{\beta}t^{\beta-1}\lambda^{\frac{1}{\beta}(1-\beta)}t^{1-\beta}e^{\lambda^{1/\beta}t} + \cdots$$

$$= \frac{1}{\beta}\lambda^{\frac{1}{\beta}-1}e^{\lambda^{1/\beta}t} + \cdots$$

$$\Gamma(\gamma)t^{\gamma+\beta}E_{1/\beta}(\lambda t^{\beta}; \beta + \gamma + 1) = \Gamma(\gamma)\frac{1}{\beta}t^{\gamma+\beta}\lambda^{\frac{1}{\beta}(1-(\gamma+\beta+1))}$$

$$\times t^{1-(\gamma+\beta+1)}e^{\lambda^{1/\beta}t} + \cdots$$

$$= \frac{\Gamma(\gamma)}{\beta}\lambda^{-\frac{\gamma+\beta}{\beta}}e^{\lambda^{1/\beta}t} + \cdots$$

$$E_{1/\beta}(\lambda t^{\beta}; 1) = \frac{1}{\beta}\lambda^{\frac{1}{\beta}(1-1)}t^{(1-1)}e^{\lambda^{1/\beta}t} + \cdots$$

$$= \frac{1}{\beta}e^{\lambda^{1/\beta}t} + \cdots . \tag{2.34}$$

From formulas (2.32) and asymptotic representations (2.34), we obtain the following equalities

$$\lim_{t\to\infty} \Phi_{11}(t) = \frac{\frac{1}{\beta}\lambda^{-1}}{\frac{1}{\beta}\lambda^{\frac{1}{\beta}-1}} = \lambda^{-\frac{1}{\beta}},$$

$$\lim_{t\to\infty} \Phi_{12}(t) = \frac{\frac{\Gamma(\gamma)}{\beta}\lambda^{-\frac{\gamma+\beta}{\beta}}}{\frac{1}{\beta}\lambda^{\frac{1}{\beta}-1}} = \Gamma(\gamma)\lambda^{-\frac{\gamma+1}{\beta}},$$

$$\lim_{t\to\infty} \Phi_{21}(t) = \frac{\frac{1}{\beta}\lambda^{-1}}{\frac{1}{\beta}} = \lambda^{-1},$$

$$\lim_{t\to\infty} \Phi_{22}(t) = \frac{\frac{\Gamma(\gamma)}{\beta}\lambda^{-\frac{\gamma+\beta}{\beta}}}{\frac{1}{\beta}} = \Gamma(\gamma)\lambda^{-\frac{\gamma+\beta}{\beta}}.$$

Thus, when $\lambda > 0$, the times under study are finite if the inequalities hold, respectively:

$$T_{11}(z_{11}^0, 0) \quad \text{if } \lambda^{-\frac{1}{\beta}} > \frac{\|z_{11}^0\|}{a-1},$$

$$T_{12}(z_{12}^0, 0) \quad \text{if } \Gamma(\gamma)\lambda^{-\frac{\gamma+1}{\beta}} > \frac{\|z_{12}^0\|}{a-1},$$

$$T_{21}(z_{21}^0, 0) \quad \text{if } \lambda^{-1} > \frac{\|z_{21}^0\|}{a-1},$$

and, finally,

$$T_{22}(z_{22}^0, 0) \quad \text{if } \Gamma(\gamma)\lambda^{-\frac{\gamma+\beta}{\beta}} > \frac{\|z_{22}^0\|}{a-1}.$$

Let us consider the case when $\lambda < 0$. Set in formula (2.31) $\rho = \frac{1}{\beta}$, $x = \lambda t^\beta$. Then

$$E_{1/\beta}(\lambda t^\beta; \mu) = -\sum_{k=1}^{\rho} \frac{\lambda^{-k}t^{-k\beta}}{\Gamma(\mu - k\beta)} + O(t^{-(1+\rho)\beta}).$$

Using this asymptotic representation, we deduce

$$t^\beta E_{1/\beta}(\lambda t^\beta; \beta + 1)$$

$$= t^\beta \left[-\sum_{k=1}^{\rho} \frac{\lambda^{-k}t^{-k\beta}}{\Gamma(\beta + 1 - k\beta)} + O(t^{-\beta(1+\rho)}) \right]$$

$$= t^\beta \left[-\frac{\lambda^{-1}t^{-\beta}}{\Gamma(1)} - \frac{\lambda^{-2}t^{-2\beta}}{\Gamma(1-\beta)} - \cdots \right] = -\lambda^{-1} - \frac{\lambda^{-2}t^{-\beta}}{\Gamma(1-\beta)} - \cdots$$

$t^{\beta-1} E_{1/\beta}(\lambda t^{\beta}; \beta)$

$$= t^{\beta-1} \left[-\sum_{k=1}^{\rho} \frac{\lambda^{-k} t^{-k\beta}}{\Gamma(\beta - k\beta)} + \cdots \right]$$

$$= t^{\beta-1} \left[-\sum_{k=2}^{\rho} \frac{\lambda^{-k} t^{-k\beta}}{\Gamma(\beta - k\beta)} + \cdots \right] = t^{\beta-1} \left[-\frac{\lambda^{-2} t^{-2\beta}}{\Gamma(-\beta)} - \cdots \right]$$

$$= -\frac{\lambda^{-2} t^{-2\beta}}{\Gamma(-\beta)} - \cdots$$

$\Gamma(\gamma) t^{\gamma+\beta} E_{1/\beta}(\lambda t^{\beta}; \gamma + \beta + 1)$

$$= \Gamma(\gamma) t^{\gamma+\beta} \left[-\sum_{k=1}^{\rho} \frac{\lambda^{-k} t^{-k\beta}}{\Gamma(\gamma + \beta + 1 - k\beta)} + \cdots \right]$$

$$= -\Gamma(\gamma) \frac{\lambda^{-1} t^{-\gamma}}{\Gamma(\gamma + 1)} - \cdots$$

$E_{1/\beta}(\lambda t^{\beta}; 1)$

$$= -\sum_{k=1}^{\rho} \frac{\lambda^{-k} t^{-k\beta}}{\Gamma(1 - k\beta)} + O(t^{-\beta(1+\rho)}) = -\frac{\lambda^{-1} t^{-\beta}}{\Gamma(1 - \beta)} - \cdots \qquad (2.35)$$

Let us analyze an asymptotic behavior of functions $\Phi_{ij}(t)$, given by formulas (2.32) in the case when $\lambda < 0$. Then functions (2.35) are not of necessity positive. However, using the inequality

$$\left| \int_0^t f(\tau) d\tau \right| \leq \int_0^t |f(\tau)| d\tau$$

for an arbitrary summable function $f(\tau)$, the asymptotic representations (2.35) and formulas (2.32), one can easily infer that

$$\Phi_{ij}(t) \underset{t \to \infty}{\to} \infty, \forall i, j = 1, 2.$$

Thus, the times $T_{ij}(z_{ij}^0, 0)$ given by formula (2.29) are finite for any $z_{ij}^0, i, j = 1, 2$. This means that in the case when $\lambda < 0$, the process under study is completely conflict controllable as far as each of the problems (2.11)–(2.12), (2.13)–(2.14), (2.22)–(2.23), (2.24)–(2.25) is concerned.

Let $\lambda = 0$. Then, taking into account formulas (2.32) for the functions $\Phi_{ij}(t), i, j = 1, 2$, together with expression (2.29), one can calculate the precise values of the termination times for the games under study, namely:

$$T_{11}(z_{11}^0, 0) = \beta \frac{\|z_{11}^0\|}{a - 1},$$

$$T_{21}(z_{21}^0, 0) = \left[\Gamma(\beta + 1)\frac{\|z_{21}^0\|}{a - 1}\right]^{\frac{1}{\beta}},$$

$$T_{12}(z_{12}^0, 0) = \left[\frac{\beta + \gamma}{B(\gamma + \beta)} \frac{\|z_{12}^0\|}{a - 1}\right]^{\frac{1}{\gamma + 1}},$$

$$T_{22}(z_{22}^0, 0) = \left[\frac{\Gamma(\beta + \gamma + 1)}{\Gamma(\gamma)} \frac{\|z_{22}^0\|}{a - 1}\right]^{\frac{1}{\gamma + \beta}}.$$

2.6 Conclusion

On the basis of the method of resolving functions, we derive general sufficient conditions for a quasi-linear conflict-controlled process trajectory approaching a cylindrical terminal set in a finite guaranteed time. To do this, the apparatus of set-valued mappings and topological theorems on measurable choice are applied. The result is specified for the systems with fractional Riemann-Liouville derivatives and regularized derivatives. In so doing, the matrix functions of Mittag-Leffler are used for solution presentation.

Theoretical results are applied for solving the game problem with simple matrix and the ball-shaped control domains. Using the Mittag-Leffler scalar function representation, an explicit form of resolving function is found and conditions for the guaranteed time finiteness are deduced, the latter expressed through the process parameters, the initial states and the order of fractional derivative.

References

[1] L. S. Pontryagin, 'Selected scientific papers', M.: Nauka, 1988 (in Russian).
[2] N. N. Krasovskii, 'Game Problems on the Encounter of Motions', M.: Nauka, 1970.
[3] O. Hajek, 'Pursuit games', New York: Academic Press, 1975, v. 12.

[4] A. A. Chikrii, 'Conflict-controlled processes', Boston; London; Dordrecht: Springer Science and Busines Media, 2013.

[5] N. N. Krasovskii, A. I. Subbotin, 'Positional differential Games', Nauka, Moscow, 1974.

[6] A. A. Chikrii, 'Quasilinear controlled processes under conflict dynamical systems', 2, J. Math. Sci., vol. 80, no. 1, 1996.

[7] A. A. Chikrii, 'Differential games with many pursuers', Mathematical contol theory, Banach center publ., PWN, Warsaw, vol.14.

[8] N. L. Grigorenko, 'Mathematical methods of control for several dynamic processes', Izdat. Gos. Univ., Moscow, 1990.

[9] B. N. Pschenitchnyi, A. A. Chikrii, J. S. Rappoport, 'Group pursuit in Differential games', Opt. invest. Stat., Germany, 1982.

[10] A. A. Chikrii, 'Minkowskii functionals in pursuit theory', Russian Akad. Sci. Dokl. Math., vol. 47, no. 2, 1993.

[11] A. A. Chikrii, 'An Analitical Method in Dynamic Pursuit Games', Proceedings of the Steklov Institute of Mathematics, vol. 271, 2010.

[12] A. A. Chikrii, 'Optimization of Game Interaction of Fractional-Order Controlled Systems', J. "Optimization Methods and Software", Taylor and Francis, Oxfordshire, UK, vol. 3, no. 1, 2008.

[13] A. A Chikrii, V. K. Chikrii, 'Image Structure of Multivalued Mapping in Game Problems of motion control', Journal of Automation and Information Sciences, 48, no. 3, 2016.

[14] A. D. Joffe, V. M. Tikhomirov, 'Theory of extremal problems', North Holland, Amsterdam, 1979.

[15] F. H. Clarke, 'Optimization and nonsmooth analysis', J. Wiley, New York, 1983.

[16] J.-P. Aubin, H. Frankowska, 'Set-valued analysis', Boston; Basel; Berlin: Birkhauser, 1990.

[17] S. G. Samko, A. A. Kilbas, O. I. Marychev, 'Integrals and derivatives of fractional order and some their applications', Minsk, 1987.

[18] M. M. Dzhrbashyan, A. B. Nersesyan, 'Fractional derivatives and Cauchy problem for differential games of fractional order', Izv. Akad. Nauk. Arm. SSR, vol. 3, no. 1, 1968.

[19] M. Caputo, 'Linear model of dissipation whose Q is almost frequency independent', Geophys. J. R. Astr. Soc., 13, 1967.

[20] A. A. Kilbas, H. M. Srivastava, J. J. Trujillo, 'Theory and applications of fractional differential equations', Amsterdam: Elsevier, 2006.

[21] J. Podlubny, 'Fractional differential equations', New York: Acad. Press, 1999.

[22] G. T. Chikrii, 'Principle of time stretching in evolutionary games of approach', Journal of Automation and Information Science, 48, no. 5, 2016.

[23] G. T. Chikrii, 'Search for a fixed target by a moving object', Journal of Applied Mathematics and Mechanics, 48, no. 4, 1984.

[24] T. Rockafeller, 'Convex analysis', Princeton Univ. Press, New York, 1970.

[25] M. M. Dzhrbashyan, 'Integral transformations and representations of functions in complex domain', Nauka, Moscow, 1996.

3

Identification and Control Automation of Cognitive Maps in Impulse Process Mode

Vyacheslav Gubarev[1,*], Victor Romanenko[2] and Yurii Miliavskyi[2]

[1] Space research institute, National academy of sciences of Ukraine
and state space agency of Ukraine, 40 Glushkov av., Kyiv, 03680, Ukraine
[2] "Institute for Applied System Analysis" of National Technical University
of Ukraine "Igor Sikorsky Kyiv Polytechnic Institute", 37a Peremohy av.,
Kyiv, 03056, Ukraine
*Corresponding Author: v.f.gubarev@gmail.com

This chapter considers identification and control problems of cognitive maps. Identification problem is solved in several statements depending on whether the CM nodes coordinates are measured precisely or with noise, whether measurement matrix is well-conditioned or ill-conditioned, whether there is additional information about relations between nodes. For each of the cases, identification method is developed and simulated for the CM of a commercial bank. Control problem is solved for the case when two types of controls are available – via direct varying CM nodes' resources and via CM edges' weights varying. A closed-loop control system for the control law developed for this case is simulated for the CM of IT company HR management.

3.1 Introduction

Cognitive maps (CM) are used for modeling of high dimensional complex systems of different nature. CM is a structural scheme of cause-and-effect relations between complex system's components (coordinates, concepts). From a mathematical standpoint, CM is a weighted directed graph with nodes (vertices) representing a complex system's coordinates and edges describing relations between these coordinates [1, 2]. CMs are usually designed by experts. Weighted edges of the directed graph allow qualitative description of interrelations between concepts of a complex system and quantitative reflection of impact of each coordinate on all others.

When the complex system evolves, CM coordinates vary in time affected by different disturbances. Let each CM node l_i has a value $Y_i(k)$ at discrete time moment $k = 0, 1, 2, \ldots$. Value $Y_i(k+1)$ at the next sampling moment is determined by the value $Y_i(k)$ and information about other coordinates l_j adjacent to l_i increasing or decreasing their values at time k. Increment $P_j(k)$ of the coordinate l_j at time k given as difference $P_j(k) = Y_j(k) - Y_j(k-1)$ is called "impulse" according to [1]. Impulse $P_j(k)$ that entered node l_j is propagated to other nodes of the CM, being either increased or decreased. Disturbances propagation rule in CM chains is defined by the difference equation [2, 5]

$$Y_i(k+1) = Y_i(k) + \sum_{j=1}^{n} a_{ij} P_j(k), \quad i = 1, 2, \ldots, n, \tag{3.1}$$

where a_{ij} is weight of the digraph's edge connecting the j-th node with the i-th one. If there is no edge from l_j to l_i, the respective coefficient $a_{ij} = 0$.

The rule (3.1) which describes an impulse process in a CM is usually formulated as a first-order difference equation in nodes coordinates increments [1]

$$\Delta Y_i(k+1) = \sum_{j=1}^{n} a_{ij} \Delta Y_j(k), \tag{3.2}$$

where $\Delta Y_i(k) = Y_i(k) - Y_i(k-1)$, $i = 1, 2, \ldots, n$. In a vectorized form, (3.2) is written as

$$\Delta \bar{Y}(k+1) = A \Delta \bar{Y}(k), \tag{3.3}$$

where A is an incidence matrix of the CM, $\Delta \bar{Y}(k)$ is a vector of increments of nodes Y_i, $i = 1, 2, \ldots, n$.

Representing CM model as dynamic equations in the state space (3.3) allows intentional influencing of the processes using external controls affecting system modes. For example, if matrix A in (3.3) has an unstable eigenvalues state, feedback can stabilize an underlying process. But control is effective if the model structure and its parameters provide adequate description of the processes in real system for all inputs. Objects described by CM are usually such that there are no other ways of building models except analyzing data obtained during observation or experimental research.

CM structure and weights a_{ij} are usually specified by experts based on cause-and-effects relations' analysis of a complex system. But it is always doubted whether a model based on expert estimates is adequate. Moreover, when a complex system evolves coefficients a_{ij} can greatly vary depending on changing the degree of interrelations between the nodes, critical and conflict cases, human factor, changes in social and political relations etc. In [3, 4], recursive least squares method is utilized to estimate coefficients a_{ij} of the matrix A, given the disturbances in (3.1) are discrete white noise.

In [3–7], CM impulse process control is automated by means of external control vector design, which varies CM node coordinates in closed-loop systems online based on known control theory methods. For this purpose, equation of forced motion in CM impulse process was formulated as follows:

$$\Delta Y_i(k+1) \quad \sum_{j=1}^{n} a_{ij}\Delta Y_j(k) + b_i\Delta u_i(k), \quad i \quad 1,\ldots,n, \qquad (3.4)$$

where $\Delta u_i(k) \quad u_i(k) - u_i(k-1)$ are control input increments that affect CM nodes directly and are implemented through varying available resources of $Y_i(k)$ coordinates. In the vector form, (3.4) can be written as

$$\Delta\bar{Y}(k+1) \quad A\Delta\bar{Y}(k) + B\Delta\bar{u}(k), \qquad (3.5)$$

where elements of control matrix B corresponding to controls in vector $\bar{u}$ are equal to 1.

In [8], a new principle of CM impulse process control design in a closed-loop system is considered. It is based on varying weights a_{ij} for implementing control inputs. This is possible when the degree of influence of one CM node on another $a_{ij}(k)$ can be changed at the k-th sampling period. For this purpose, forced motion CM impulse process model is presented as follows:

$$\Delta\bar{Y}(k+1) \quad A\Delta\bar{Y}(k) + L(k)\Delta\bar{a}(k), \qquad (3.6)$$

where matrix $L(k)$ is composed of measured coordinates $Y_j(k)$, which affect coordinates $Y_i(k+1)$ via edges with variable weights $\Delta a_{ij}(k) \quad a_{ij}(k) - a_{ij}(k-1)$, $\Delta\bar{a}(k)$ is a vector composed of all non-zero $\Delta a_{ij}(k)$.

Control inputs designed as discrete control laws in [3–7] are implemented by a decision-maker who operates as an actuating mechanism in a closed-loop control system.

The present chapter develops new approaches and methods of solving both problems – identification and control of CM. Identification method to be considered is based on the subspace method (4SID) widely used for identification of multivariable systems in state space. The 4SID methods have their origin in state-space realization theory as developed in 1960's; a classic contribution is by Ho and Kalman in 1966 [9]. Here, the main idea and subspace method technique will be adapted to CM conditions and peculiarities. Results and effectiveness of the suggested identification method are illustrated by numerous simulation experiments. Full information case, when all nodes $Y_i, i = 1, 2, \ldots, n$ are measurable, is considered. The incidence matrix A is initially completely filled with unknown coefficients.

After having solved the identification problem, an automated control system for CM impulse process should be designed. The number of possible controls $\dim \Delta \bar{u}$ in (3.5) or $\dim \Delta \bar{a}$ in (3.6) can be much smaller than number of nodes $\dim \Delta \bar{Y}$ that impairs accuracy and response speed of the control system. Thus, in the present chapter, we consider combined control vector in a closed-loop system, i.e., we will vary simultaneously resources of some of the CM nodes $\Delta \bar{u}$ and weights of some of the edges $\Delta \bar{a}$. Hence, a combined control vector should be composed of two parts $\Delta \bar{U}(k) = [\Delta \bar{u}^{\mathrm{T}}(k) \, \Delta \bar{a}^{\mathrm{T}}(k)]^{\mathrm{T}}$.

3.2 Cognitive Maps Identification with Full Information

CM identification with full information signifies that model dimension, testing signal, and measured outputs $Y_i(k)$ for all nodes are known at each time moment k. Difference equation (3.4) can be written as follows:

$$\sum_{j=1}^{n} a_{ij} \Delta Y_j(k) = \Delta q_i(k), \quad i = 1, \ldots, n \qquad (3.7)$$

or in the vector form

$$A \Delta \bar{Y}(k) = \Delta \bar{Q}(k), \qquad (3.8)$$

where $\Delta q_i(k) = \Delta Y_i(k+1) - b_i \Delta u_i(k)$, $\Delta \bar{Y}(k)$ is a vector of nodes increments $Y_i, i = 1, \ldots, n$; $\Delta \bar{Q}(k) = \Delta \bar{Y}(k+1) - B \Delta \bar{u}(k)$.

3.2.1 Parametric Identification in Deterministic Case

Impulse process (3.8) can be represented as the following sequence of equations sets for time moments k $0, 1, \ldots, K - 1$

$$A \cdot \Delta \bar{Y}(0) \quad \Delta \bar{Q}(0),$$
$$A \cdot \Delta \bar{Y}(1) \quad \Delta \bar{Q}(1),$$

$$\vdots$$

$$A \cdot \Delta \bar{Y}(K - 1) \quad \Delta \bar{Q}(K - 1). \tag{3.9}$$

For known outputs $\Delta \bar{Y}$ and inputs $\Delta \bar{Q}$, identification problem for matrix A can be solved from (3.9). In a deterministic case, when signals $\{\Delta \bar{Y}(k)\}$ and $\{\Delta \bar{Q}(k)\}$ are given precisely, we can select any n independent equations sets, e.g., the first n sets in the sequence (3.9). It is easy to obtain n independent linear algebraic equations sets for finding the matrix A elements in the first, second, $\ldots$, n-th rows. Assuming n K, these equations sets are represented as follows:

$$\Delta Y_1(0)a_{11} + \Delta Y_2(0)a_{12} + \ldots + \Delta Y_n(0)a_{1n} \quad \Delta q_1(0),$$
$$\Delta Y_1(1)a_{11} + \Delta Y_2(1)a_{12} + \ldots + \Delta Y_n(1)a_{1n} \quad \Delta q_1(1),$$

$$\vdots \tag{3.10}$$

$$\Delta Y_1(n-1)a_{11} + \Delta Y_2(n-1)a_{12} + \ldots + \Delta Y_n(n-1)a_{1n}$$
$$\Delta q_1(n-1);$$

$$\Delta Y_1(0)a_{21} + \Delta Y_2(0)a_{22} + \ldots + \Delta Y_n(0)a_{2n} \quad \Delta q_2(0),$$
$$\Delta Y_1(1)a_{21} + \Delta Y_2(1)a_{22} + \ldots + \Delta Y_n(1)a_{2n} \quad \Delta q_2(1),$$

$$\vdots \tag{3.11}$$

$$\Delta Y_1(n-1)a_{21} + \Delta Y_2(n-1)a_{22} + \ldots + \Delta Y_n(n-1)a_{2n}$$
$$\Delta q_2(n-1);$$

$$\cdots\cdots\cdots\cdots\cdots\cdots\cdots\cdots\cdots\cdots\cdots\cdots$$

$$\Delta Y_1(0)a_{n1} + \Delta Y_2(0)a_{n2} + \ldots + \Delta Y_n(0)a_{nn} \quad \Delta q_n(0),$$
$$\Delta Y_1(1)a_{n1} + \Delta Y_2(1)a_{n2} + \ldots + \Delta Y_n(1)a_{nn} \quad \Delta q_n(1),$$

$$\vdots \tag{3.12}$$

$$\Delta Y_1(n-1)a_{n1} + \Delta Y_2(n-1)a_{n2} + \ldots + \Delta Y_n(n-1)a_{nn}$$
$$\Delta q_n(n-1).$$

The systems (3.10), (3.11), (3.12) can be unified and written as

$$\Delta Y \bar{a}_i \quad \Delta \bar{q}_i, \tag{3.13}$$

where vector $\Delta\bar{q}_i$ is given as

$$\Delta\bar{q}_i \quad (\Delta q_i(0),\ \Delta q_i(1),\ \ldots,\ \Delta q_i(n-1)) \tag{3.14}$$

and $\bar{a}_i^{\mathrm{T}}\quad(a_{i1}\ a_{i2}\ \ldots\ a_{in})$ is the vector containing elements of the i-th row of matrix A. Matrix ΔY is composed of the CM nodes' discrete measurements from zero to $(n-1)$-st sampling periods as follows:

$$\Delta Y \quad \begin{pmatrix} \Delta Y_1(0) & \Delta Y_2(0) & \cdots & \Delta Y_n(0) \\ \Delta Y_1(1) & \Delta Y_2(1) & \cdots & \Delta Y_n(1) \\ \cdots & \cdots & & \cdots \\ \Delta Y_1(n-1) & \Delta Y_2(n-1) & \cdots & \Delta Y_n(n-1) \end{pmatrix}. \tag{3.15}$$

So, weighting coefficients of matrix A are calculated according to expression

$$\bar{a}_i \quad \Delta Y^{-1}\cdot\Delta\bar{q}_i. \tag{3.16}$$

Formula (3.16) makes sense when $\det\Delta Y\ /\ 0$. A Solution similar to (3.16) can be obtained also with any other way of building square systems (3.13) out of (3.9). But they should be selected in such a way that matrix A elements are constant during the respective time period.

3.2.2 Identification in Case with Noise

In this case, instead of accurate values of matrix ΔY and vector $\Delta\bar{q}_i$, we have information about approximate values of matrix $\Delta\tilde{Y}$ and vector $\Delta\tilde{\bar{q}}_i$. So, after solving (3.16), we will get an estimate $\hat{\bar{a}}_i$. For identification, it is very important to know how estimates differ from precise values. Note that this depends significantly on the condition number of matrix ΔY. If this matrix is ill-conditioned or close to degenerate, even small disturbances or inaccuracy in measurements of node coordinates can make the problem improperly posed. Ill-conditioning depends also on parameters of the data-generating system. It is easy to show that the condition number grows quickly with an increase in CM dimensionality. That is why, for big n, it is always necessary to estimate discrepancy between an obtained approximate solution and a precise one.

3.2.2.1 Combinatorial method of solution

Consider a procedure of estimating membership intervals of precise values $\bar{a}_i$ (per component) to predefined constrains on measurement error of CM node coordinates. To accomplish this, we need well-known concepts of matrix norm, matrix inequality, and absolute value of a matrix (or vector as a particular case). For matrix A, having size $n \times n$:

1. $\|A\|_\infty \quad \max\limits_{1 \le j \le n} \sum\limits_{i=1}^{m} |a_{ij}|,$

2. $B \le A \Rightarrow b_{ij} \le a_{ij}, \quad i \quad 1,\ldots,m, \quad j \quad 1,\ldots,n,$ (3.17)

3. $B \quad |A| \Rightarrow b_{ij} \quad |a|_{ij}, i \quad 1,\ldots,m, j \quad 1,\ldots,n.$

Let measurement error be in the form

$$\Delta \tilde{Y}_i(k) \quad \Delta Y_i(k) + \xi_i(k) \quad \text{and} \quad |\xi_i(k)| \le \varepsilon \quad \text{for all } i \text{ and } k, \quad (3.18)$$

where $\xi_i(k)$ – is an additive noise of the i-th node coordinate measurement at time moment k, and ε is a sufficiently small value.

Consider the system of linear square equations

$$\Delta \tilde{Y} \cdot \hat{\bar{a}}_i \quad \Delta \tilde{\bar{q}}_i, \quad (3.19)$$

where $\Delta \tilde{\bar{q}}_i \quad \Delta \bar{q}_i + \bar{\xi}_i$ and $\bar{\xi}_i \quad (\xi_i(1) \; \xi_i(2) \; \ldots \; \xi_i(n))$. In the system (3.19), matrix $\Delta \tilde{Y}$ and vector $\Delta \tilde{\bar{q}}_i$ are given with error, and

$$(\Delta \tilde{Y} + \Xi)\bar{a}_i \quad \Delta \tilde{\bar{q}}_i - \bar{\xi}_i \quad (3.20)$$

corresponds to the precise equation, where Ξ is additive matrix with elements equal to measurement errors of CM nodes coordinates (with "minus" sign). To obtain membership intervals of the precise values (component-wise), we can apply the following theorem [10].

Theorem 1. *Let us have a nondegenerate approximate system (3.19) and a precise system (3.20) such that*

$$\|\Xi\|_\infty \le \delta \|\Delta \tilde{Y}\|_\infty, \quad \|\bar{\xi}_i\|_\infty \le \delta \|\Delta \tilde{\bar{q}}_i\|_\infty. \quad (3.21)$$

If the condition number $\rho_\infty(\Delta \tilde{Y})$ meets the requirement

$$\delta \rho_\infty(\Delta \tilde{Y}) \quad r < 1, \quad (3.22)$$

then $\Delta \tilde{Y} + \Xi$ is also nondegenerate and

$$\left\|\hat{\bar{a}}_i - \bar{a}_i\right\|_\infty \le \frac{2\delta}{1-r} \left\||\Delta \tilde{Y}^{-1}||\Delta \tilde{Y}|\right\|_\infty \left\|\hat{\bar{a}}_i\right\|_\infty. \quad (3.23)$$

Value $\rho_\infty(\Delta\tilde{Y})$ $\|\Delta\tilde{Y}\|_\infty\|\Delta\tilde{Y}^{-1}\|_\infty$ in (3.22) is condition number of matrix $\Delta\tilde{Y}$ in ∞-norm.

In this theorem, specific realizations of Ξ and $\bar{\xi}_i$ are unknown; hence, it is impossible to find δ and r out of (3.21), (3.22). But if we consider the most unfavorable realization (despite its low probability), we can obtain a guaranteed membership interval from (3.21) which is true for all realizations subject to the condition (3.18). For this realization, we have

$$\Xi \quad \varepsilon\Xi_1, \quad \bar{\xi}_i \quad \varepsilon\bar{e}, \tag{3.24}$$

where Ξ_1 is matrix composed of unit values and $\bar{e}$ is vector with all unit components. Then

$$\|\Xi\|_\infty \quad n\varepsilon, \quad \|\bar{\xi}_i\|_\infty \quad \varepsilon. \tag{3.25}$$

The result obtained allows applying combinatorial method for identification problem with approximate data. It is applicable when condition (3.22) of the theorem is met. So, initially, we need to check it. For this purpose, we find the smallest δ satisfying non-strict inequalities

$$\varepsilon \le \delta\frac{\|\Delta\tilde{Y}\|_\infty}{n}, \varepsilon \le \delta\|\delta\tilde{q}_i\|_\infty. \tag{3.26}$$

Then we check strict inequality (3.22). It is expectable that if matrix $\Delta\tilde{Y}$ is ill-conditioned, r under the smallest δ won't be less than 1. In fact, this can indicate big range of the obtained solutions based on the square equations followed from (3.9), i.e., the method based on guaranteed result cannot be applied. In case, when (3.22) is held for this set of square systems, a combinatorial method will be effective. The description of the method is given in [11]. Its advantage is that when there are multiple estimates, it allows significant shrinking of guaranteed membership intervals for the precise values of vector's $\bar{a}_i$ components (under some statistical conditions). The solution algorithm according to the combinatorial method is the following. Create a set of square systems out of the overdetermined system (3.9) via random omitting other equations. There are C_K^n such combinations. Leave only those systems for which (3.22) is met, i.e., nondegenerate and well-conditioned ones. Let the number of such systems be S. Solve these square systems and find approximations $\hat{\bar{a}}_{ij}$ of the parameters $\bar{a}_{ij}$. Each s-th square system, $s \quad 1, \ldots, S$, has a guaranteed component-wise error according to (3.23)

$$\varepsilon_{is} \quad \frac{2\delta}{1-r}\left\|\,|\Delta\tilde{Y}_1^{-1}||\Delta\tilde{Y}_1|\,\right\|_\infty\left\|\hat{\bar{a}}_i\right\|_\infty.$$

Inequality (3.23) then helps obtain an element-wise estimate for $\bar{a}_{ij}$ as

$$|\hat{a}_{ij}^s - \bar{a}_{ij}^s| \le \varepsilon_{is} \quad \text{or} \quad \hat{a}_{ij}^s - \varepsilon_{is} \le \bar{a}_{ij} \le \hat{a}_{ij}^s + \varepsilon_{is}. \qquad (3.27)$$

For different S and fixed i, j (3.27) produces a system of guaranteed estimates which leads to the fact that precise values should belong to a minimal truncated interval

$$\max_s(\hat{a}_{ij}^s - \varepsilon_{is}) \le \bar{a}_{ij} \le \min_s(\hat{a}_{ij}^s + \varepsilon_{is}). \qquad (3.28)$$

Inequality (3.28) implies that with an increase of S and range of realizations, $\hat{a}_{ij}^s$ estimation accuracy should be higher. It makes sense to consider midpoint of the interval $[\max_s(\hat{a}_{ij}^s - \varepsilon_{is}); \min_s(\hat{a}_{ij}^s + \varepsilon_{is})]$ as an estimate of $\bar{a}_{ij}$, and half of its range characterizes estimation accuracy.

3.2.2.2 Identification problem solution with least squares method

As mentioned above, combinatorial method is not applicable when condition (3.22) is not fulfilled because estimate (3.23) will not be correct. In this case we suggest using ordinary or weighted least squares method (LSM). Equations sequence (3.9) allows building overdetermined equations system (3.19) with the same vector $\hat{a}_i$ to be found where $\Delta\tilde{q}_i$ $(\Delta\tilde{q}_i(0) \; \Delta\tilde{q}_i(1) \; \dots \; \Delta\tilde{q}_i(K-1))^{\mathrm{T}}$,

$$\Delta\tilde{Y} \quad \begin{pmatrix} \Delta\tilde{y}_1(0) & \Delta\tilde{y}_2(0) & \dots & \Delta\tilde{y}_n(0) \\ \Delta\tilde{y}_1(1) & \Delta\tilde{y}_2(1) & \dots & \Delta\tilde{y}_n(1) \\ \dots & \dots & \dots & \dots \\ \Delta\tilde{y}_1(K-1) & \Delta\tilde{y}_2(K-1) & \dots & \Delta\tilde{y}_n(K-1) \end{pmatrix}.$$

If we use LSM to solve an overdetermined system of linear algebraic equations with an approximate right-hand side $\Delta\tilde{q}_i$ and matrix $\Delta\tilde{Y}$, then estimate of the vector $\hat{a}_i$ is calculated as

$$\hat{a}_i \quad (\Delta\tilde{Y}^{\mathrm{T}} \cdot \Delta\tilde{Y})^{-1} \cdot \Delta\tilde{Y}^{\mathrm{T}} \cdot \Delta\tilde{q}_i. \qquad (3.29)$$

In case of using weighted LSM where weights are given by diagonal matrix D of size K, formula (3.29) is modified:

$$\hat{a}_i \quad (\Delta\tilde{Y}^{\mathrm{T}} \cdot D \cdot \Delta\tilde{Y})^{-1} \cdot \Delta\tilde{Y}^{\mathrm{T}} \cdot D \cdot \Delta\tilde{q}_i. \qquad (3.30)$$

Both these methods are the most effective, if $\Delta\bar{u}(k)$ is a persistently exciting input impulse [12]. In our case, they are defined as follows. Let us have a sequence of input impulses

$$\Delta\bar{u}^{\mathrm{T}} \quad (\Delta u_1(0)\ldots\Delta u_n(0)\Delta u_1(1)\ldots\Delta u_n(1)\ldots\Delta u_1(K-1)\ldots \\ \Delta u_n(K-1)).$$

Sequence $\Delta\bar{u}$ is persistently exciting of K-th order, if

$$\mathrm{rank}(\Delta\bar{u}\cdot\Delta\bar{u}^{\mathrm{T}}) \quad nK. \tag{3.31}$$

If signal $\Delta\bar{u}$ is stationary white noise with zero mean, the sequence $\Delta\bar{u}(j)$ has the following statistical property:

$$E\left(\left(\begin{array}{c} \Delta\bar{u}(j) \\ \Delta\bar{u}(j+1) \\ \ldots \\ \Delta\bar{u}(j+k-1) \end{array}\right)\left(\Delta\bar{u}^{\mathrm{T}}(j)\ \Delta\bar{u}^{\mathrm{T}}(j+1)\ \ldots\ \Delta\bar{u}^{\mathrm{T}}(j+k-1)\right)\right)$$

$$\sigma^2\left(\begin{array}{cccc} I_n & 0 & \ldots & 0 \\ 0 & I_n & \ldots & 0 \\ \ldots & \ldots & \ldots & \ldots \\ 0 & 0 & \ldots & I_n \end{array}\right),$$

where $\Delta\bar{u}(j)$ $(\Delta u_1(j)\ \ldots\ \Delta u_n(j))^{\mathrm{T}}$, E is expectation of the stationary random sequence and σ^2 is its variance.

3.2.2.3 Regularized solution of the identification problem

Under some conditions, the identification problem can become severely ill-posed. This happens when information matrix $\Delta\tilde{Y}$ is ill-conditioned (for example at big n). This leads to high sensitivity of the solution to the available errors in data. In such cases, it is reasonable to utilize additional information about the solution and introduce a regularization procedure into the algorithm based on this information. This allows finding solutions which are stable with respect to errors' variances. Additional information about incidence matrix, i.e. about interrelations between CM nodes, can be extremely useful. Absence of some edges means that respective coefficients a_{ij} of the matrix A are zero. This implies reducing size of vector $\bar{a}_i$ and information matrices in (3.29). If this is insufficient to get a stable solution and there is no other additional information, Tikhonov regularization method [13] can be applied to solve ill-conditioned systems of linear algebraic equations with approximate right-hand side and main information matrix.

Consider a stabilizer

$$\Omega(\bar{a}_i) \quad \|\bar{a}_i\|_2^2, \tag{3.32}$$

with Euclid norm. Smoothing functional according to [13] is

$$M^\alpha(\bar{a}_i, \Delta\tilde{\tilde{q}}_i, \Delta\tilde{Y}) \quad \|\Delta\tilde{Y}\bar{a}_i - \Delta\tilde{\tilde{q}}_i\|_2^2 + \alpha\|\bar{a}_i\|_2^2. \tag{3.33}$$

One needs to build a decreasing sequence $\{\alpha_j\}$, e.g., decreasing geometrical progression, and to solve the sequence of the minimization problems with these α_j. Regularized solution is the solution that corresponds to the minimal α_j found according to the residual principal, i.e., minimal α_j satisfying inequality

$$\|\Delta\tilde{Y}\bar{a}_i - \Delta\tilde{\tilde{q}}_i\|_2 \le \varepsilon(\|\Xi_1\|_2 + \|\bar{e}\|_2). \tag{3.34}$$

For α_{j-1}, (3.34) should not be held. That $\hat{\bar{a}}_i$ which guarantees minimum of (3.33) with α found according to (3.34) will get a regularized solution of the parametric identification problem, i.e., the solution stable to errors in data.

The result can be improved if a regularization parameter α is found as quasi-optimal [13]. Starting from some α_j, we decrease it unless infinum of the functional

$$f(\alpha) \quad \|\alpha\frac{d\bar{a}_i(\alpha)}{d\alpha}\|_2 \tag{3.35}$$

is reached, where $\bar{a}_i(\alpha)$ is a minimizing vector for given α.

3.3 Control Automation in Cognitive Maps in Impulse Process Mode

To design an automated control algorithm in CM impulse process by means of synthesizing combined control vector $\Delta\bar{U}(k) \quad [\Delta\bar{u}^T(k) \; \Delta\bar{a}^T(k)]^T$, the following model of CM nodes' forced motion is utilized:

$$\Delta\bar{Y}(k+1) \quad A\Delta\bar{Y}(k) + B\Delta\bar{u}(k) + L(k)\Delta\bar{a}(k) + \Delta\bar{\xi}(k). \tag{3.36}$$

It is obtained by combining (3.5) and (3.6). In [8], the following rules of composing weights increments vector $\Delta\bar{a}(k)$ and control matrix $L(k)$ are formulated:

(a) Vector of weights increments $\Delta\bar{a}(k)$ has dimension $p < n$ — it includes only non-zero elements $\Delta a_{ij}(k) \; / \; 0$. If some weight in CM cannot be varied, then increment $\Delta a_{ij}(k) \quad 0$ is not included into $\Delta\bar{a}(k)$.

(b) Matrix $L(k)$ has dimension $n \times p$ and contains no more than one element in each i-th row equal to Y_j which affects node Y_i via weight increment Δa_{ij}. This element's column number is equal to number of element Δa_{ij} in the vector $\Delta \bar{a}(k)$.

(c) If any node Y_i does not have any incoming edges that may be varied, then all elements of the i-th row are zero.

An incidence matrix A has size $n \times n$. If vector $\Delta \bar{u}(k)$ has length $m < n$, then control matrix B has size $n \times m$ and is composed in the same way as it is in the model (3.5). Uncontrollable random disturbances $\Delta \bar{\xi}(k)$ have zero mean $E \Delta \xi_i(k)$ 0.

Forced motion dynamics of the CM impulse process (3.36) can be written as

$$\bar{Y}(k+1) \quad (I + A - Aq^{-1})\bar{Y}(k) + [B \; L(k)] \begin{bmatrix} \Delta \bar{u}(k) \\ \Delta \bar{a}(k) \end{bmatrix} + \Delta \bar{\xi}(k), \quad (3.37)$$

where q^{-1} is inverse shift operator (by one sampling period).

Combined optimal control vector $\Delta \bar{U}(k)$ $[\Delta \bar{u}^T(k) \; \Delta \bar{a}^T(k)]^T$ with total length $m + p \leq n$ is synthesized to minimize the following quadratic optimality criterion:

$$J(k+1) \quad E\Big\{ [\bar{Y}(k+1) - \bar{G}]^T [\bar{Y}(k+1) - \bar{G}]$$

$$+ \big(\Delta \bar{u}^T(k) \; \Delta \bar{a}^T(k) \big) \begin{pmatrix} R_1 & 0 \\ 0 & R_2 \end{pmatrix} \begin{pmatrix} \Delta \bar{u}(k) \\ \Delta \bar{a}(k) \end{pmatrix} \Big\}, \quad (3.38)$$

where $\bar{G}$ is a reference-input vector that defines desirable behavior of the CM nodes coordinates, R_1, R_2 are diagonal positive-definite matrices, E is conditional expectation operator as of time moment k. Having minimized criterion (3.38) w.r.t. vector $[\Delta \bar{u}^T(k) \; \Delta \bar{a}^T(k)]^T$ based on model (3.37) as restriction, we find the combined control vector

$$\begin{pmatrix} \Delta \bar{u}(k) \\ \Delta \bar{a}(k) \end{pmatrix} \quad -\Big\{ \begin{pmatrix} B^T \\ L^T(k) \end{pmatrix} (B \; L(k)) + \begin{pmatrix} R_1 & 0 \\ 0 & R_2 \end{pmatrix} \Big\}^{-1} \begin{pmatrix} B^T \\ L^T(k) \end{pmatrix}$$

$$\{(I + A - Aq^{-1})\bar{Y}(k) + \Delta \bar{\xi}(k) - \bar{G}\}. \quad (3.39)$$

Thus, at each time moment k, the designed controller generates a combined control vector. The first part of this vector $\Delta \bar{u}(k)$ is fed into the closed-loop system as a direct impact on controllable nodes (resources varying) and the second part $\Delta \bar{a}(k)$ is implemented via corresponding edges' weights varying, i.e., via increase or decrease of some of the weights.

After substituting (3.39) into (3.37), the closed-loop control system of a CM impulse process can be written as:

$$\bar{Y}(k+1) \quad \left\{ I - (B \, L(k)) \left[\left(\begin{array}{c} B^T \\ L^T(k) \end{array} \right) (B \, L(k)) + \left(\begin{array}{cc} R_1 & 0 \\ 0 & R_2 \end{array} \right) \right]^{-1} \right.$$
$$\left. \left(\begin{array}{c} B^T \\ L^T(k) \end{array} \right) \right\} \cdot (I + A - Aq^{-1}) \bar{Y}(k)$$

$$+ (B \, L(k)) \left[\left(\begin{array}{c} B^T \\ L^T(k) \end{array} \right) (B \, L(k)) + \left(\begin{array}{cc} R_1 & 0 \\ 0 & R_2 \end{array} \right) \right]^{-1}$$
$$\cdot \left(\begin{array}{c} B^T \\ L^T(k) \end{array} \right) (\bar{G} - \Delta \bar{\xi}(k)) + \Delta \bar{\xi}(k). \tag{3.40}$$

3.4 Experimental Research

3.4.1 Research of the Algorithms of CM Incidence Matrix Identification

The problem of identifying coefficients $\hat{a}_{ij}$ is researched through an example of a CM of a commercial bank (Figure 3.1) described in [14].

CM incidence matrix in model (3.8) is written as

$$A \quad \left(\begin{array}{ccccccc} 0 & 0.15 & 0 & 0 & 0.1 & 0 & 0 \\ -0.2 & 0 & 0.13 & -0.2 & 0.03 & 0 & 0 \\ 2 & 0 & 0.75 & 0 & 0 & 0 & 0 \\ 1 & 0 & 0 & 0.8 & 0 & -0.5 & 0 \\ 0 & 0.85 & -0.95 & 0.9 & 0 & 0 & -0.7 \\ 0 & -0.5 & 0.3 & 0 & 0 & 0.7 & 0 \\ 0 & 0 & 0.1 & -0.2 & 0.8 & 0 & 0 \end{array} \right).$$

CM nodes have the following meaning: 1 – regional network, 2 – capital, 3 – loans, 4 – deposits, 5 – liquid assets, 6 – stability risk measurement, 7 – liquidity risk measurement. Eigenvalues of the matrix A are less than one in absolute values, and so the system is stable.

The impulse process has been simulated based on the CM model (3.8) where testing signal $\Delta \bar{u}$ is a sequence of independent standard normally distributed vectors, i.e., white noise (unless otherwise stated). Maximal simulation interval has been selected equal to K 40.

First, a deterministic case was considered when all the nodes are measured precisely. In this case, it has appeared to be sufficient to take the first n 7 measurements of the output vector and use formula (3.16) to obtain almost perfectly accurate estimates of the matrix A elements (with accuracy comparable to the computer precision). It is sufficient to feed test signals only at the initial time moment and even not to all nodes because this almost does not affect estimation accuracy.

The problem becomes significantly more difficult when the coordinates are measured with errors. The following approach is used for simulation in this case. We take $\Delta Y_{\max}$ – maximal absolute value among all $\Delta \bar{Y}$ (not for each of the coordinates but total maximum for all of them because common constraint for all noises will be used in the algorithms). Some accuracy ζ is set for experimental purposes; it is changed from 0 to 0.1 with step 0.01. For each ζ, we calculate ε $\zeta \cdot \Delta Y_{\max}$ and generate random vector $\bar{\xi}$ with

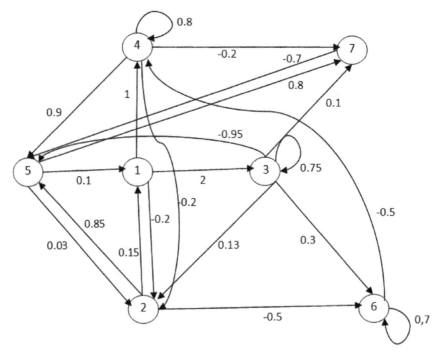

Figure 3.1 CM of a commercial bank.

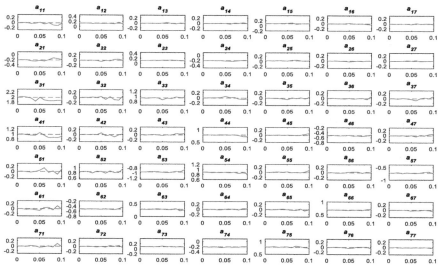

Figure 3.2   Identification of CM weights.

independent coordinates in the interval $[-\varepsilon, \varepsilon]$. In case under consideration, each coordinate is either uniformly distributed in this interval or normally distributed with zero mean and standard deviation $\varepsilon/3$ that guarantees 99% of all values in this interval. Note that in the second case, all obtained estimates are more accurate because most of random values are concentrated closer to zero. Then vector of noisy measurements is calculated according to formula (3.18).

Applicability limits of the combinatorial method to the given problem have been researched. They are determined mostly by inequalities (3.22) and (3.26). In our case, they are met under $\varepsilon < 0.01$, i.e., under minor noise. A combinatorial method provides very good results, comparable to those in deterministic case. But specificity of the given CM is that under bigger noise the method appears inapplicable.

The next method researched here is based on applying LSM (3.29). The best results have been obtained under K 40 and test signals $\Delta \bar{Q}$ fed to all CM nodes at each time moment. Figure 3.2 shows estimates of all edges' weights of CM vs. accuracy ζ (dotted line is for real values).

As a result of simulation, three important regularities are deduced.

1. The quality of coefficients $\hat{a}_{ij}$ estimation significantly depends on number of nodes' measurements K $(n \le k \le K)$.

2. The estimation accuracy depends (even more significantly) on duration of feeding test exciting inputs $\Delta\bar{Q}$ (obviously long-term testing is not always possible in practice).

3. Reducing number of non-zero elements in vector $\Delta\bar{Q}$ is in some sense "equivalent" to reducing interval of non-zero testing signal. Note that in practice testing inputs, Δq_i usually cannot be fed to all nodes because it is physically impossible or practically undesirable in real complex system. In this case, we suggest "compensating" reduced number of unexcited nodes by increasing exciting period for other nodes.

Consider a practically wide-spread case when information about the absence of some edges between CM nodes is available before the identification procedure. To account for this information and to obtain guaranteed stable solution, Tikhonov's regularization is applied. Here instead of smoothing functional (3.33), the following modified functional is utilized:

$$M^\alpha(\bar{a}_i, \Delta\tilde{\bar{Y}}_i, \Delta\tilde{Y} + \Delta Q) \qquad \|(\Delta\tilde{Y} + \Delta Q)\bar{a}_i - \Delta\tilde{\bar{Y}}_i\|_2^2$$

$$+\alpha \sum_{j=1}^{n} w_{ij} a_{ij}^2,$$

where

$$w_{ij} \quad \begin{cases} 100, & \text{if } a_{ij} \quad 0, \\ 0, & \text{if } a_{ij} \,/\, 0. \end{cases}$$

Thus, additional information about relations between CM nodes is accounted for in the stabilizer. As a result, zero coefficients estimates become more accurate while accuracy of others remains the same, as shown in Figure 3.3. For simulation, 40 measurements have been used.

3.4.2 Automating Control of Impulse Processes in CM of Human Resources Management in IT Company

CM for human resources (HR) management of IT company [15] is shown on Figure 3.4. CM nodes represent the following coordinates of control system: Y_1 – "career and staff management", Y_2 – "staff certification", Y_3– "bonuses for early completion of tasks", Y_4 – "bonuses for new skills development", Y_5– "level of monitoring", Y_6 – "planning of staff training process", Y_7 – "average salary", Y_8 – "company finance per employee", Y_9 – "employees' satisfaction", Y_{10} – "promotion perspectives", Y_{11} – "spending on sports", Y_{12} – "staff training without specialization change",

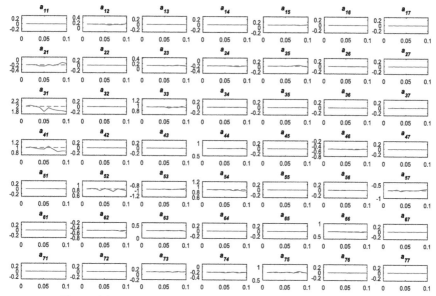

Figure 3.3 Identification of CM weights with using regularization.

Y_{13} – "retraining with change of main specialization", Y_{14} – "human resources professional skills", Y_{15} – "products innovativeness", Y_{16} – "supporting staff training", Y_{17} – "spending on Research and development", Y_{18} – "graduate school effectiveness". All coordinates are measured in 0 to 10 scale.

Incidence matrix A of the CM, where weighting coefficients $a_{ij}(i, j$ $1, 2, \ldots, 18)$ are found based on cause-and-effect relations, in the IT company is as follows:

For simulation CM nodes $Y_1, Y_2, Y_5, Y_6, Y_{13}$ are selected as ones that are controlled through varying of resources. Then control vector $\Delta \bar{u}$ in model (3.36) is written as

$$\Delta \bar{u} \quad (\Delta u_1 \; \Delta u_2 \; \Delta u_5 \; \Delta u_6 \; \Delta u_{13})^{\mathrm{T}}.$$

Nodes $Y_3, Y_4, Y_7, Y_{10}, Y_{11}, Y_{12}, Y_{14}, Y_{16}$ can be controlled via weights varying. Then vector $\Delta \bar{a}$ in model (3.36) can be written as

$$\Delta \bar{a} \quad (\Delta a_{3,5} \; \Delta a_{4,5} \; \Delta a_{7,8} \; \Delta a_{10,1} \; \Delta a_{11,8} \; \Delta a_{12,6} \; \Delta a_{14,2} \; \Delta a_{16,6})^{\mathrm{T}},$$

where coefficients a_{ij} can be varied when controlling the i-th CM node by means of changing degree of influence of the j-th node on it. In this case

$$
A =
\begin{pmatrix}
0 & 0 & 0 & 0 & 0 & 0 & 0 & 0 & 0 & 0.4 & 0 & 0 & 0 & 0.4 & 0 & 0 & 0 & 0.4 \\
0 & 0 & 0 & 0 & 0 & 0 & 0 & 0 & 0 & 0 & 0 & 0 & 0 & 0.4 & 0.25 & 0 & 0 & 0.3 \\
0 & 0 & 0 & 0 & 0 & 0 & 0.1 & 0 & 0.4 & 0 & 0 & 0 & 0.1 & 0 & 0.3 & 0 & 0 \\
0.5 & 0 & 0 & 0 & 0 & 0 & 0.2 & 0.4 & 0 & 0.4 & 0 & 0 & 0 & 0.3 & 0 & 0 & 0 \\
0 & 0 & 0 & 0.5 & 0.3 & 0 & 0 & 0 & 0 & 0 & 0 & 0 & 0.4 & 0 & 0 & 0 & 0 \\
0 & 0.5 & 0 & 0 & 0 & 0 & 0 & 0 & 0 & 0 & 0 & 0 & 0 & 0 & 0 & 0 & 0 \\
0 & 0.2 & 0 & 0 & 0 & 0 & 0 & 0 & 0 & 0 & 0 & 0 & 0 & 0 & 0 & 0 & 0 \\
0 & 0 & 0 & 0 & 0 & 0 & 0 & 0 & 0.3 & 0 & 0 & 0 & 0 & 0 & 0 & 0 & 0 \\
0 & 0 & 0 & 0 & 0 & 0 & 0.05 & 0 & 0.2 & 0.5 & 0 & 0 & 0 & 0 & 0 & 0 & 0 \\
0 & 0 & 0 & 0 & 0 & 0 & 0 & 0 & 0 & 0 & 0 & 0 & 0 & 0.15 & 0 & 0 & 0 \\
0 & 0 & 0 & 0 & 0 & 0 & 0.5 & 0.4 & 0 & 0 & 0.4 & 0.5 & 0.5 & 0 & 0 & 0.3 & 0 \\
0 & 0 & 0 & 0 & 0 & 0 & 0 & 0 & 0.4 & 0 & 0 & 0 & 0 & 0 & 0 & 0 & 0 \\
0 & 0 & 0 & 0 & 0 & 0 & 0 & 0 & 0 & 0 & 0.4 & 0.4 & 0 & 0 & 0.8 & 0 & 0 \\
0 & 0 & -0.8 & -0.8 & 0 & 0 & 0 & 0 & 0 & 0 & 0 & 0 & 0 & 0.3 & 0 & 0 & 0 \\
0 & 0 & 0 & 0 & 0 & 0.2 & 0 & 0 & 0 & 0 & 0 & 0 & 0 & 0 & 0 & 0 & 0 \\
0 & 0 & 0 & 0 & 0 & 0.2 & 0 & 0 & 0 & 0 & 0 & 0 & 0 & 0 & 0 & 0 & 0 \\
0 & 0 & 0 & 0 & 0 & 0.4 & 0 & 0 & 0.4 & 0 & 0 & 0 & 0.7 & 0 & 0 & 0 & 0 \\
0 & 0.4 & 0 & 0 & 0.5 & 0 & 0 & 0 & 0.5 & 0 & 0.4 & 0.4 & 0 & 0 & 0 & 0 & 0
\end{pmatrix}
$$

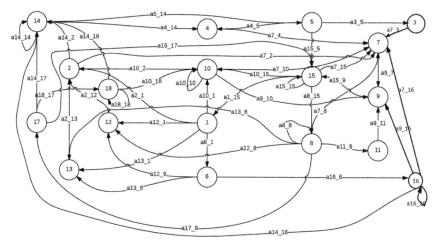

Figure 3.4 CM of HR management in IT company.

dynamic control matrix $L(k)$ in model (3.36) is composed as follows:

$$L(k) \begin{pmatrix}
0 & 0 & 0 & 0 & 0 & 0 & 0 & 0 \\
0 & 0 & 0 & 0 & 0 & 0 & 0 & 0 \\
Y_5(k) & 0 & 0 & 0 & 0 & 0 & 0 & 0 \\
0 & Y_5(k) & 0 & 0 & 0 & 0 & 0 & 0 \\
0 & 0 & 0 & 0 & 0 & 0 & 0 & 0 \\
0 & 0 & 0 & 0 & 0 & 0 & 0 & 0 \\
0 & 0 & Y_8(k) & 0 & 0 & 0 & 0 & 0 \\
0 & 0 & 0 & 0 & 0 & 0 & 0 & 0 \\
0 & 0 & 0 & 0 & 0 & 0 & 0 & 0 \\
0 & 0 & 0 & Y_1(k) & 0 & 0 & 0 & 0 \\
0 & 0 & 0 & 0 & Y_8(k) & 0 & 0 & 0 \\
0 & 0 & 0 & 0 & 0 & Y_6(k) & 0 & 0 \\
0 & 0 & 0 & 0 & 0 & 0 & 0 & 0 \\
0 & 0 & 0 & 0 & 0 & 0 & Y_2(k) & 0 \\
0 & 0 & 0 & 0 & 0 & 0 & 0 & 0 \\
0 & 0 & 0 & 0 & 0 & 0 & 0 & Y_6(k) \\
0 & 0 & 0 & 0 & 0 & 0 & 0 & 0 \\
0 & 0 & 0 & 0 & 0 & 0 & 0 & 0
\end{pmatrix}.$$

For simulation of the closed-loop control system (3.40) we assume that CM nodes of different physical nature are measured in 10 scores scale and are at medium level 5 at the initial time moment.

We have investigated performance of the closed-loop automated control system for the problem of shifting nodes $Y_8, Y_{10}, Y_{14}, Y_{15}$ from level 5 to level 7 through respective step increase of the components $G_8, G_{10}, G_{14}, G_{15}$

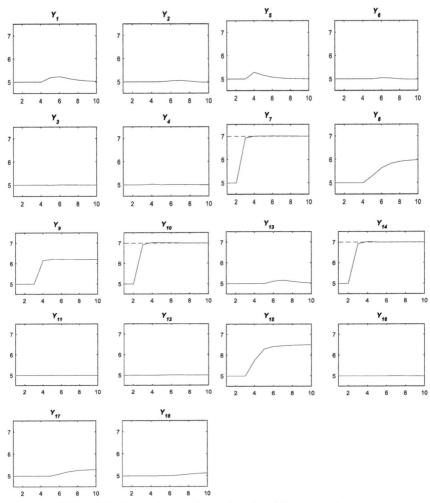

Figure 3.5   CM nodes coordinates in a closed-loop control system.

of the reference vector in control law (3.39). Resulting dynamics of CM impulse process are shown on Figure 3.5.

3.5 Conclusion

Two major problems of cognitive modeling have been considered: identification of CM edges' weights and control using these weights' varying.

The parametric identification problem deals with the necessity of finding weighting coefficients of CM edges relevant for further analysis, simulation, and control. The solutions suggested here are applicable for the case when all CM nodes are measurable, although these measurements can be inaccurate. Several methods are proposed depending on properties of the specific problem: method for deterministic case, combinatorial method for the case when measurements errors are small and matrices are well-conditioned, a method based on least square method (LSM) for a more general case of noisy measurements, regularized method for the case when the problem is ill-posed and additional information is available. All these methods were simulated using the example of CM of a commercial bank. The simulation allowed formulating important and practically useful regularities about accuracy of estimation depending on the number of observations and test inputs.

The control problem solved in the present chapter deals with the popular case when the number of CM nodes that can be directly affected by a decision-maker is small. It is suggested to add another type of control inputs in the form of CM weights varying. Both types of controls are then combined into one control vector, and control law is developed based on the optimality quadratic criterion. The method has been simulated using the example of an IT company's human resources management system described by CM. Efficiency of the method has been demonstrated for this example.

References

[1] F. Roberts, 'Discrete Mathematical Models with Applications to Social, Biological, and Environmental Problems', Englewood Cliffs: Prentice-Hall, 1976.

[2] R. Axelrod, 'The Structure of Decision: Cognitive Maps of Political Elites', Princeton University Press, 1976.

[3] V. Romanenko, Y. Milyavskiy, A. Reutov, 'Adaptive Control Method for Unstable Impulse Processes in Cognitive Maps Based on Reference Model', Journal of Automation and Information Sciences, vol. 47(3), pp. 11–23, 2015.

[4] M. Zgurovsky, V. Romanenko, Y. Milyavsky, 'Adaptive Control of Impulse Processes in Complex Systems Cognitive Maps with Multirate Coordinates Sampling', Advances in Dynamical Systems and Control, Studies in Systems, Decision and Control, Springer International Publishing, vol. 69, pp. 363–374, Switzerland, 2016.

[5] V. Romanenko, Y. Milyavsky, 'Stabilizing of Impulse Processes in Cognitive Maps Based on State-Space Models', System Research and Information Technologies, vol. 1, pp. 26–42, 2014 (in Russian).

[6] M. Zgurowsky, V. Romanenko, Y. Milyavskiy, 'Principles and Methods of Impulse Processes Control in Cognitive Maps of Complex Systems. Part 1', Journal of Automation and Information Sciences, vol. 48(3), pp. 36–45, 2016.

[7] M. Zgurowsky, V. Romanenko, Y. Milyavskiy, 'Principles and Methods of Impulse Processes Control in Cognitive Maps of Complex Systems. Part 2', Journal of Automation and Information Sciences, vol. 48(7), pp. 4–16, 2016.

[8] V. Romanenko, Y. Milyavsky, 'Control method in cognitive maps based on weights increments', Cybernetics and Computer Engineering, vol. 184, pp. 44–55, 2016.

[9] B. Ho, R. Kalman, 'Efficient Construction of Linear State Variable Models from Input/output Functions', Regulungstechnik, vol. 14, pp. 545–548, 1966.

[10] G. Golub, C. VanLoan, 'Matrix Computations', Baltimore: John Hopkins University Press, 1989.

[11] V. Gubarev, S. Melnichuk, 'Guaranteed State Estimation Algorithms for Linear Systems in the Presence of Bounded Noise', Journal of Automation and Information Sciences, vol. 47(3), pp. 1–10, 2015.

[12] M. Verhaegen, P. Dewilde, 'Subspace model identification. Part 1. The output-error state-space model identification class of algorithms', Int. J. of Control, vol. 56(3), pp. 1187–1210, 1992.

[13] A. Tikhonov, V. Arsenin, 'Methods of solving ill-posed problems', Moscow: Nauka, 1979 (in Russian).

[14] V. Romanenko, Y. Milyavsky, 'Impulse Processes Stabilization in Cognitive Maps of Complex Systems Based on Modal State-space Controllers', Cybernetics and Computer Engineering, vol. 179, pp. 43–55, 2015 (in Russian).

[15] V. Romanenko, Y. Milyavsky, 'Decision-making in cognitive maps impulse processes models based on synthesis of increments of weights and vertices coordinates', Abstracts of the II International Scientific and Practical Forum "Science and Business", Dnipro, pp. 247–254, 2016 (in Ukrainian).

4

Decentralized Guaranteed Cost Inventory Control of Supply Networks with Uncertain Delays

Leonid M. Lyubchyk and Yuri I. Dorofieiev*

National Technical University "Kharkiv Polytechnic Institute",
Kharkiv, Ukraine
*Corresponding Author: dorofeev@kpi.kharkiv.edu

The problem of guaranteed cost inventory control of supply networks with uncertain transport delays in the presence of disturbances modeling the changing demand is considered. As a model of uncertainty of demand, a model of "unknown but bounded" disturbance was adopted. The supply network model is adopted in the form of a graph with nodes described by discrete models in the state space, in which resource stock levels are used. An equivalent model of the system, taking into account the presence of a transport delay, is obtained using the descriptor transformation and delay-dependent Lyapunov-Krasovskii functional used for stability analysis is constructed. The solution of the problem of decentralized guaranteed cost inventory control under conditions of uncertain demand is obtained on the basis of the method of invariant ellipsoids. Using the technique of linear matrix inequalities, the problem of controller's synthesis, which minimizes the upper boundary value of the quadratic quality criterion, is reduced to the problem of semidefinite programming. Equations of local stabilizing controllers for nodes of supply network are obtained taking into account the uncertain delay. Stability analysis of large-scale controlled supply network, taking into account the interconnections between the nodes, is obtained on the basis of Lyapunov vector functions method using the dynamic aggregated comparison system. The results of computer simulation of supply network with decentralized guaranteed cost control are also presented.

4.1 Introduction

Among the modern control problems of complex technical and economic systems, the problem of supply networks control attracts considerable attention. Supply network is an integrated production process, including production nodes, warehouses, stores, and distribution channels that extract raw materials, carry out its processing, transport and store semi-finished and finished products, and also supply it to retail network or end-users. The goal of each network node is to effectively add value to its products as it moves through the supply network to distributed markets in required quantities, in the right combination at the required time and at a competitive price [1].

The term "supply network" is used as a generalization of the term "supply chain" [2], which is used for systems characterized by linear or tree-like structure of interconnections between nodes. At present, the most widespread are the production-storage-distribution systems with hierarchical multi-level structure, which is characteristic for complex industrial plants, network retailers, transnational companies, where deliveries can be made between any nodes.

The supply network model is usually represented as an oriented graph, the vertices of which corresponds to network nodes, determine the types and sizes of controlled stocks, and the arcs represent controlled and uncontrolled flows in the network. Controlled flows describe the processes of conversion and redistribution of resources between network nodes, as well as the processes of supplying raw materials from outside. Uncontrolled flows describe the demand for resources, which is formed by external consumers.

A characteristic feature of production-storage-distribution systems is availability of transport delay between the formation of an order for the stocks replenishment and the receipt of ordered resources to the warehouse. Since the presence of delay is the reason for the decrease in the quality of the system operation, and possibly of loss of stability, considerable attention is paid to the problems of stability analysis and the controller's synthesis for systems with an uncertain delay (see [3] and references therein).

The synthesis of control systems for objects with delay in control requires consideration of the delay influencing the stability and quality of transient processes in a closed system. To solve the problems of stabilizing dynamic systems with delay, two approaches have become most widespread: i) using standard schemes for constructing feedback by state or output; ii) using various methods for compensating delays by including in scheme special "predictors". The main disadvantage of systems built using predictors is high sensitivity to model uncertainty and errors in estimating delay value.

Another approach to solving control problems of linear objects with delay, based on the transformation of models of initial systems to equivalent models without delay, was first proposed by R. R. Bate [4], and later independently by G. L. Slater and W. R. Wells [5]. The transformation is carried out by extending the state space. The disadvantage of this approach is a significant increase in the dimensionality of the model, and in the case of uncertainty of the delay value, such a transformation is inapplicable.

In last decade, the method of Model Predictive Control (MPC) has been widely used to control objects with delay [6]. The method began to develop in the early 60s of the XX century to control processes and equipment in petrochemical and power generation, for which the use of traditional synthesis methods was difficult due to the complexity of mathematical models. Currently, the application area of MPC has expanded significantly, including inventory control problems [7]. However, the results of modeling and practical application have shown that the quality of MPC directly depends on the chosen model of external disturbances prediction. Therefore, to use this method, it is necessary to have reliable information for the synthesis of an adequate disturbance model.

In practice, there is usually no information for constructing an adequate model of external demand for resources, which is considered as external disturbances. One of the approaches to solving the inventory control problem in conditions of demand uncertainty is the use of the concept of "unknown but bounded" disturbances [8], in accordance with which the uncertainty of demand is described by a set of intervals within which the components of the vector function describing demand take on values in an arbitrary way. The boundaries of the intervals are determined based on the available statistics.

At present, significant attention is paid to the problem of stability analysis of linear discrete systems with an unknown but bounded delay, at that, stability analysis methods in the time domain, which are based on the Lyapunov stability theory, are usually used.

It should be noted that while for systems without delay stability analysis methods in the time domain based on the existence of a strictly decreasing positive definite Lyapunov function (LF), the classic Lyapunov theory is not directly applicable to systems with delay. This is due to the fact that the influence of delayed states can cause a violation of the monotonic decrease condition, which the standard LF satisfies.

The first Lyapunov method for linear systems with delay was developed in the works of V. I. Zubov and R. Bellman. The second (direct) Lyapunov method admits two generalizations to systems with delay. The first of these was proposed by N. N. Krasovskii in 1956 [9]. As an analogue of LF, the author used functional whose argument, along with the components of the system state vector, is the components of the state vector with delay. The use of Lyapunov-Krasovskii functional (LKF) made it possible to obtain a criterion for the exponential stability of linear discrete stationary systems with delay – this is the existence of a positive definite functional having a negative definite first difference along all trajectories of the system.

The second generalization of the direct Lyapunov method to systems with delay, also proposed in 1956, belongs to B. S. Razumikhin [10]. The LF is used for the stability analysis, but the negative definiteness of their first derivative is checked only on the set of functions that satisfy the special constraint – the Razumikhin condition. Although this approach is generally easier to use, it is based on the "small-gain condition" [11] and, as a result, is inherently "conservative". In fact, Razumikhin's approach can be considered as a special case of Krasovskii's approach [12].

The main idea of both generalizations is to obtain sufficient conditions for the stability of systems with delay by constructing the corresponding LF or LKF. However, for a long time, there were no algorithms for constructing these functions. Only after the researchers began to apply the technique of linear matrix inequalities (LMI), and also computational methods based on the ideas of convex optimization were developed [13] and corresponding algorithms and software were designed for their implementation, among which it is necessary to single out freely distributed software SeDuMi [14] and CVX [15] for the MATLAB, it was possible to simplify the process of constructing LF and LKF, which contributed to the development and application of these methods. At present, new results continue to appear one after another (see the review [16] and references therein).

A set of stability criteria for systems with delay can be divided into two classes. The first includes criteria in which the stability conditions do not depend on the delay, that is, information on the delay value is not taken into account. Similar results in the literature on the control theory are called "conservative". Conservatism is manifested in the fact that the results which are obtained suggest implementation of the worst possible version of uncertainty, the probability of which, from a practical point of view, is very small. The second class includes criteria that take into account information about the delay value. Usually the delay-dependent criteria are less conservative, because they offer conditions that ensure the stability of closed systems only when the delay amount does not exceed a certain value. Therefore, in recent years, the attention of researchers has mainly been drawn to the study of delay-dependent stability criteria for systems with an uncertain delay, which value belongs to some interval. It is worth noting that most of the results are obtained for continuous time systems, while much less attention has been paid to discrete systems.

It is known that the existence of a complete quadratic LKF (CQLKF) for a system with delay is a necessary and sufficient condition for its asymptotic stability [17]. Since the construction of CQLKF leads to complex infinite-dimensional LMI, many authors used special forms of LKF, which make it possible to obtain LMI of finite dimension. The results which obtained are based only on sufficient stability conditions, so that they lead to conservatism. Another reason for conservatism is the use of various ways of transforming the model describing the system with delay.

If additional state variables, which are algebraically related to the main variables, are introduced in the construction of control object models, the systems are called descriptor ones. The search for the most "convenient" transformation led to the use of the descriptor representation of the system [18], which is equivalent to the original system and minimizes conservatism. Most of the papers using the descriptor approach are devoted to the analysis and synthesis of systems in continuous time. Among the publications in which the descriptor transformation is applied to discrete systems, it should be noted [19].

These circumstances make it necessary to develop an approach to solving the synthesis problem of a decentralized guaranteed cost inventory control system for discrete supply networks with uncertain delays in conditions of unknown but bounded external demand based on the descriptor transformation of the model and the construction of a delay-dependent LKF.

4.2 Problem Statement

Mathematical models of supply networks are formed on the basis of a dynamic description of the networks nodes and the processes of interaction between them. Almost all the work on supply networks modeling based on ordinary differential equations (in the case of continuous time) or difference equations (in the case of discrete time) to describe "production dynamics", that is, changes in the values of the supply network characteristics with time. The use of dynamic models is extremely important for understanding the dynamic nature of changes in such indicators as resource stock levels, output levels, deficit volumes, etc. Static models cannot adequately describe dynamic systems, the characteristic feature of which is the presence of delays. Therefore, as emphasized in [20], the natural choice for modeling and researching the dynamics of supply networks is the application of dynamic models and methods of control theory. This is confirmed by the results presented in the review [21], which presents an extensive analysis of the classic approaches of control theory to the design and control of supply networks.

For the mathematical description of a controlled supply network, in this chapter, discrete dynamic models in the state space are used. Discreteness in the supply network model arises from the fact that obtaining information about external demand, measuring available levels of resource stocks, as well as forming the orders' size for resources replenishment occurs at discrete time instants. The following assumptions are used in constructing the model: i) the sampling period is selected and all time intervals are considered as a multiple of the selected period; ii) the time is incremented, the current time is denoted k $0, 1, 2, ...$, at the end of each period, the state of the system is calculated using the model equations; iii) the system state is characterized by the stocks level of each type of resources during this period.

The state variables of the model are the available levels of resource stocks. The orders' sizes for resources replenishment, which are formed by network nodes in the current period, are considered as control actions, and the volumes of external demand that comes to the network nodes from outside are disturbances.

It is assumed that the structure of the supply network is known and the levels of resource stocks are available for direct measurement. When building a model, the network nodes are numbered and grouped according to the stages of processing raw materials and semi-finished products, starting with those for which external demand is received. In this case, any node of the layer l is a resource supplier for nodes belonging to layers with numbers less than l, or nodes of the layer l with numbers less than this. It is also assumed that the oriented graph representing the supply network model has no cycles.

Each of the n nodes of the supply network is described by a difference equation with delay

$$x_i(k+1) \quad x_i(k) + B_i u_i(k - h_i(k)) + G_i w_i(k), \quad i \quad \overline{1,n}, \quad (4.1)$$

where $x_i(k) \in \mathrm{R}^{n_i}$, $u_i(k) \in \mathrm{R}^{m_i}$, and $w_i(k) \in \mathrm{R}^{n_i}$ are state, control actions and external disturbances vectors of the node i; $B_i \in \mathrm{R}^{n_i \times m_i}$, $G_i \in \mathrm{R}^{n_i \times n_i}$ are control and disturbances matrices; $h_i(k)$ is a positive integer multiple of the selected sampling period Δt that determines the replenishment delay and is assumed to be unknown but satisfying the inequality

$$0 \leq h_i(k) \leq h_i^{\max}, \quad (4.2)$$

where $h_i^{\max}$ is known.

The composite vector $x(k) \quad \mathrm{col}\{x_1(k), x_2(k), \ldots, x_n(k)\}$ with dimension $N \quad \sum_{i=1}^{n} n_i$, which is constructed from the state vectors of individual nodes, is a state vector of the supply network. External disturbances of each node include functions of external demand, as well as demand generated by network nodes, for which this node is the resources supplier

$$w_i(k) \quad \sum_{j=1, j \neq i}^{n} \Pi_{ij} u_j(k) + H_i d(k), \quad (4.3)$$

where $u_j(k) \in \mathrm{R}^{m_j}$ is control actions vector of the node j; $d(k) \in \mathrm{R}^q$ is external disturbances vector of the network; $\Pi_{ij} \in \mathrm{R}^{n_i \times m_j}$, $i, j \quad \overline{1,n}$ are productive matrices, which are formed on the basis of the description of the technological process provided by the supply network: the value of the element $[\Pi_{ij}]_{pr}$ is equal to the number of resource units $p \quad \overline{1, n_i}$ of the node i, which is necessary for the production of resource unit $r \quad \overline{1, n_j}$ by node j; $H_i \in \mathrm{R}^{n_i \times q}$ is matrix of the influence of external disturbances on the state vector of the node i.

Block matrix

$$\Pi \quad \begin{bmatrix} \Pi_{11} & \cdots & \Pi_{1n} \\ \vdots & \ddots & \vdots \\ \Pi_{n1} & \cdots & \Pi_{nn} \end{bmatrix} \quad (4.4)$$

with dimension $N \times N$ completely characterizes the interaction of nodes, which is determined by the supply network structure and the technological process being implemented. The selected method of numbering the network nodes ensures that the matrix Π is the lower triangular; on the main diagonal

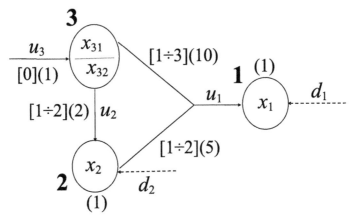

Figure 4.1 Graphical representation of the supply network model.

of the matrix, there are zeros, and the values of the elements below the main diagonal are nonnegative.

Let us illustrate the principle of forming the model matrices, for example, described in the paper [22]. Consider the supply network of three nodes, which is shown in Figure 4.1, and is described by a graph G $(\{1, 2, 3\}, \{(2, 1), (3, 1), (3, 2)\})$.

Let the controlled flow u_1 describe the process of transportation and assembly, as a result of which from 10 resource units of the first type of node 3, the transportation time of which varies from 1 to 3 sampling periods, and five resource units of node 2, the transportation time of which varies from 1 to 2 sampling periods, the unit of production of node 1 is received, the lead time is equal to one sampling period. The flow u_1 is represented on the graph by a hyper-arc connecting nodes 2 and 3 with node 1. The controlled flow u_2 describe the process, as a result of which from 2 resource units of the second type of node 3, the transportation time of which varies from 1 to 2 sampling periods, the unit of production of node 2 is received, the lead time is equal to one sampling period. The arcs d_1, d_2, which are represented by dashed lines, describe an external demand. The value of the transportation time and the number of resource units that are required in accordance with the technological process are indicated for each controlled flow in square brackets and in parentheses, respectively. The lead time value is indicated near each node in parentheses.

The dimensions of the model are equal: n_1 n_2 1, n_3 2, m_1 m_2 1, m_3 2, q 2, N 3. We compute the upper boundary value of the delay of the controlled flows u_1 and u_2: $h_1^{\max} \quad \max\{3+1, \ 2+1\} \quad 4$, $h_2^{\max} \quad 2+1 \quad 3$. The model matrices are equal: $B_1 \quad B_2 \quad 1, B_3 \quad I_2$, $G_1 \quad G_2 \quad -1, G_3 \quad -I_2$, where I_n is the unit matrix of the corresponding dimension; $H_1 \quad [1\ 0], H_2 \quad [0\ 1], H_3 \quad 0_{2\times 2}$, where $0_{n\times m}$ is the zero matrix of the corresponding dimension. The block matrix has the form

$$\Pi \quad \begin{bmatrix} \Pi_{11} & 0 & \Pi_{12} & 0 & \Pi_{13} & 0_{1\times 2} \\ \Pi_{21} & 5 & \Pi_{22} & 0 & \Pi_{23} & 0_{1\times 2} \\ \Pi_{31} & \begin{bmatrix} 10 \\ 0 \end{bmatrix} & \Pi_{32} & \begin{bmatrix} 0 \\ 2 \end{bmatrix} & \Pi_{33} & 0_{2\times 2} \end{bmatrix}.$$

The external disturbances of the network model satisfy the constraints

$$d(k) \in D \quad \{d \in \mathbb{R}^q : 0 \le d^{\min} \le d \le d^{\max}\}, \tag{4.5}$$

where vectors $d^{\min}$ and $d^{\max}$, which determine the boundary values of demand, are considered known.

The additional constraints on the state values in supply network control problems are imposed, as a rule, which are due to the maximum allowable capacity of storage facilities and the requirement of nonnegativity of the stock sizes values. The following constraint must be holds at each time

$$x(k) \in X \quad \{x \in \mathbb{R}^n : 0 \le x \le x^{\max}\}, \tag{4.6}$$

where vector $x^{\max}$ is considered given.

Many approaches have been developed for inventory control in supply networks [23]. It should be noted that most publications are devoted to the development of heuristic approaches or methods of mathematical programming. However, it has now become clear that the optimization of the material resources flows in supply network is impossible without the use of a system approach and the control theory methodology.

Inventory control is the process of forming and regulating of inventory levels that consists in determining of the orders formation time and the sizes of orders for stock replenishment. We use a model of periodic check, which involves monitoring the stock levels and the orders formation in each period.

The formation of the safety stock levels is a traditional control method at demand uncertainty. The expedient safety stock level depends on a large number of conditions related both to the specifics of the production process and the supply network structure, as well as to external factors.

The sizes of safety stock levels for the whole network is calculated using the Leontief productive model [24] on the basis of the upper boundary values of demand

$$\bar{x}^* \quad (I_N - \Pi)^{-1} \begin{bmatrix} H_1 \\ \vdots \\ H_n \end{bmatrix} d^{\max}. \tag{4.7}$$

Equation (4.7) makes sense if the matrix Π is productive, which is the case if and only if the matrix that is the inverse for matrix $(I_N - \Pi)$ exists and its elements are nonnegative.

Statement. A sufficient condition for the productivity of a matrix Π is that it is lower triangular, there are zeros on the main diagonal, and the values of the elements below the main diagonal are nonnegative.

Proof. Consider a supply network consisting of two consecutively connected nodes whose dimensions are 1. Then N 2, and the matrix Π $\begin{bmatrix} 0 & 0 \\ 1 & 0 \end{bmatrix}$ satisfies the conditions of the statement. The matrix $(I_N - \Pi)$ $\begin{bmatrix} 1 & 0 \\ -1 & 1 \end{bmatrix}$ is unitriangular, and the values of the elements below the main diagonal are nonpositive. As a result, the inverse matrix is equal to $(I_N - \Pi)^{-1}$ $\begin{bmatrix} 1 & 0 \\ 1 & 1 \end{bmatrix}$, i.e., the matrix Π is productive.

Suppose that for N n, where n is an arbitrary natural number, the elements values of the matrix $(I_N - \Pi)^{-1}$ are nonnegative. Let us prove that at N $n + 1$ the elements of the matrix $\begin{bmatrix} (I_N - \Pi) & 0_{N \times 1} \\ V_{1 \times N} & 1 \end{bmatrix}^{-1}$ are also nonnegative, where $V_{1 \times N}$ is a row vector with nonpositive elements. In accordance with the Frobenius formula for inversion of a nondegenerate block matrix, we obtain $\begin{bmatrix} (I_N - \Pi) & 0_{N \times 1} \\ V_{1 \times N} & 1 \end{bmatrix}^{-1}$ $\begin{bmatrix} (I_N - \Pi)^{-1} & 0_{N \times 1} \\ -V(I_N - \Pi)^{-1} & 1 \end{bmatrix}$.
It is easy to verify that the values of all elements of the resulting matrix are nonnegative. Thus, according to the method of mathematical induction, this statement is true for any natural number N. *End of proof.*

The safety stock levels are recalculated taking into account delays at stock replenishment

$$x_i^* \quad h_i^{\max} \bar{x}_i^*, \quad i \quad \overline{1, n}. \tag{4.8}$$

The decentralized control law for a node i is constructed in the form of feedback on the deviation between available and safety stock levels

$$u_i(k) \quad K_i(x_i(k) - x_i^*), \tag{4.9}$$

where $K_i \in \mathrm{R}^{m_i \times n_i}$ is a feedback gain matrix. The equation of a node that is closed by feedback (4.9) has the form

$$x_i(k+1) \quad x_i(k) + B_i K_i \left(x_i(k - h_i(k)) - x_i^*\right) + G_i w_i(k). \qquad (4.10)$$

The local quality criterion in the case of an infinite time horizon is chosen in the form

$$J_i^\infty(k) \quad \sum_{t=k}^{\infty} \beta^t \left((x_i(t) - x_i^*)^\mathrm{T} W_{xi}(x_i(t) - x_i^*) + u_i^\mathrm{T}(t) W_{ui} u_i(t)\right),$$

$$(4.11)$$

where $W_{xi} \in \mathrm{R}^{n_i \times n_i}$ and $W_{ui} \in \mathrm{R}^{m_i \times m_i}$ are positive definite diagonal weight matrices; $0 < \beta < 1$ is discount coefficient. The first term in (4.11) determines the sizes of penalty for deviating the available resource levels from safety levels, and the second is the cost of production and storage of resources; the presence of a multiplier β^t ensures the boundedness of criterion on infinite time interval.

For a supply network consisting of n nodes, each of which is controlled by local feedback (4.9) and described by (4.10) with constraints (4.2), (4.5), and (4.6), and the interconnections between nodes are described by the matrix (4.4), it is necessary solve the problem of local controllers synthesis, which for any permissible external demand and permissible delays ensure:

i) complete and timely satisfaction of external and internal demand for resources, that is, the fulfillment of the condition of states nonnegativity

$$x_i(k) \in X_i \quad \{x_i \in \mathrm{R}^{n_i} : 0 \leq x_i\}, \quad i \quad \overline{1, n}; \qquad (4.12)$$

ii) stability of closed systems (4.10), as well as the stability of the entire supply network;

iii) guaranteed cost control, which means that the values of local quality criteria (4.11) for closed nodes of the network will not exceed the corresponding boundary values J_i^* for any implementation of model uncertainties.

4.3 Descriptor Transformation of Supply Network Model

The approach based on the descriptor transformation of the model has two important advantages: first, it can be applied not only to analyze the stability of a system with delay, but also to solve the controller synthesis problem;

second, the introduction of additional state variables makes it possible to reduce the conservatism of the control results.

To simplify the notation, in the following, we omit the index "i" defining the node number and introduce an additional state variable

$$y(k) \quad x(k+1) - x(k). \tag{4.13}$$

Consider descriptor transformation of the closed-loop system model (4.10)

$$\begin{bmatrix} x(k+1) \\ 0_{n\times 1} \end{bmatrix} \quad \begin{bmatrix} x(k) + y(k) \\ -y(k) + x(k) + BK(x(k-h(k)) - x^*) \\ +Gw(k) - x(k) \end{bmatrix}. \tag{4.14}$$

It is easy to verify that equality holds

$$x(k-h(k)) \quad x(k) - \sum_{j=k-h(k)}^{k-1} y(j). \tag{4.15}$$

The system (4.14), taking into account (4.15), is represented in the form

$$\begin{bmatrix} x(k+1) \\ 0_{n\times 1} \end{bmatrix} \quad \begin{bmatrix} x(k) + y(k) \\ -y(k) + BKx(k) - BK\sum_{j=k-h(k)}^{k-1} y(j) \\ -BKx^* + Gw(k) \end{bmatrix}.$$

We construct a composite state vector of the descriptor model $\xi(k)$ $\mathrm{col}\{x(k) - x^*, y(k)\}$ and introduce the notation: E $\begin{bmatrix} I_n & 0_{n\times n} \\ 0_{n\times n} & 0_{n\times n} \end{bmatrix}$,

A $\begin{bmatrix} I_n & I_n \\ BK & -I_n \end{bmatrix}$, $\overline{B}$ $\begin{bmatrix} 0_{n\times n} \\ BK \end{bmatrix}$, $\overline{G}$ $\begin{bmatrix} 0_{n\times n} \\ G \end{bmatrix}$. Finally, we obtain the descriptor model of the network node, closed by control (4.9)

$$E\xi(k+1) \quad A\xi(k) - \overline{B}\sum_{j=k-h(k)}^{k-1} y(j) + \overline{G}(w(k) - w^*) + \overline{G}w^*, \tag{4.16}$$

where $w^* \in \mathrm{R}^n$ is a vector that determines the coordinates of center of the ellipsoid, which will be constructed to approximate the set of values of the node external actions.

Thus, if the sequence $x(k), k \quad 0, 1, \ldots$ is a solution of system (4.10), then the sequence $\mathrm{col}\{x(k) - x^*, y(k)\}, k \quad 0, 1, \ldots$, where $y(k)$ is determined in accordance with (4.13), is a solution of system (4.16) and vice versa.

4.4 Construction of Delay-Dependent Lyapunov-Krasovskii Functional

The formulation of new stabilization problems of processes with delays, as well as the absence of a universal method for constructing Lyapunov functional satisfying the conditions of general stability theorems, leads to the need to study methods for their modification and development. Krasovskii's approach is based on sufficient stability conditions and consists in constructing functional possessing properties analogous to the properties of Lyapunov functions for systems without delay.

The descriptor system (4.16) is asymptotically stable [25] if there exist a functional $V(k) : \underbrace{\mathrm{R}^n \times \ldots \times \mathrm{R}^n}_{h^{\max}} \to \mathrm{R}$ and such scalar parameters $\alpha, \beta > 0$ that for any $x(k)$ and $y(k)$ satisfying (4.16) at any $k \geq 0$ conditions holds true:

$$0 \leq V(k) \leq \beta\{ \max_{j \in [k-h^{\max}, k]} |x(j)|^2, \max_{j \in [k-h^{\max}, k-1]} |y(j)|^2 \}; \qquad (4.17)$$

$$V(k+1) - V(k) \leq -\alpha |x(k)|^2 . \qquad (4.18)$$

If the properties (4.17) and (4.18) hold, then the functional $V(k)$ is called the Lyapunov-Krasovskii functional (LKF) for system (4.16) with uncertain delay (4.2). Control (4.9) must ensure that the properties (4.17) and (4.18) are satisfied along any trajectory of the closed system (4.16). In addition, the required guaranteed cost control should ensure minimization of the upper boundary value of the criterion (4.11).

In the synthesis of control systems for objects with delay, the efforts of developers are aimed at constructing controllers that not only ensure the stability of a closed system, but also guarantee a sufficient level of performance. One of the approaches to solving this problem is the synthesis of the so-called guaranteed cost control, first proposed in [26]. The problem of choosing the optimal control is formulated as a game task and the optimal control strategy is defined as a strategy guaranteeing the achievement of the best result under the worst (most unfavorable) combinations of uncertainty factors. If the control results satisfy predetermined quality requirements, then the resulting control is called the guaranteed cost control.

In [27], control actions are called guaranteed cost control, if the minimization of the upper boundary value of the quality criterion is guaranteed for any permissible disturbances and any variant of realization of the model uncertainty, that is, the condition holds

$$J^* \quad \inf_{u(k) \in U} \quad \sup_{d(k) \in D, \, 0 \leq h(k) \leq h^{\max}} J_\infty(k).$$

The main advantage of this approach is that it allows determining the upper boundary value of a quality criterion; as a result the developer has the ability to assess the system performance degradation caused by the presence of delay. On the basis of this idea, a lot of results were obtained for linear discrete systems with delay [28]. The development of LMI technique makes it possible to apply this approach for the optimal inventory control synthesis in supply networks with uncertain delays in stocks replenishment.

We define the block matrix $P \quad \begin{bmatrix} P_1 & P_2 \\ P_2^T & P_3 \end{bmatrix}$, where $0 \prec P_j \in \mathbb{R}^{n \times n}$,

$j \quad 1, 2, 3, P_1 \quad P_1^T, P_3 \quad P_3^T$, and construct the LKF for the descriptor system (4.16) in the form

$$V(k) \quad V_2(k) + V_2(k), \tag{4.19}$$

$$V_1(k) \quad \beta^k \xi^T(k) E P E \xi(k) \quad \beta^k (x(k) - x^*)^T P_1 (x(k) - x^*), \tag{4.20}$$

$$V_2(k) \quad \beta^k \sum_{i=-h^{\max}}^{-1} \sum_{j=k+i}^{k-1} y^T(j) Z y(j), \quad 0 \prec Z \quad Z^T \in \mathbb{R}^{n \times n}. \tag{4.21}$$

Compute the first difference in k of the LKF (4.19–4.21) in virtue of the system (4.16):

$$\begin{aligned}
\Delta V_1(k) \quad & V_1(k+1) - V_1(k) \\
& \beta^k [(x(k+1) - x^*)^T (\beta - 1) P_1 (x(k+1) - x^*) \\
& + (x(k+1) - x^*)^T \beta P_1 B K (x(k - h(k)) - x^*) \\
& + (x(k+1) - x^*)^T \beta P_1 G (w(k) - w^*) \\
& + (x(k+1) - x^*)^T \beta P_1 G w^* \\
& + (x(k - h(k)) - x^*)^T \beta K^T B^T P_1 (x(k+1) - x^*) \\
& + (x(k - h(k)) - x^*)^T \beta K^T B^T P_1 B K (x(k - h(k)) - x^*) \\
& + (x(k - h(k)) - x^*)^T \beta K^T B^T P_1 G (w(k) - w^*) \\
& + (x(k - h(k)) - x^*)^T \beta K^T B^T P_1 G w^*
\end{aligned}$$

$$+(w(k) - w^*)^\mathrm{T} \beta G^\mathrm{T} P_1 (x(k+1) - x^*)$$
$$+(w(k) - w^*)^\mathrm{T} \beta G^\mathrm{T} P_1 BK(x(k - h(k)) - x^*)$$
$$+(w(k) - w^*)^\mathrm{T} \beta G^\mathrm{T} P_1 G(w(k) - w^*)$$
$$+(w(k) - w^*)^\mathrm{T} \beta G^\mathrm{T} P_1 Gd^* + w^{*\mathrm{T}} \beta G^\mathrm{T} P_1 (x(k+1) - x^*)$$
$$+w^{*\mathrm{T}} \beta G^\mathrm{T} P_1 BK(x(k - h(k)) - x^*)$$
$$+w^{*\mathrm{T}} \beta G^\mathrm{T} P_1 G(w(k) - w^*) + w^{*\mathrm{T}} \beta G^\mathrm{T} P_1 Gw^*],$$

$$\Delta V_2(k) \quad V_2(k+1) - V_2(k)$$

$$\beta^k \left[h^{\max} y^\mathrm{T}(k) Z\, y(k) - \sum_{j=k-h_m}^{k-1} y^\mathrm{T}(j) Z\, y(j) \right]$$

$$\beta^k [h^{\max} y^\mathrm{T}(k) Z\, y(k) + (x(k) - x^*)^\mathrm{T} Z(x(k) - x^*)$$
$$-(x(k) - x^*)^\mathrm{T} Z(x(k - h^{\max}) - x^*)$$
$$+(x(k - h^{\max}) - x^*)^\mathrm{T} Z(x(k - h^{\max}) - x^*)].$$

As a result, we obtain

$$\Delta V(k) \quad \Delta V_1(k) + \Delta V_2(k) \quad s^\mathrm{T}(k) M(h^{\max}) s(k), \tag{4.22}$$

where

$$s(k) \quad \mathrm{col}\{\beta^{k/2}(x(k) - x^*), \beta^{k/2} K(x(k - h(k)) - x^*),$$
$$\beta^{k/2}(x(k - h^{\max}) - x^*), \beta^{k/2} y(k),$$
$$\beta^{k/2}(w(k) - w^*), \beta^{k/2} w^*\}; M(h^{\max})$$

$$\begin{bmatrix}
(\beta - 1)P_1 + Z & \beta P_1 B & -Z & 0_{n \times n} & \beta P_1 G & \beta P_1 G \\
* & \beta B^\mathrm{T} P_1 B & 0_{m \times n} & 0_{m \times n} & \beta B^\mathrm{T} P_1 G & \beta B^\mathrm{T} P_1 G \\
* & * & Z & 0_{n \times n} & 0_{n \times n} & 0_{n \times n} \\
* & * & * & h^{\max} Z & 0_{n \times n} & 0_{n \times n} \\
* & * & * & * & \beta G^\mathrm{T} P_1 G & \beta G^\mathrm{T} P_1 G \\
* & * & * & * & * & \beta G^\mathrm{T} P_1 G
\end{bmatrix},$$

the symbol "*" denotes the corresponding block in the symmetric matrix of inequality.

Stability condition (4.18) requires that the value of the LKF (4.19–4.21) decreases with time at some guaranteed rate determined by the current value of the criterion (4.11)

$$\Delta V(k) \leq -\beta^k (x(k) - x^*)^{\mathrm{T}} (W_x + K^{\mathrm{T}} W_u K)(x(k) - x^*). \quad (4.23)$$

The fulfillment of inequality (4.23) implies that condition holds, which is a sufficient condition for the asymptotic stability of closed-loop system (4.16)

$$\Delta V(k) \leq -\beta^k \lambda_{\min}(W_x + K^{\mathrm{T}} W_u K)|x(k) - x^*|^2,$$

where $\lambda_{\min}(\cdot)(\lambda_{\max}(\cdot))$ denotes the minimal (maximal) eigenvalue of matrix $(\cdot)$.

Summing up the left and right sides of (4.23) with respect to k from 0 to ∞, we obtain

$$J^\infty(k) \leq (x(0) - x^*)^{\mathrm{T}} P_1 (x(0) - x^*) + h^{\max} y^{\mathrm{T}}(0) Z\, y(0)$$

$$- \sum_{j=-h^{\max}}^{-1} y^{\mathrm{T}}(j) Z\, y(j), \quad (4.24)$$

that is, the LKF (4.19–4.21), computed at the time $k \quad 0$, determines the upper boundary value of the criterion (4.11). Then the guaranteed cost control synthesis problem is equivalent to the problem

$$u(k) \quad \arg \min_{u(k) \in U} V(0). \quad (4.25)$$

The classic approach to solving the problem of control synthesis minimizing the quadratic quality criterion is based on the solution of the algebraic Riccati equation [29] and guarantees the optimal solution for arbitrary initial conditions. The control actions obtained as a result of solving the problem (4.25) differs in that it guarantees an optimal solution only for the given initial conditions.

4.5 Guaranteed Cost Inventory Control Based on Invariant Ellipsoids Method

An effective method for solving guaranteed cost control problems under uncertainty is invariant ellipsoids method [30]. We will use ellipsoid description in the form

$$E(x^*, Q) \quad \{x \in \mathbb{R}^n : (x(k) - x^*)^{\mathrm{T}} Q^{-1}(x(k) - x^*) \leq 1\}, \quad (4.26)$$

where $x^* \in \mathrm{R}^n$ is a vector whose components determine the coordinates of ellipsoid center; $0 \prec Q \in \mathrm{R}^{n \times n}$ is an ellipsoid matrix.

The representation (4.26) is the most convenient way of describing the ellipsoid from the point of view of LKF constructing; however, it is non-linear in the totality x^* and Q. We reduce it to linear with the help of a Schur lemma modification for non-strict matrix inequalities [30]

$$E(x^*, Q) \quad \left\{ x \in \mathrm{R}^n : \begin{bmatrix} 1 & (x(k) - x^*)^{\mathrm{T}} \\ x(k) - x^* & Q \end{bmatrix} \succeq 0 \right\}.$$

An ellipsoid (4.26) is said to be invariant with respect to the state for the system (4.10), if from the condition $x(0) \in E(x^*, Q)$ follows that $x(k) \in E(x^*, Q) \ \forall k \geq 0$. In other words, any trajectory of the system, starting in invariant ellipsoid, remains in it at any time. A trajectory that starts outside the invariant ellipsoid converges to it (asymptotically or in finite time), that is, the ellipsoid (4.26) is attractive [30].

Invariant ellipsoids can be considered as an approximation of the reachability set [31] of a closed system, that is, they allow one to characterize the effect of external disturbances and the uncertainty of the model parameters on the trajectory of a closed system. Then the minimization (in some sense) of the invariant ellipsoid (4.26) ensures the robust control of the system (4.10).

We will approximate the set of values of external actions for each network node by the least volume ellipsoid

$$E(w^*, Q_w) \quad \{w \in \mathrm{R}^n : (w(k) - w^*)^{\mathrm{T}} Q_w^{-1} (w(k) - w^*) \leq 1\}. \quad (4.27)$$

Let us introduce the following notations:

$$W \quad Q_w^{-1/2} \succ 0, \quad z \quad Q_w^{-1/2} w^* \quad (4.28)$$

and represent the ellipsoid (4.27) in the form $E \quad \{w \in \mathrm{R}^n : \|Ww - z\| \leq 1\}$. The inequality $\|Ww - z\| \leq 1$ is equivalent to an inequality $(Ww - z)^{\mathrm{T}}(Ww - z) \leq 1$ which, using Schur lemma, can be represented in the LMI form

$$\begin{bmatrix} 1 & (Ww - z)^{\mathrm{T}} \\ Ww - z & I_n \end{bmatrix} \succeq 0.$$

The volume of the ellipsoid (4.27) is equal to $c_n \sqrt{\det Q_w}$, where c_n is the volume of the unit ball in the n-dimensional space. The function $f(Q_w)$ $\sqrt{\det Q_w}$ is nonlinear. In order to represent the problem of finding an approximating ellipsoid in the form of a convex optimization problem, we choose as an objective function $f(Q_w)$ $-\lg \det Q_w$ that is convex and linear for $Q_w \succ 0$. According to (4.28) Q_w W^{-2}, then $\lg \det Q_w$ $\lg \det W^{-2}$ $-2\lg \det W$. Finally, we obtain semidefinite programming problem (SDP) [31]

$$-\lg \det W \to \min \qquad (4.29)$$

under constraints on the matrix W $W^{\mathrm{T}} \in \mathrm{R}^{n \times n}$ and vector $z \in \mathrm{R}^n$ variables:

$$W \succ 0, \quad \begin{bmatrix} 1 & (Ww_j - z)^{\mathrm{T}} \\ Ww_j - z & I_n \end{bmatrix} \succeq 0, \quad j \quad \overline{1, 2^n}, \qquad (4.30)$$

where $w_j \in \mathrm{R}^n$ are vectors containing the coordinates of the vertices of a convex polyhedron that determines the set of admissible values of the external actions of the network node, that is, all possible combinations of the components of the boundary value vectors $w^{\max}$ and $w^{\min}$. In this case, the upper boundary values are calculated using the Leontief productive model in accordance with (4.7), that is, $w^{\max}$ $\bar{x}^*$, and the lower ones are analogous with the replacement of the vector $d^{\max}$ by the vector $d^{\min}$.

The next step in solving the problem is to approximate the admissible set of state values of each network node by an ellipsoid of least volume

$$E(\tilde{x}^*, Q_x) \quad \{x \in \mathrm{R}^n : (x(k) - \tilde{x}^*)^{\mathrm{T}} Q_x^{-1}(x(k) - \tilde{x}^*) \leq 1\}. \qquad (4.31)$$

As there is no need for levels of available stocks $x(k)$ to exceed the size of safety stocks x^*, the constraint (4.12) is represented as

$$x(k) \in X \quad \{x \in \mathrm{R}^n : 0 \leq x \leq x^*\}. \qquad (4.32)$$

Then vector $\tilde{x}^* \in \mathrm{R}^n$, which determine the coordinates of the center of the ellipsoid (4.31), is determined using safety stock vector

$$\tilde{x}^* \quad \frac{1}{2}x^*, \qquad (4.33)$$

and the ellipsoid matrix $Q_x \in \mathrm{R}^{n \times n}$ is calculated as a result of solving the SDP problem

$$\mathrm{tr}\, Q_x \to \min \qquad (4.34)$$

under constraints on the matrix variable:

$$Q_x \succ 0, \quad \begin{bmatrix} 1 & (x_j - \tilde{x}^*)^{\mathrm{T}} \\ x_j - \tilde{x}^* & Q_x \end{bmatrix} \succeq 0, \quad j \quad \overline{1, 2^n}, \qquad (4.35)$$

where $x_j \in \mathbb{R}^n$ are vectors, containing coordinates of vertices of a convex polyhedron X, that is, all possible combinations of the components of the vector x^* and the vector of minimal admissible values of states consisting of zeros.

4.6 Synthesis of Guaranteed Cost Inventory Control

By analogy with (4.26), we define a family of ellipsoids that are invariant for system (4.10) with respect to the state with delay

$$E_j(x^*, R) \quad \{x \in \mathbb{R}^n : (x(k-j) - x^*)^T R^{-1}(x(k-j) - x^*) \le 1\},$$

$$j \quad \overline{1, h^{\max}}. \tag{4.36}$$

We understand the sum and difference of ellipsoids in the sense of the sum and difference of sets on Minkowski [32]. Then the sum of the ellipsoid (4.26) and the family of ellipsoids (4.36) can be considered as an approximation of the reachability set of the descriptor system (4.16). The second term in the LKF (4.19–4.21) is represented in the form

$$V_2(k) \quad \beta^k \left[h^{\max}(x(k) - x^*)^T Z(x(k) - x^*) \right.$$

$$\left. - \sum_{i=-h^{\max}}^{-1} (x(k-i) - x^*)^T Z(x(k-i) - x^*) \right]. \tag{4.37}$$

Comparison of expressions (4.20) and (4.26) and also (4.37) and (4.36) allows to assert that if conditions holds

$$P_1 \quad Q^{-1}, \quad Z \quad R^{-1}, \tag{4.38}$$

then the sum of the ellipsoid (4.26) and the family of ellipsoids (4.36) represents a set that is an approximation of the level set of the LKF (4.19–4.21).

Thus, the synthesis of guaranteed cost control for system (4.16) reduces to constructing a controller that ensures minimization by a certain criterion of the sum of these invariant ellipsoids under given constraints. The sum of the squares of the semi-axes of the ellipsoids is chosen as the criterion, that is, the sum of the trace of the matrix Q and the trace of matrix R.

Inequality (4.23), taking into account (4.22), may be written as

$$\Delta V(k) \le s^T(k) M_0(h^{\max}) s(k), \tag{4.39}$$

where

$$M_0(h^{\max}) \begin{bmatrix} (\beta-1)P_1 & \beta P_1 B & -Z & 0_{n\times n} & \beta P_1 G & \beta P_1 G \\ +Z+W_x & & & & & \\ +K^\mathrm{T} W_u K & & & & & \\ * & \beta B^\mathrm{T} P_1 B & 0_{m\times n} & 0_{m\times n} & \beta B^\mathrm{T} P_1 G & \beta B^\mathrm{T} P_1 G \\ * & * & Z & 0_{n\times n} & 0_{n\times q} & 0_{n\times q} \\ * & * & * & h^{\max} Z & 0_{n\times q} & 0_{n\times q} \\ * & * & * & * & \beta G^\mathrm{T} P_1 G & \beta G^\mathrm{T} P_1 G \\ * & * & * & * & * & \beta G^\mathrm{T} P_1 G \end{bmatrix}.$$

We introduce the notations: $f_j(s) \quad s^\mathrm{T} M_j s$, $M_j \in \mathrm{R}^{(5n+m)\times(5n+m)}$, $j \quad 0, 1$, M_1 block diag$\{0_{n\times n}, 0_{m\times m}, 0_{n\times n}, 0_{n\times n}, Q_w^{-1}, 0_{n\times n}\}$.

Then the inequality (4.39) ensuring the decrease of the LKF (4.19–4.21) along any trajectory of a closed-loop system (4.16), and also the inequality (4.27) describing the ellipsoid approximating the set of values of external actions of the network node, we write in the form $f_0(s) \le 0 \; \forall s : f_1(s) \le 1$.

Taking into account the lossless of the S-procedure under one constraint [29], the sufficient condition for the sign-definiteness of the written quadratic forms is the fulfillment for some scalar parameter $\tau > 0$ of matrix inequality

$$M_0(h^{\max}) \preceq \tau M_1,$$

which can be represent in the form

$$\begin{bmatrix} \beta^{1/2} \\ \beta^{1/2} B^\mathrm{T} \\ 0_{1\times n} \\ 0_{1\times n} \\ \beta^{1/2} G^\mathrm{T} \\ \beta^{1/2} G^\mathrm{T} \end{bmatrix} P_1 \begin{bmatrix} \beta^{1/2} & \beta^{1/2} B & 0_{n\times 1} & 0_{n\times 1} & \beta^{1/2} G & \beta^{1/2} G \end{bmatrix}$$

$$+ \begin{bmatrix} \Psi & 0_{m\times m} & -Z & 0_{n\times n} & 0_{n\times n} & 0_{n\times n} \\ * & 0_{m\times m} & 0_{m\times n} & 0_{m\times n} & 0_{m\times n} & 0_{m\times n} \\ * & * & Z & 0_{n\times n} & 0_{n\times n} & 0_{n\times n} \\ * & * & * & h^{\max} Z & 0_{n\times n} & 0_{n\times n} \\ * & * & * & * & -\tau Q_w^{-1} & 0_{n\times n} \\ * & * & * & * & * & 0_{n\times n} \end{bmatrix} \preceq 0,$$

where $\Psi \quad -P_1 + Z + W_x + K^\mathrm{T} W_u K$.

Using the Schur lemma modification for non-strict matrix inequalities, we write the last inequality in the form

$$
\begin{bmatrix}
-P_1^{-1} & \beta^{1/2} & \beta^{1/2}B & 0_{n\times n} & 0_{n\times n} & \beta^{1/2}G & \beta^{1/2}G \\
* & -P_1 + Z + W_x + K^\mathrm{T}W_u K & 0_{n\times m} & -Z & 0_{n\times n} & 0_{n\times n} & 0_{n\times n} \\
* & * & 0_{m\times m} & 0_{m\times n} & 0_{m\times n} & 0_{m\times n} & 0_{m\times n} \\
* & * & * & Z & 0_{n\times n} & 0_{n\times n} & 0_{n\times n} \\
* & * & * & * & h^{\max}Z & 0_{n\times n} & 0_{n\times n} \\
* & * & * & * & * & -\tau Q_w^{-1} & 0_{n\times n} \\
* & * & * & * & * & * & 0_{n\times n}
\end{bmatrix} \preceq 0.
$$

We introduce the matrix variable $Y \quad KQ$, from where by virtue of $Q \succ 0$ gain matrix K recover uniquely

$$K \quad YQ^{-1}. \tag{4.40}$$

Three times applying Schur lemma, and then applying to the resulting matrix of inequality a congruent transformation with matrix block $\operatorname{diag}\{I_n, P_1^{-1}, I_n, -Z^{-1}, -Z^{-1}, I_n, I_n, I_n, I_n, I_n\}$ and using the matrix variables (4.38), we obtain LMI

$$
\begin{bmatrix}
-Q & \beta^{1/2}Q & \beta^{1/2}B & 0_{n\times n} & 0_{n\times n} & \beta^{1/2}G & \beta^{1/2}G & 0_{n\times n} & 0_{n\times n} & 0_{n\times m} \\
* & -Q & 0_{n\times m} & Q & 0_{n\times n} & 0_{n\times n} & 0_{n\times n} & Q & QW_x^{1/2} & Y^\mathrm{T}W_u^{1/2} \\
* & * & 0_{m\times m} & 0_{m\times n} & 0_{m\times n} & 0_{m\times n} & 0_{m\times n} & 0_{m\times n} & 0_{m\times n} & 0_{m\times m} \\
* & * & * & -R & 0_{n\times n} & 0_{n\times n} & 0_{n\times n} & 0_{n\times n} & 0_{n\times n} & 0_{n\times m} \\
* & * & * & * & -h^{\max}R & 0_{n\times n} & 0_{n\times n} & 0_{n\times n} & 0_{n\times n} & 0_{n\times m} \\
* & * & * & * & * & -\tau Q_w^{-1} & 0_{n\times n} & 0_{n\times n} & 0_{n\times n} & 0_{n\times m} \\
* & * & * & * & * & * & 0_{n\times n} & 0_{n\times n} & 0_{n\times n} & 0_{n\times m} \\
* & * & * & * & * & * & * & -R & 0_{n\times n} & 0_{n\times m} \\
* & * & * & * & * & * & * & * & -I_n & 0_{n\times m} \\
* & * & * & * & * & * & * & * & * & -I_m
\end{bmatrix} \preceq 0.
$$
$$\tag{4.41}$$

The upper boundary value of the quality criterion (4.11) according to (4.24) depends on the initial conditions of the system (4.16). In accordance with the approach proposed in [28], we assume that the initial state of each network node belongs to the ellipsoid

$$E(\tilde{x}^*, Q_x) \quad \{x \in \mathrm{R}^n : (x(-j) - \tilde{x}^*)^\mathrm{T}Q_x^{-1}(x(-j) - \tilde{x}^*) \le 1,$$
$$j \quad 0, 1, \ldots, h^{\max}\},$$

whose matrix Q_x is calculated as a result of solving the problem (4.34) under constraints (4.35). Then the estimate (4.24) leads to the inequality

$$J^\infty(k) \le \lambda_{\max}(Q_x^\mathrm{T}P_1 Q_x) + h^{\max}\lambda_{\max}(Q_x^\mathrm{T}Z Q_x),$$

whence, taking into account the notations (4.38), an inequality follows for estimating the upper boundary value J^* of the quality index of a node closed by feed-back control (4.9)

$$J^\infty(k) \leq \lambda_{\max}(Q_x^\mathrm{T} Q^{-1} Q_x) + h^{\max} \lambda_{\max}(Q_x^\mathrm{T} R^{-1} Q_x) \quad J^*. \quad (4.42)$$

Thus, if for system (4.16) with an uncertain delay (4.2) and constraints (4.5), (4.32) matrices $Q \in \mathrm{R}^{n \times n}$, $R \in \mathrm{R}^{n \times n}$ and $Y \in \mathrm{R}^{m \times n}$ are obtained as a solution of the optimization problem

$$\mathrm{tr}\, Q + \mathrm{tr}\, R \to \min \quad (4.43)$$

with constraint (4.41) on matrix variables $Q \quad Q^\mathrm{T} \succ 0$, $R \quad R^\mathrm{T} \succ 0$, Y, and scalar parameter $\tau > 0$, then:

i) for any initial state $x(-j) \in E(\tilde{x}^*, Q_x), j \quad 0, 1, \ldots, h^{\max}$, any admissible external demand $d(k) \in D$, and any delay value $0 \leq h(k) \leq h^{\max}$, the closed-loop system (4.16) is stable;

ii) the control

$$u(k) \quad Y\, Q^{-1}(x(k) - x^*) \quad (4.44)$$

provides the minimum of matrix trace criterion for sum of invariant ellipsoid (4.26) and the family of ellipsoids (4.36) among all linear control (4.9);

iii) the value of the local quality criterion (4.11) of a node, closed by feedback (4.44), satisfies inequality (4.42).

The problem of minimizing the linear function (4.43) under constraint represented in the form of LMI (4.41) is a SDP problem and can be solved numerically. Since the descriptor model (4.16) is equivalent to the model (4.10) of the supply network node, the control law (4.44) provides an optimal decentralized guaranteed cost inventory control for the local network node with an uncertain delay (4.2).

4.7 Stability Analysis of Controlled Supply Network

Controlled supply networks can be considered as large-scale systems for which the problem of local controllers design should be supplemented by the problem of ensuring the stability of the entire system in the presence of interconnections. Vector Lyapunov functions method (VLF) as well as a comparison system approach [33] are used to analyze the stability of controlled supply network, consisting of a set of interconnected nodes with local feedback controllers.

The equations of closed-loop nodes descriptor models, taking into account the interconnections (4.3), take the form

$$
E_i \xi_i(k+1) \quad A_i \xi_i(k) - \tilde{B}_i \sum_{j=k-h(k)}^{k-1} \xi_i(j) + \sum_{j=1, j \neq i}^{n} B_{ij} u_j(k) + F_i d(k),
$$

(4.45)

where $\tilde{B}_i \quad [\, 0_{2n_i \times n_i} \mid \overline{B}_i \,]$, $B_{ij} \quad \overline{G}_i \Pi_{ij}$, $F_i \quad \overline{G}_i H_i$.

After substitution of control (4.44) in (4.45), we obtain

$$
E_i \xi_i(k+1) \quad A_i \xi_i(k) - \tilde{B}_i \sum_{j=k-h(k)}^{k-1} \xi_i(j)
$$

$$
+ \sum_{j=1, j \neq i}^{n} B_{ij} Y_j Q_j^{-1}(x_j(k) - x_j^*) + F_i d(k).
$$

We extend state-space descriptor models of nodes with composite state vector $\overline{\xi}_i(k) \quad \mathrm{col}\{\xi_i(k), \xi_i(k-1), \ldots, \xi_i(k-h_i^{\max})\} \in \mathrm{R}^{\overline{N}_i}$, where $\overline{N}_i \quad 2n_i(1+h_i^{\max})$, and introduce block matrices:

$\overline{E}_i \quad$ block diag$\{E_i, E_i, \ldots, E_i\} \in \mathrm{R}^{\overline{N}_i \times \overline{N}_i}$,

$$
\overline{A}_i \quad
\begin{bmatrix}
A_i & -\tilde{B}_i & -\tilde{B}_i & \cdots & -\tilde{B}_i \\
0_{2n_i \times 2n_i} & I_{2n_i} & -\tilde{B}_i & \cdots & -\tilde{B}_i \\
0_{2n_i \times 2n_i} & 0_{2n_i \times 2n_i} & I_{2n_i} & \cdots & -\tilde{B}_i \\
\vdots & \vdots & \vdots & \ddots & \vdots \\
0_{2n_i \times 2n_i} & 0_{2n_i \times 2n_i} & 0_{2n_i \times 2n_i} & \cdots & I_{2n_i}
\end{bmatrix}
\in \mathrm{R}^{\overline{N}_i \times \overline{N}_i},
$$

$\overline{F}_{ij} \quad$ block diag$\{B_{ij} Y_j Q_j^{-1}, 0_{2n_i \times 2n_j}, \ldots, 0_{2n_i \times 2n_j}\} \in \mathrm{R}^{\overline{N}_i \times \overline{N}_j}$,

$\overline{F}_i \quad [\, F_i^{\mathrm{T}} \mid 0_{2n_i \times q}^{\mathrm{T}} \mid \cdots \mid 0_{2n_i \times q}^{\mathrm{T}} \,]^{\mathrm{T}} \in \mathrm{R}^{\overline{N}_i \times q}$,

$\overline{P}_i \quad$ block diag$\left\{ E_i^{\mathrm{T}} \begin{bmatrix} P_{1i} & P_{2i} \\ P_{2i}^{\mathrm{T}} & P_{3i} \end{bmatrix} E_i, \begin{bmatrix} 0_{n_i} & Z_i \\ 0_{n_i} & 0_{n_i} \end{bmatrix}, \ldots, \begin{bmatrix} 0_{n_i} & Z_i \\ 0_{n_i} & 0_{n_i} \end{bmatrix} \right\}$

$\in \mathrm{R}^{\overline{N}_i \times \overline{N}_i}$.

The matrices P_{1i} of block matrices P_i are calculated in accordance with (4.38) after solving the problems (4.43), the matrices P_{2i} and P_{3i} are chosen arbitrarily, since equality holds

$$E_i^{\mathrm{T}} \begin{bmatrix} P_{1i} & P_{2i} \\ P_{2i}^{\mathrm{T}} & P_{3i} \end{bmatrix} E_i \quad \begin{bmatrix} P_{1i} & 0_{n\times n} \\ 0_{n\times n} & 0_{n\times n} \end{bmatrix}, \quad i \quad \overline{1, n}.$$

The extended descriptor model of the node is represented in the form

$$\overline{E_i}\xi_i(k+1) \quad \overline{A_i}\,\overline{\xi_i}(k) + \sum_{j=1, j\neq i}^{n} \overline{F_{ij}}\,\overline{\xi_j}(k) + \overline{F_i}d(k). \tag{4.46}$$

For a supply network consisting of interconnected nodes described by (4.46), consider VLF

$$\Psi(k) \quad \mathrm{col}\{V_1(k), \dots, V_n(k)\}, \tag{4.47}$$

the components of which are partial LKF of local nodes in the form (4.19–4.21). The general Lyapunov function is constructed on the basis of VLF (4.47)

$$V_0(k) \quad \sum_{i=1}^{n} l_{0i} V_i(k) \quad L_0 \Psi(k), \tag{4.48}$$

where $L_0 \quad [l_{01}, \dots, l_{0n}] \in \mathrm{R}^{1\times n}, l_{0i} > 0, i \quad \overline{1, n}.$

We collate the set of controlled local nodes of network with a linear comparison system, which is determined by difference equations:

$$\begin{aligned} v(k+1) &\quad \Lambda v(k), \quad v(0) \quad \Psi(0), \\ \eta(k) &\quad P_0 v(k), \end{aligned} \tag{4.49}$$

where $v(k) \in \mathrm{R}^n$ is a state vector of the comparison system; $\eta(k)$ is a scalar function that is the output of a comparison system; $\Lambda \in \mathrm{R}^{n\times n}$ is a dynamics matrix of the comparison system with nonnegative elements.

Quadratic forms $V_i(\overline{A_i}(k)) \quad \beta^k \overline{\xi_i}^{\mathrm{T}}(k) \overline{A_i}^{\mathrm{T}} \overline{P_i}\,\overline{A_i}\,\overline{\xi_i}(k)$ and $V_j(k)$ $\beta^k \overline{\xi_j}^{\mathrm{T}}(k) \overline{P_j}\,\overline{\xi_j}(k)$ define a pencil of quadratic forms $V_i(\overline{A_i}(k)) - \mu V_j(k)$, where μ is a some scalar parameter. Similarly, quadratic forms $V_i(\overline{F_{ij}}(k))$ $\beta^k \overline{\xi_i}^{\mathrm{T}}(k) \overline{F_{ij}}^{\mathrm{T}} \overline{P_i}\,\overline{F_{ij}}\,\overline{\xi_i}(k)$ and $V_j(k)$ define a pencil of forms $V_i(\overline{F_{ij}}(k)) - \mu V_j(k)$. The calculation of the elements of matrix Λ is carried out according to the characteristic equations of quadratic forms pencils [34]:

$$\begin{aligned} \det(\overline{A_i}^{\mathrm{T}} \overline{P_i A_i} - \mu_{ii}\overline{P_i}) &\quad 0, \quad i \quad \overline{1, n}, \\ \det(\overline{F_{ij}}^{\mathrm{T}} \overline{P_i F_{ij}} - \mu_{ij}\overline{P_j}) &\quad 0, \quad i, j \quad \overline{1, n}, \, j \, / \, i, \end{aligned}$$

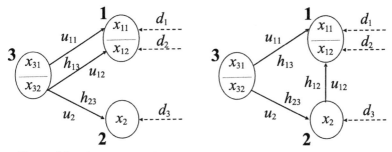

Figure 4.2 The structure of the supply network: *a* – existing; *b* – proposed.

where $[\Lambda]_{ij}$ $(\mu_{ij}^{\max})^{1/2}$, $\mu_{ij}^{\max}$ is a maximum value of the root of corresponding equation.

As a result, for the vector (4.48) and general (4.49) Lyapunov functions, the following inequalities hold:

$$\Psi(k) \leq v(k), \quad V_0(k) \leq \eta(k).$$

Thus, the comparison system (4.49) component-wise majorizes the vector and general Lyapunov functions constructed for a controlled supply network. As a result, the stability analysis of the supply network is reduced to a stability analysis of the linear positive comparison system (4.49). Therefore, to ensure the stability of a controlled supply network, the nodes of which are closed by feedbacks (4.44), it is necessary to choose such values of weight matrices W_{xi} and W_{ui} of local quality criteria (4.11) for which the dynamics matrix of the comparison system is a Shur's matrix, that is, condition holds

$$\sigma(\Lambda) < 1, \tag{4.50}$$

where $\sigma(\cdot)$ is a spectral radius of the matrix $(\cdot)$.

4.8 Numerical Example

As an example, we study a company engaged in the sale of alcoholic beverages. Data are provided by Cloud Works LTD (http://www.cloudwk.com/), which develops software for automated inventory control systems. The two most popular types of products are selected for modeling, sales of which are carried out independently of each other. The structure of supply network is shown in Figure 4.2.

At present, node 1 sells two types of products x_{11} and x_{12} and replenishes their stocks directly from the warehouse 3. The sampling period is 1 day.

The lead time h_{13} is from 2 to 6 days. Node 2 sells product x_2 and replenishes stocks from the warehouse 3. The lead time h_{23} is from 2 to 4 days. The arcs d_1, d_2, and d_3, shown in dotted lines, denote an external demand (see Figure 4.2(a)).

The company is interested in the issue of optimizing the transportation process by sending product x_{12} from warehouse 3 to node 2, for which node 1 will form orders for its delivery (see Figure 4.2(b)). The lead time h_{12} is from 1 to 2 days.

The dimensions of the model are equal: n_1　m_1　2, n_2　m_2　1, n_3　m_3　2, N　5. The following values of the matrices describing the interconnections between the nodes correspond to the variant a: Π_{21}

$\begin{bmatrix} 0 & 0 \end{bmatrix}$, Π_{31} 　$\begin{bmatrix} 1 & 0 \\ 0 & 1 \end{bmatrix}$, Π_{32} 　$\begin{bmatrix} 0 \\ 1 \end{bmatrix}$; and the variant b: Π_{21}

$\begin{bmatrix} 0 & 1 \end{bmatrix}$, Π_{31} 　$\begin{bmatrix} 1 & 0 \\ 0 & 0 \end{bmatrix}$, Π_{32} 　$\begin{bmatrix} 0 \\ 1 \end{bmatrix}$. The matrices of external demand

are equal: H_1 　$\begin{bmatrix} 1 & 0 & 0 \\ 0 & 1 & 0 \end{bmatrix}$, H_2 　$\begin{bmatrix} 0 & 0 & 1 \end{bmatrix}$, H_3 　$\begin{bmatrix} 0 & 0 & 0 \\ 0 & 0 & 0 \end{bmatrix}$.

The boundary values of demand were determined on the basis of information on sales volumes for the first 50 days of 2018 year: $d^{\min}$ col$\{6; 19; 22\}$, $d^{\max}$ 　col$\{112; 326; 263\}$. In accordance with (4.4) and (4.5), the size of the safety resources stocks for both options are determined, which are selected as the initial state of the system: $x_a(0)$ 　x_a^* 　col$\{672; 1956; 1052; 112; 589\}$, $x_b(0)$ 　x_b^* col$\{672; 652; 2356; 112; 589\}$. It is easy to see that the total sizes of safety stocks in both variants are the same; they are only distributed in different ways among the nodes of the network.

The numerical solution of the corresponding SDP problems is performed in the MATLAB using the freely distributed software CVX [15]. As a result, the parameters of ellipsoids (4.27) approximating the sets of values of external actions for each network node are determined: for node 1, variant a: Q_{w_1} 　diag$\{5, 616 \cdot 10^3; 4, 713 \cdot 10^4\}$, w_1^* 　col$\{59; 172, 5\}$; variant b: the problems are solved independently for different types of products (the lead time is different); as a result, the ellipsoids degenerate into segments: $Q_{w_{11}}$ 　2, 809 $\cdot$ 10^3, w_{11}^* 　59, $Q_{w_{12}}$ 　2, 356 $\cdot$ 10^4, w_{12}^* 　172, 5; for node 2, the ellipsoid also degenerates into segment: variant a: Q_{w_2} 1, 452 $\cdot$ 10^4, w_2^* 　142, 5; variant b: Q_{w_2} 　7, 508 $\cdot$ 10^4, w_2^* 　315; for node 3, the ellipsoid parameters in both variants are the same: Q_{w_3} diag$\{5, 616 \cdot 10^3; 1, 502 \cdot 10^5\}$, w_3^* 　col$\{59; 315\}$.

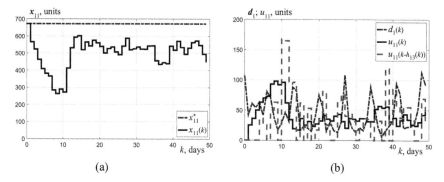

Figure 4.3 Changes in stock levels of product x_{11}.

The values of the delays $h_{13}(k) \in [2; 6]$, $h_{23}(k) \in [2; 4]$ and $h_{12}(k) \in [1; 2]$ in each period were randomly generated, β $0, 8$.

The controller matrices are calculated in accordance with (4.40) using the results of the solution for each node of problem (4.43) under constraint (4.41): variant *a*: K_1 $\begin{bmatrix} 0,139 & -0,167 \\ -0,033 & -0,192 \end{bmatrix}$, K_2 $-0,243$,

K_3 $\begin{bmatrix} -0,143 & -0,243 \\ 0,341 & -0,699 \end{bmatrix}$; variant *b*: K_{11} $-0,242$, K_{12} $-0,420$,

K_2 $-0,222$, K_3 $\begin{bmatrix} -0,119 & -0,201 \\ 0,341 & -0,699 \end{bmatrix}$.

The comparison system for a controlled supply network is built. The dynamics matrix of the comparison system for variant *b* is

$$\Lambda \quad \begin{bmatrix} 5,145 \cdot 10^{-4} & 0 & 0 & 0 \\ 0 & 5,050 \cdot 10^{-4} & 0 & 0 \\ 0 & 4,164 \cdot 10^{-1} & 5,010 \cdot 10^{-4} & 0 \\ 3,353 \cdot 10^{-1} & 0 & 3,333 \cdot 10^{-1} & 7,646 \cdot 10^{-4} \end{bmatrix}.$$

It is easy to verify that the stability condition (4.50) is satisfied. Thus, the comparison system, and, consequently, the controlled supply network are stable.

The simulation results for variant *b* are shown in Figures 4.3–4.5, where a – values of safety and available inventory levels and b – the values of demand, nominal and real (with delay) control actions.

It can be seen from the graphs that the presence of delays leads to losses in deliveries and unevenness in the sizes of deliveries due to the combination of orders made in different periods. Nevertheless, the proposed inventory control

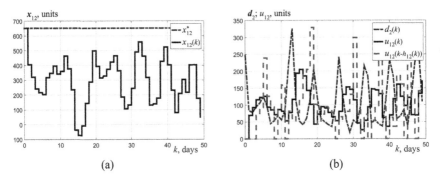

Figure 4.4 Changes in stock levels of product x_{12}.

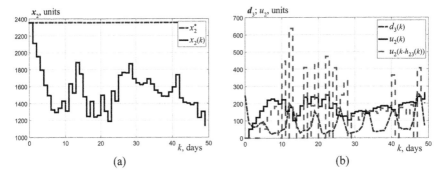

Figure 4.5 Changes in stock levels of product x_2.

strategy ensures that there is no deficit of resources in the network nodes, as well as guaranteed costs for production and storage of resources.

The value of the quality criterion obtained by summing the values of the local criteria calculated in accordance with (4.11) for variant b is equal $J_\Sigma^\infty(k) \quad 2,375 \cdot 10^6$, which is 40.8 % less than the value obtained for variant a. Thus, the resource transport scheme shown in Figure 4.2(b), is more advantageous. The upper boundary value of the criterion obtained by summing the local boundary values calculated in accordance with (4.42) is equal $J_\Sigma^* \quad 9,874 \cdot 10^6$.

4.9 Conclusion

The approach for solving the decentralized guaranteed cost inventory control synthesis problem in supply networks with uncertain transport delays under the conditions of the "unknown but bounded" demand is proposed. Using the descriptor transformation of the discrete controlled models of network nodes, the delay-dependent Lyapunov-Krasovskii functional is constructed. The condition of existence of local stabilizing controller that implements a control law in the form of linear feedback with respect to deviation between available and safety resource stock levels is obtained. The problem of the controller synthesis, which minimizes the upper boundary value of the quadratic quality criterion, is reduced to the semidefinite programming problem on the basis of the invariant ellipsoids method using the technique of linear matrix inequalities. The method of vector LFs and the method of constructing a comparison system are used to stability analysis of supply network taking into account the interconnections between controlled nodes.

References

[1] J. Shapiro, 'Modelling the supply chain', 2nd ed., Duxbury Press, 2007.

[2] C. Daganzo, 'A theory of supply chains', Springer, Heidelberg, 2003.

[3] X. Zhu, G. Yang, 'New results of stability analysis for systems with time-varying delay', Int. J. Robust Nonlinear Control, vol. 20, pp. 596–606, 2010, doi: 10.1002/rnc.1456.

[4] R. Bate, 'The optimal control of systems with transport lags', in Advances in Control Systems, ed. by C. Leondes, vol. 7, pp. 165–224, Academic Press, 1969.

[5] G. Slater, W. Well, 'On the reduction optimal time delay systems to ordinary ones', IEEE Trans. Automat. Contr., vol. AC-17, pp. 154–155, 1972.

[6] E. Camacho, C. Bordons, 'Model predictive control', Springer-Verlag, 2004.

[7] E. Aggelogiannaki, P. Doganis, H. Sarimveis, 'An adaptive model predictive control configuration for production-inventory systems', Int. J. Prod. Econ., vol. 114, pp. 165–178, 2008.

[8] D. Bertsekas, I. Rhodes, 'Recursive state estimation for a set-membership description of uncertainty', IEEE Trans. Automat. Control, vol. 16, pp. 117–128, 1971.

[9] N. Krasovskii, 'On the application of the second Lyapunov method for equations with time delays', Applied Mathematics and Mechanics, vol. 20, no. 3, pp. 315–327, 1956 (in Russian).

[10] B. Razumikhin, 'On the stability of systems with delay', Applied Mathematics and Mechanics, vol. 20, no. 4, pp. 500–512, 1956 (in Russian).

[11] A. Teel, 'Connections between Razumikhin-type theorems and the ISS nonlinear small gain theorem', IEEE Trans. Automat. Control, vol. 43, no. 7, pp. 960–964, 1998.

[12] V. Kolmanovskii, A. Myshkis, 'Introduction to the theory and applications of functional differential equations', Kluwer Academic Publishers, 1999.

[13] Y. Nesterov, A. Nemirovsky, 'Interior-point polynomial algorithms in convex programming', SIAM, 1994.

[14] J. Sturm, 'Using SeDuMi 1.02, a MATLAB toolbox for optimization over symmetric cones', Optimization Methods and Software, vol. 11, no. 12, pp. 625–653, 1999, doi: 10.1080/10556789908805766.

[15] M. Grant, S. Boyd, 'CVX: MATLAB software for disciplined convex programming, version 2.0', Mode of access: URL: http://cvxr.com/cvx, last access: 12.02.18.

[16] R. Gielen, 'Stability analysis and control of discrete-time systems with delay', Technische Universiteit Eindhoven, 2013, doi: 10.6100/IR747896.

[17] V. Wu, Y. He, J.-H. She, 'Stability analysis and robust control of time-delay systems', Springer, 2010.

[18] E. Fridman, U. Shaked, 'A descriptor system approach to H_∞ control of linear time-delay systems', IEEE Trans. Automat. Control, vol. 47, pp. 253–270, 2002.

[19] W. Zhang et al., 'Robust stability test for uncertain discrete-time systems: a descriptor system approach', Lat. Am. Appl. Res., vol. 41, no. 4, pp. 359–364, 2011.

[20] K. Subramanian et al., 'Integration of control theory and scheduling methods for supply chain management', Comput. Chem. Eng., vol. 51, pp. 4–20, 2013, doi: 10.1016/j.compchemeng.2012.06.012.

[21] M. Ortega, L. Lin, 'Control theory applications to the production-inventory problem: a review', Int. J. Prod. Res., vol. 42, pp. 2303–2322, 2004, doi: 10.1080/00207540410001666260.

[22] Y. Dorofieiev, A. Nikulchenko, 'Construction of mathematical models of controlled supply networks with allowance for flow delays', System Research and Information Technologies, vol. 1, pp. 16–27, 2013 (in Russian).

[23] V. Lototskii, A. Mandel', 'Models and methods of inventory management', Science, Moscow, 1991 (in Russian).

[24] A. Solodovnikov, V. Babaitsev, A. Brailov, 'Mathematics in economics', Finance and Statistics, Moscow, 1999 (in Russian).

[25] E. Fridman, U. Shaked, 'Robust stability of uncertain discrete systems with time-varying delay', Proc. 16-th IFAC World Congress, Prague, Elsevier, vol. 38, no. 1, pp. 54–59, 2005.

[26] S. Chang, T. Peng, 'Adaptive guaranteed cost control of systems with uncertain parameters', IEEE Trans. Autom. Contr., vol. AC-17 (4), pp. 474–483, 1972.

[27] V. Afanasiev, 'Guaranteed control of nonlinear objects', MIEM, Moscow, 2012 (in Russian).

[28] I. Petersen, D. McFarlane, M. Rotea, 'Optimal guaranteed cost control of discretetime uncertain linear systems', Int. J. Robust Nonlinear Control, vol. 8, pp. 649–657, 1998.

[29] B. Polyak, P. Shcherbakov, 'Robust stability and control', Science, Moscow, 2002 (in Russian).

[30] A. Poznyak, A. Polyakov, V. Azhmyakov, 'Attractive ellipsoids in robust control', Springer International Publishing, 2014.

[31] B. Polyak, P. Shcherbakov, 'Attainability and attraction sets of linear systems with bounded control: description with the help of invariant ellipsoids', In Stochastic optimization in computer science, ed. by O. Granichin, Issue 4, St. Petersburg State University, pp. 3–23, 2008 (in Russian).

[32] E. Polovinkin, M. Balashov, 'Elements of convex and strongly convex analysis', FIZMATLIT, Moscow, 2004 (in Russian).

[33] A. Martynyuk, 'Stability by Liapunov's matrix function method with applications', Marcel Dekker, Inc, 1998.

[34] A. Bobtsov, G. Boltunov, S. Bystrov, V. Grigoriev, 'Control of continuous and discrete processes', St. Petersburg State University, 2010 (in Russian).

5

Application of a Special Method of Nondimensionization in the Solution of Nonlinear Dynamics Problems

**Maksym V. Maksymov[1,*], Olexander I. Brunetkin[1]
and Oksana B. Maksymova[2]**

[1]Odesa National Polytechnic University, Odesa, Ukraine
[2]Odesa National Academy of Food Technologies, Odesa, Ukraine
*Corresponding Author: prof.maksimov@gmail.com

A possibility of reducing the number of criteria compared to their number predicted according to the dimension theory is shown. In a number of cases, it is possible to define the view of normalizing quantities equating all similarity criteria in a nondimensionized model to one. In other words, it is possible to achieve self-similarity by all similarity criteria. On the one hand, it allows to perform distorted modeling. On the other hand, a decrease in model parameters by a number of reduced criteria makes it possible to obtain approximate analytic solutions or approximation of experimental data in solving dynamic tasks. Possibilities of this approach are demonstrated in the examples of:

- approximation of experimental data in finding natural frequencies of oscillation of a fluid with free surface in containers of various forms;
- an approximate analytic solution of a nonlinear task of the motion of a mathematical pendulum, both mathematical and in the presence of dissipative forces;
- an approximate analytic solution of the task of non-stationary heat transfer through a flat wall.

5.1 Introduction

A comprehensive study of any process is closely related to modeling. The variety of areas of science and technology in which modeling is used, as well as our desire for a model to best fit the characteristics of a problem, generates a large number of specific models and types of modeling. Often, in their diversity, it is difficult to choose a path leading to the creation of the most appropriate model in each particular case. Consequently, along with the exact approach that research requires, an element of creativity appears, a heuristic approach in the process of developing an adequate model.

One of the methods that allows to adjust the magnitude of adequacy is the ability to bring a model to a nondimensionized kind. The similarity theory is closely related to this method. In the indicated field of science, there is a basic pi theorem (in the English-language literature – Buckingham's theorem, in the French-language one – Vaschy's theorem), fixing a possible number of nondimensionized values in the models being converted. Nevertheless, there are still attempts to develop methods which make it possible to obtain a smaller number of nondimensionized quantities than prescribed by the theorem.

Certain progress has been made on the way of the further development of nondimensionization methods. But methods used by researchers are the result of their developers' intuition and do not indicate the limits of the theory of nondimensionization development. In the scientific approach, it is necessary to speak of the method as a harmonious logical system. This gives grounds for further work in this direction and justifies relevance of research.

5.2 Models and Modeling

5.2.1 Method of Nondimensionization of Mathematical Models

5.2.1.1 Formulation of the problem

Decreasing (due to nondimensionization) the number of values taken into account in a model helps analyze the available solutions and the causes of possible errors, provides analytical solutions to new problems, and also significantly reduces the number of necessary experimental studies (physical and numerical). When developing new methods, the number of nondimensionized variables predicted on the basis of the pi theorem is adopted as a reference point. In relation to this value, the degree of reduction in the number of nondimensionized values obtained using methods proposed by different researchers is determined. Thus, the possibility of such procedures is demonstrated in [1, 2]. It is noted that the greatest possible reduction in the number of nondimensionized quantities has been achieved. But it is not specified how it was determined that deeper transformations of mathematical models (MM) in this direction are impossible.

Nondimensionization of models can be used to analyze the solutions obtained and their reliability. In [3], on the basis of the analysis of nondimensionized properties and models, an attempt to determine the reasons for the inconsistency of the results obtained in the intensification of the phenomena under study has been made.

In a number of works, nondimensionization of models in combination with other methods is used to obtain new solutions. Thus, in [4], the application of the Laplace transform to nondimensionized equations is used to simplify the algebraic equations obtained. And in [5], based on the application of the group theory and nondimensionized differential equations, new solutions are sought. Moreover, nondimensionization operations can serve to identify a group of homogeneous stretches. From this point of view, in [6], the question is considered: what is the goal – to nondimensionize variables or to reduce the number of model parameters? An unbiased view on this question suggests that the goal is precisely to reduce the number of parameters, and nondimensionization is just a means which in some cases makes it possible to achieve exactly this result. Within the framework of these studies, no attempts were made to develop new methods of bringing models to a nondimensionized form. But it follows from their context that such a procedure would help expand the boundaries of the application of other methods of solving.

Representation of MMs in a nondimensionized form allows using their properties for modeling processes that are difficult to realize under experimental conditions [7]. This also contributes to generalization of the results obtained in numerical and physical experiments [8]. It can be assumed that development of nondimensionization methods will further improve modeling processes. In a number of cases [9], nondimensionization of models is called "a problem of reduction to a minimal parametric form". But in this case, the question of achieving self-similarity in parameters is not considered.

Systematization of the results of the above-mentioned studies allows us to conclude that the method of analysis, solutions, and generalization of the results obtained using a MM reduced to a nondimensionized form is effective. The results of works of a number of authors indicate a possibility of reducing the number of nondimensionized variables to values smaller than those established by the pi theorem. A deterrent factor for further reduction of the number of nondimensionized quantities in models is a lack of a general method for such transformations and, as a consequence, uncertainty of the lower bound of a possible number of such quantities. The need to develop such a method determines relevance of the work.

5.2.1.2 Purpose and objectives of the study

The objective of this chapter is to develop a method that minimizes the number of nondimensionized variables for the model in question.

Within the framework of achieving this goal, the following tasks should be solved:

- development of an algorithm that formalizes the process of measuring the MM values to minimize their number in comparison with the results prescribed in the pi theorem;
- development of a procedure to determine the lower bound of a possible number of nondimensionized values for the relevant models;
- ensuring reproducibility (coverage, inclusion in its composition) of the results of MM nondimensionization performed by other methods.

5.2.1.3 Scheme for ensuring self-similarity by criteria for models

Let us consider the pi theorem. "Any equation connecting N physical and geometric quantities the dimension of which is expressed in terms of n basic units of measurement can be transformed into the similarity equation $\pi = N - n$". According to this theorem, due to nondimensionization, a decrease in the number of quantities entering a model can be made only by the value "n". Thus, within the SI system, mechanical quantities can only be described by means of three units of measurement: mass [M], length [L], and time [T]. So, in the respective models, the number of quantities included in them can only be reduced to three units. But on the other hand, in this definition, we can consider the way to further reducing the number of quantities included in a model.

It is known that the current SI system or the previously used CGS and other similar systems are based not on any physical sense, but on metrological convenience. Applying quantities from this system to models recorded in a dimensional form, one has to agree on the use of uniform scales of dimensional quantities. But physical laws which are taken into account while compiling models reflect the relationship between the quantities entering them irrespective of their scale. As a result, a record of physical laws using dimensional quantities leads to formation of dimensional physical constants which take into account the scale of the measurement system currently in use. These constants in different combinations as well as variables constitute a set of "N" physical and geometric quantities used in the formulation of the pi theorem. The number of variables without changing a model itself cannot be reduced. Therefore, it is necessary to reduce the number of constants.

A standard procedure of nondimensionization reflected in the pi theorem is aimed at minimizing the number of constants. When we normalize dimensional variables by any characteristic values of the same nature, the standard scales of dimensional quantities are reduced, e.g., from the SI system by taking into account the internal scales of the processes described by the model being nondimensionized. In this case, the resulting criteria are complexes formed by different scales. And they are also the scales of the studied processes. A positive side of this procedure is taking into account the scales of ongoing processes in each model individually. In fact, this is the reason for reducing the number of quantities in a model due to a combination of the internal scales of processes under study. A disadvantage is that quantities of the same nature as the ones being nondimensionized are used as normalizing. As noted above, the dimensional values used are not selected from physical considerations. For this reason, it is not possible to reduce the number of scales (criteria) to the minimum until they are eliminated completely.

Another possible way to simplify expressions due to the reduction of the quantities taken into account is elimination of a number of physical constants from consideration. This effect is manifested when using natural units of measurement. In these systems, the basic units of measurement are chosen not from metrological considerations, but using physical constants themselves. The constants selected as basic units are equated to one and then all other quantities are expressed from this. The systems of units of M. Planck, G. Lewis, D. Hartree, P. Dirac, etc. were built in such a way. For example, consider the expression of the Coulomb law in various units of measurement. In the SI system, the electric constant in Coulomb's law looks like: $\varepsilon_0 \quad 8.99 \cdot 10^9 \ [\text{H} \cdot \text{M}^2 \cdot \text{K}\pi^{-2}]$. Bearing this in mind, it is written as follows:

$$F \quad \varepsilon_0 \frac{q_1 \cdot q_2}{R^2} \quad 8.99 \cdot 10^9 \cdot \frac{q_1 \cdot q_2}{R^2} \tag{5.1}$$

In CGS system, where $\varepsilon_0 \quad 1$ is considered as one of the basic units, this law has a simpler form:

$$F \quad 1 \cdot \frac{q_1 \cdot q_2}{R^2}. \tag{5.2}$$

A positive side in this kind of a procedure is the equality of the physical constants to one. Since they are components of similarity criteria, a number of criteria become equal to one. In other words, self-similarity is achieved by the corresponding criteria. But the procedure described has a drawback. A specific natural system of units is convenient for a particular model. In other models, the values determined on its basis usually have values inconvenient for use: very large or very small. In addition, physical constants are determined with an error. As a result, values determined on their basis (for example, masses and process times) will have errors inadmissible for practical use.

In the proposed method of nondimensionization, positive aspects of the pi theorem (taking into account the scales of the processes occurring in each model individually) are combined with the ones from the introduction of natural units of measurement (equality of physical constants to one).

At the first stage, we express

$$p_q \qquad \bar{p}_q \cdot p_q^\Delta, \quad \forall q \in J_k. \tag{5.3}$$

Here $p, \bar{p}_q$-dimensional and nondimensionized MM values, respectively, p_q^Δ – a normalizing value (scale), q – the number of variables in a MM, k – the number of dimensional quantities.

In a usual nondimensionization procedure, a value of the same type as the one being nondimensionized is chosen as a scale. Thus, the geometric characteristics of the research space are normalized by a value corresponding to some characteristic dimension, temperature – a characteristic temperature, etc. This is a significant and unjustified restriction. Normalization can be made by a quantity of the same nature as the normalized quantity, or of the same dimension, but not necessarily of the same kind. For example, for geometric characteristics of space, the normalizing quantities having the dimension of length [L], depending on the quantities entering a MM, can take the form:

$$x^\Delta \qquad \nu \cdot \sqrt{\frac{\rho}{\Delta P}} \quad \text{or} \quad x^\Delta \quad \sqrt[3]{\frac{\nu^2}{g}}, \tag{5.4}$$

for speed, having the dimension $[LT^{-1}]$:

$$u^\Delta \qquad \sqrt{\frac{\Delta P}{\rho}} \quad \text{or} \quad u^\Delta \quad \sqrt[3]{\nu \cdot g}. \tag{5.5}$$

Here ΔP – pressure drop, ν – kinematic viscosity, ρ – density, and g – acceleration of gravity. With this approach, the value x is a scale for geometric characteristics, but not a characteristic size, and similarly, for u as a scale for velocity and other normalizing quantities.

Further, the procedure of nondimensionization is similar to the standard:

- by removing the normalizing quantities and physical constants beyond the sign of the operator and forming complexes with the same dimension;
- nondimensionization of complexes dividing by one of them.

As a result, nondimensionized complexes appear that correspond to similarity criteria externally but differ from them in nature. Similarity criteria are formed from physical constants and scales of variables, which are unchanging characteristic values for the process under consideration: characteristic size, time, velocity, pressure, temperature, etc. For this reason, the criteria have an unchanging view. In the proposed method of nondimensionization, normalizing quantities are not chosen at this stage of transformations. Expressions (5.4) and (5.5) are given to demonstrate the ability of their wider representation. It remains possible to vary them to represent the obtained nondimensionized complexes of the required kind.

At the second stage, the condition of equality of all the obtained nondimensionized complexes to one is set. An analogous result, but only for certain complexes and for particular models, due to their simplification, is achieved by introducing for them a natural reference system. In this case, as in the case of a standard nondimensionization procedure, the basic units of measurement are constant. They only change the form: a transition is made from characteristic values of the process to physical constants corresponding to this process. As a result, there is no possibility of changing the type of nondimensionized complexes which are criteria.

In the proposed method of nondimensionization, the procedure is constructed from the reverse. The desired kind of nondimensionized complexes (in the case under consideration equal to one) is given and this is achieved by varying the type of normalizing quantities.

5.2.1.4 Concept of the procedure of ensuring self-similarity by criteria

Let the size value q_i be determined by measure (numerical value) p_i and the unit of measurement e_i pertaining to the i-th comparison class Q_i. Quantities q_i, q_j belong to the corresponding comparison classes $q_i \in Q_i$, $q_j \in Q_j$. When the units of measurement change, the corresponding measure value is determined in terms of the scale of change of units μ. For two different classes, there are no scales that would establish the correspondence between measures in these classes:

$$\forall q_i, \quad q_i \in Q_i, q_i \quad p_i \cdot e_i, q_i' \quad p_i' \cdot e_i' \quad \exists \mu : p_i' \quad \mu_i \cdot p_i,$$
$$\forall q_i, \quad q_j : q_i \in Q_i, q_j \in Q_j \quad Q_i \cap Q_j \quad \emptyset \bar{\exists} \mu : p_i' \quad \mu_i \cdot p_i.$$

It can be shown that in physical laws that are independent of the choice of a measurement system, the relationships between the scales of dimensional quantities when the units of measurement change are:

$$\mu_i \quad \prod_j \mu_j^{\alpha_{ij}}, \quad i, j \in J_u, \tag{5.6}$$

where α_{ij} – a corresponding exponent,
u – the number of dimensional quantities.

The value of the exponents in (5.6) is determined by the structure of a MM.

It is useful to establish a relationship between the scale ratios and measures of dimensional quantities, which is induced by MM structure. If the scales of all (u) dimensional quantities included in a MM are in relation $\mu_i \quad \prod_j \mu_j^{\alpha_{ij}}, \forall i, j \in J_u$, then the measures of these quantities are in this relation:

$$p_i \quad \prod_j p_j^{\alpha_{ij}}, \quad \forall i, j \in J_u.$$

This follows from $p_i \quad \prod_j p_j^{\alpha_{ij}} \Rightarrow \mu_i \quad \prod_j \mu_j^{\alpha_{ij}}$, which is determined by the homogeneity of the original dependence on scales and is verified by direct substitution $p_i' \quad \mu_i \cdot p_i$. The statement $\mu_i \quad \prod_j \mu_j^{\alpha_{ij}} \Rightarrow p_i \quad \prod_j p_j^{\alpha_{ij}}$ is proved by contradiction. Let $\mu_i \quad \prod_j \mu_j^{\alpha_{ij}} \Rightarrow p_i \quad \prod_j p_j^{\beta_{ij}}$, but then $\prod_j p_j^{\alpha_{ij}}$
$\prod_j p_j^{\beta_{ij}}$. Comparison of values p_i, p_j can only be in its class, from where $\alpha_{ij} \quad \beta_{ij}$. So then:

$$\mu_i \quad \prod_j \mu_j^{\alpha_{ij}} \Leftrightarrow p_i \quad \prod_j p_j^{\alpha_{ij}}. \tag{5.7}$$

By the Fourier rule in the formula $p_i \quad \prod_j p_j^{\alpha_{ij}}$, the dimensions of the quantities in the right and left parts are equal to $E_i \quad \prod_j E_j^{\alpha_{ij}}$, which allows to present nondimensionized complexes $\prod_j p_j^{\alpha_{ij}} / p_i$ together with results (5.7) for the transformation of MM space.

An MM can be written in the following form:

$$
\left.
\begin{array}{c}
\displaystyle\sum_{j=1}^{i=n_j} A_{ij}(\vec{p}) \quad 0; \\[3mm]
\vec{p} \in P, \quad P \quad Y \otimes S \otimes W, \\[3mm]
\vec{y} \in Y, \quad \vec{s} \in S, \quad \vec{w} \in W,
\end{array}
\right\}
$$

where A_{ij} – operator of the i-th term, in j-th equation or MM formula;

$\vec{p}$ $(p_1, p, \ldots, p_n)$ – tuple of all dimensions in a MM;

$\vec{y}$ $(y_1, y, \ldots, y_{ny})$ – tuple of quantities that determine the behavior of the system;

$\vec{s}$ (s_1, s_2, s_3, s_4) – tuple of variables that determine the coordinates of geometric space and time;

$\vec{w}$ $(w_1, w_2, \ldots, y_{nw})$ – tuple of quantities, which complements $\vec{s}$ and $\vec{y}$ to $\vec{p}$

n_j – number of terms in j-th equation or MM formula;

m – number of equations and formulas in a MM.

According to the theory of similarity, the establishment of conditions for physical modeling is performed on an MM reduced to a nondimensionized form. This can be done in various ways. However, the most effective is a procedure performed in a general way. For this, we normalize all dimensional variables. At the same time, the normalizing quantities are not yet determined. It is convenient to do this later to get the results we need. After the normalization of all dimensional quantities, the normalizing quantities are removed beyond the sign of the operator. The product of the normalizing quantities and other constants form coefficients whose dimension in each equation or formula, according to the Fourier rule, is the same. Nondimensionized complexes are formed by dividing all the others by any coefficient. These complexes and normalized variables form a list of quantities that determine the process described by the original MM. A formalized record of the procedure for reducing to a nondimensionized form and in this case obtaining nondimensionized variables and complexes can be represented in the following form:

$$\varphi : \sum_{i=1}^{i=n_j} A_{ij}(\vec{p}) \quad 0 \rightarrow \sum_{i=1}^{i=n_j} A_{ij}(\vec{\bar{y}}, \vec{\bar{s}}, \vec{\pi}) \quad 0, \qquad (5.8)$$

$$\vec{\bar{y}} \quad (\bar{y}_1, \ldots, \bar{y}_{n_y}), \qquad \vec{\bar{s}} \quad (\bar{s}_1, \ldots, \bar{s}_{n_s}),$$

$$\bar{y}_l \quad \frac{y_l}{y_l^\Delta}, \quad \bar{s}_k \quad \frac{s_k}{s_k^\Delta}, \quad \pi_{ij} \quad \prod_{q=1}^{q=u} p_q^{\alpha_{ijq}}, \qquad (5.9)$$

where u – the number of dimensional quantities in a MM,
Δ – denomination of a normalizing quantity.

Some of π_{ij} are identically equal to one or other numbers. We exclude them from consideration as identical for all systems described by the original MM, and, therefore, do not introduce additional information within the considered MM class. Let us put one and only one element of a sequence of natural numbers to each pair (I, j) corresponding to a nondimensionized complex π_{ij}, not equal to one or to another number:

$$(i, j) \leftrightarrow h, \quad h \in J_t, \qquad (5.10)$$

where t – the number of nondimensionized complexes.

Now, each complex can be represented as follows:

$$\pi_h \quad \prod_{q=1}^{q=u} p_q^{\alpha_{nq}}$$

Thus, as a result of transformation (5.8), the MM is reduced to a nondimensionized form, the solution of which, if it exists, can be written in the following form:

$$\vec{\bar{y}} \quad f(\vec{\bar{s}}, \vec{\pi}), \qquad (5.11)$$

where $\vec{\pi} \quad (\pi_1, \pi_2, \ldots, \pi_t)$.

It is known that a necessary and sufficient condition for similarity of two processes G is the equality of the same criteria and nondimensionized variables:

$$G : \forall s_k \in \bar{S}, \quad \pi_h \in \Pi \quad \Pi \supset D^* : (\bar{s}_k \quad \bar{s}_k^*) \wedge (\pi_h \quad \pi_h^*) \qquad (5.12)$$

where * – designation of belonging to the natural system,
D – range of values defined by the technical specification for the system.

In (5.12), the similarity condition is specified so that the values of the criteria should be in the area specified by the technical specification for the original.

If the MM of the original and its model corresponds to condition (5.12) after procedure (5.8), they become indistinguishable, and the identity of the MM operators of the original and model determines the indistinguishability of the solutions (5.11) for the original and model.

Of course, further research should begin with an attempt to find transformations that reduce the dimensionality of the space in which the problem is solved. If we introduce some one-to-one conversion Ψ between the set of dimensional values of the natural system and the set of values of the model quantities, simulation conditions with allowance for the transformation Ψ can be written in the following form:

$$\exists p_q : (\Psi : p_q^* \leftrightarrow p_q) \wedge G, \quad \forall p_q^* \; q \in J_u.$$

Suppose that transformation Ψ has the following form:

$$p_q \quad \mu_q \cdot p_q^*, \quad \forall q \in J_u. \tag{5.13}$$

In this case, the first numbers of the natural sequence are assigned to the elements of the tuple $\overrightarrow{y}$, the next $-\overrightarrow{s}$ and the remaining dimensional values of the MM. The last ones are parameters entering the boundary conditions and the physical constants of the working objects participating in the process.

Substitution of expression (5.13) into formula (5.9), taking into account relation (5.10), yields the following result:

$$\prod_{q=1}^{q=u} \mu_q^{\alpha_{hq}} \quad 1, \quad \forall h \in J_t. \tag{5.14}$$

The logarithm of expression (5.13) makes it possible to obtain a system of linear homogeneous algebraic equations

$$A_1 \cdot \overrightarrow{M} \quad 0, \tag{5.15}$$

where A_1 $\begin{bmatrix} \alpha_{11} & \alpha_{12} & \cdots & \alpha_{1u} \\ \cdots & \cdots & \cdots & \cdots \\ \alpha_{t1} & \alpha_{t2} & \cdots & \alpha_{tu} \end{bmatrix}$ – matrix of exponents α_{nq},

$\overrightarrow{M}$ $[\ln(\mu_1), \ln(\mu_2), \ldots, \ln(\mu_u)]^T$ – vector-column of logarithms of scales of all dimensional quantities.

Using the Gauss Jordan algorithm, matrix A_1 can be transformed to the form:

$$A_1 \rightarrow [E\vdots B],$$

where E – a unit matrix of size $(r \times r)$,
r – rank$[A_1]$. Lines with linearly dependent elements from matrix A_1 are deleted,
   B – matrix of exponents β_{hv} of size $[r \times (u - r)]$.

In a general form, $[E\vdots B]$ can be represented in the form:

$$p_1, \ldots, p_q, \ldots, p_r, \ p_{r+1}, \ldots, p_u$$

$$y_1, \ldots, y_{n_y}, s_1, \ldots, s_{n_s}, z_1, \ldots, z_{n_z}, z_{\Omega_1}, \ldots, z_{\Omega_n}, c_1, \ldots, c_{n_c} \tag{5.16}$$

$$
\begin{array}{c}
\pi_1 \\
\cdots \\
\pi_n \\
\cdots \\
\pi_r
\end{array}
\left[
\begin{array}{ccc|ccc}
1 & \cdots & 0 & \beta_{11} & \cdots & \beta_{1(u-r)} \\
  &   &   &   &   &   \\
\cdots & 1 & \cdots & \cdots & \beta_{nv} & \cdots \\
  &   &   &   &   &   \\
0 & \cdots & 1 & \beta_{r1} & \cdots & \beta_{r(u-r)}
\end{array}
\right]
$$

Above the matrix, a tuple of all dimensional quantities of the MM is written. Matrix A_1 is composed of the powers in complexes π_h.
   Here z – process parameters,
   c – physical constants of working objects,
   Ω – quantities entering the boundary conditions of the problem.

Using matrix $[E\vdots B]$, it is easy to express solution of system (5.14) and analyze results.
   The solution of the system (5.15) for the q-th scale will have the form:

$$\mu_q \prod_{v=1}^{v=u-r} \mu_v^{-\beta_{qv}}, \quad \forall q \in J_r$$

Now, in view of (5.6), the normalizing quantities will take the form:

$$p_q^{\Delta} \prod_{v=1}^{v=u-r} p_v^{-\beta_{qv}}, \quad \forall q \in J_r \tag{5.17}$$

And the normalized ones:

$$\bar{p}_q \quad \frac{p_q}{\prod_{v=1}^{v=u-r} p_v^{-\beta_{qv}}}. \tag{5.18}$$

Nondimensionized complexes π_h after transformations will be displayed in accordance with the expression:

$$\pi'_h \quad \prod_{q=1}^{q=r} (p_q^{\triangle})^{\alpha_{nq}} \quad \prod_{q=r+1}^{q=u} (p_q^{\alpha_{hq}}). \qquad (5.19)$$

Substituting expression (5.17) into (5.19), it is possible to find:

$$\pi'_h \quad \prod_{q=1}^{q=r} (\prod_{v=1} p_v^{-\beta_{qv}})^{\alpha_{nq}} \quad \prod_{q=r+1}^{q=u} (p_q^{\alpha_{hq}}).$$

From the results of the transformations, the following two conclusions can be made:

1. If $rank\ A_1 < n_y + n_s$, a MM is reducible to a self-similar form. This follows from the fact that if $rank\ A_1 \quad r, n_y + n_s - n_r$ of independent variables enter the right-hand side of formula (5.17), and n_r of the normalizing quantities will be expressed in terms of these variables and other $u - (n_y + n_s)$ dimensional quantities. The space of independent variables after transformations (5.18) is reduced, and the MM in a new space will have a self-similar form.

2. If $rank\ A_1 < t + n_y + n_s$, the number of nondimensionized complexes will be less than that follows from the π-theorem. This follows from the statement of the π-theorem that the number of nondimensionized complexes is equal to $n - k$, where k is the number of dimensional quantities with independent dimensions. This does not take into account the structural features of a MM. Thus, $n - k \geq t + n_y + n_s$. The number of nondimensionized quantities after transformations (5.18) and (5.19) will be n_r, where $n_r \leq rank\ A_1$. Hence, taking into account the initial condition, it follows that $n_r < n - k$. In other words, after transformations (5.18) and (5.19), the number of nondimensionized complexes will be less than this follows from the π-theorem.

The transformations that have been performed make it possible to obtain all known results of reducing the dimensionality of space, up to bringing a MM to a self-similar form, and also to obtain new results due to a deep penetration into connection of the similarity theory with MM structure. The developed algorithm of transformations makes it possible to widely use reductions of the dimensionality of MM space in various studies.

The dimensions of matrix A_1 of complex processes turn out to be large, which causes considerable difficulties in transforming to a nondimensionized form, if rational fractions are taken as matrix elements in accordance with the procedure developed above. To facilitate the implementation of the algorithm on a computer in accordance with ratio (5.20), at the initial stage, it is proposed to replace rational fractions with their decimal values. After all calculations are performed, at the final stage, the results in the form of decimal fractions are replaced by rational fractions, for example, according to Euclid's algorithm. This allows us to perform all further transformations (5.17) and (5.18).

5.2.1.5 Physical modeling

Condition (5.12) is sometimes an insurmountable constraint in the conditions of physical modeling. However, here we should emphasize a limited nature of this conclusion. First, this concerns the form of recording MM (5.8) and the level of detail, which is characterized by the number of criteria included in the description of the process. The point is that the smaller the number of nondimensionized variable complexes considered in the MM is, the easier it is to ensure the condition of modeling. (this provision will be proved below). For some MMs, it is possible to find transformations induced by their structure, leading to a form with a smaller number of variables and complexes. This is one of the possible ways to improve modeling conditions and it is carried out without loss of information, since, after the solution of the problem, it is possible to uniquely find the values of all variables and parameters included in the correctly formulated original MM. The second way is connected with the transition to a MM, which is more general than the original one. Climbing the hierarchical structure of the MM, one can find the degree of idealization of the model in which physical modeling is possible due to negligence which has little influence on the process, i.e., condition (5.12) is satisfied for a new MM. However, the movement along the second path is accompanied by loss of information, in terms and definitions given by C. Shannon. Indeed, not all parameters and variables can be uniquely determined by inverse transformation after the solution of the problem. Some of them are defined as medium integral, others as random ones in a certain range of values. In general, the result is obtained as a random variable with a variance greater than it would be when obtaining the result from the original MM. As a result, the sought values will be in a larger confidence interval for a given probability. This, however, does not mean the unacceptability of a simplified model. The final judgment can be made after checking the MM for adequacy of the process.

If it is possible to find transformations leading to diminishing the number of independent variables, as is known, an MM is brought to its auto-similar form.

In physical modeling, in comparison with mathematical, the situation is complicated by the fact that it is necessary to select real substances as model working objects, whose physical constants depend on the substance selected and process parameters. These additional links included in the conditions of physical modeling are essential limitations on the possibilities of physical modeling. As a result, in complex models, physical modeling is not possible. The most significant results in physical modeling can be obtained using thermodynamic similarity of substances. In case of a multiplicative form of connection between substance physical constants and process parameters, a problem solution about the admissibility of physical modeling can be obtained analytically.

Let the dependence of physical constant C_i on other physical constants and process parameters z be of the form:

$$c_i \quad c_i(\vec{c}, \vec{z}); \quad i \in J_{n_c}. \tag{5.20}$$

In its simplest form, such a relationship can be:

$$c_i \quad const. \tag{5.21}$$

Substituting (5.13) into (5.20) or (5.21) and after logarithm of the obtained expressions, additional connections imposed on the scales of dimensional values by the previously noted physical modeling specifics are determined:

$$A_2 \cdot \vec{M} \quad \vec{F},$$

where A_2 – matrix of size $(n_c \times u)$,
$\vec{F}$ – vector-column of constants, depending on the selected model and actual working objects, properties of objects.

The scale of dimensional physical quantities is determined on the basis of the solution of the system of linear equations:

$$\begin{bmatrix} A_1 \\ A_2 \end{bmatrix} \cdot \vec{M} \quad \begin{bmatrix} 0 \\ F \end{bmatrix}. \tag{5.22}$$

Let us consider the above-mentioned expression. If the rank of the system of the matrix lines $\begin{bmatrix} A_1 \\ A_2 \end{bmatrix}$ is less than or equal to u, physical modeling of the process described by the original MM is possible. Otherwise, the process is not physically modeled. Indeed, in a case when the rank of the system of the matrix lines $[\frac{A_1}{A_2}]$ is less than or equal to u, the system of linear inhomogeneous equations (5.22) is consistent and has an infinite number of solutions or has only one if system (5.22) is a Cramer system. In this case, the scales of the simulated quantities satisfying condition (5.12) are determined from the solution of system (5.22). In a case when the rank of the system of the matrix lines $[\frac{A_1}{A_2}]$ is greater than u, the system of equations (5.22) is incompatible and there are no scales that satisfy condition (5.12), which makes physical simulation impossible.

To fulfil modeling conditions, it is necessary to go to another level of the MM hierarchical structure where it is described with greater simplifications. In research practice, the following ways in this direction are most rational:

1. In the series of nondimensionized complexes ranked by importance for the description of the process, the last complexes are cut off. In fact, this means that phenomena that are less significant for the process are not taken into account. After fulfiling the conditions of modelability due to the reduction in the number of complexes, it is necessary to check the MM for adequacy of the process.
2. In the MM, they use integral characteristics instead of some differential ones. This makes it possible to reduce the number of complexes, and consequently, the rank of matrix A_1.
3. In the MM, they allocate subtasks, which can be solved autonomously. Solving a particular problem allows us to reduce the number of complexes and thereby improve modeling conditions.

5.2.1.6 Examples of reduction of models to a nondimensionized form in problems of technical systems dynamics

To demonstrate the efficiency of the proposed method of nondimensionization, an example of hydraulic shock in a pipe is considered as an example in two versions: without taking into account and taking into account a dissipative term. In each variant, the process of nondimensionization is considered according to a conventional method and using the proposed method.

Nondimensionization of an MM written without taking into account a dissipative term

In the classic formulation, such a model has the form:

$$\begin{cases} -\dfrac{\partial P}{\partial x} & \rho \cdot \dfrac{\partial \omega}{\partial t}; \\[4mm] -\dfrac{\partial P}{\partial t} & \rho \cdot c^2 \cdot \dfrac{\partial \omega}{\partial x}. \end{cases} \qquad (5.23)$$

boundary conditions

$$initial \left\{ t \quad 0 \quad \begin{matrix} \omega & \omega_o; \\ P & 0; \end{matrix} \right. \quad border \left\{ \begin{matrix} x & 0 & P & 0; \\ x & l & \omega & 0. \end{matrix} \right. \quad (5.24)$$

Here P – pressure in the stream; ω – flow velocity; x, t – coordinate along the length of the pipe and the time of the process, respectively; ρ – density of the fluid flowing through the pipe, c – the speed of sound in the liquid; ω_o – initial flow velocity; l – length of the pipe. As the model is linear, instead of absolute pressure, we consider its deviation from the initial value P to be 0.

Conventional method of nondimensionization
When using normalizations $P^\triangle, \omega^\triangle, t^\triangle, x^\triangle$, nondimensionized values of the corresponding variables are written:

$$\bar{P} \quad \frac{P}{P^\triangle}; \quad \bar{\omega} \quad \frac{\omega}{\omega^\triangle}; \quad \bar{t} \quad \frac{t}{t^\triangle}; \quad \bar{x} \quad \frac{x}{x^\triangle}.$$

Further, when used in models (5.23) and (5.24), dimensional complexes are distinguished in front of the operators. The operators are written in a nondimensionized form. By the Fourier theorem, dimensional complexes in the same equation have the same dimension. Further within the framework of each equation, by dividing all dimensional complexes by one of them, nondimensionized complexes are formed. All equations and, accordingly, the model are nondimensionized:

$$\left\{ \begin{matrix} -\frac{\partial \bar{P}}{\partial \bar{x}} & \pi_1 \cdot \frac{\partial \bar{\omega}}{\partial \bar{t}}; \\ -\frac{\partial \bar{P}}{\partial \bar{t}} & \pi_2 \cdot \frac{\partial \bar{\omega}}{\partial \bar{x}}. \end{matrix} \right.$$

boundary conditions:

$$initial \left\{ \bar{t} \quad 0 \quad \begin{matrix} \bar{\omega} & 1; \\ \bar{P} & 0; \end{matrix} \right. \quad border \left\{ \begin{matrix} \bar{x} & 0 & \bar{P} & 0; \\ \bar{x} & 1 & \bar{\omega} & 0. \end{matrix} \right. \quad (5.25)$$

$$\pi_1 \quad \frac{\rho \cdot \omega^\triangle \cdot x^\triangle}{t^\triangle \cdot P^\triangle}; \quad \pi_2 \quad \frac{\rho \cdot c^2 \cdot \omega^\triangle \cdot t^\triangle}{x^\triangle \cdot P^\triangle}; \quad \pi_3 \quad \frac{\omega_o}{\omega^\triangle}; \quad \pi_4 \quad \frac{l}{x^\triangle}. \quad (5.26)$$

Further, on the basis of a heuristic approach, there is a decrease in the number and simplification of the type of complexes π_i. The result depends on the complexity of the model and the experience of a researcher. Assuming in the boundary conditions $\pi_3 \quad 1, \pi_4 \quad 1$, the initial quantities are determined. Here $\pi_1, \pi_2, \pi_3, \pi_4$ are nondimensionized complexes of normalizing values:

ω^Δ ω_o, x^Δ l. The normalizing value for time can be determined from the ratio of characteristic values of the process: t^Δ l/c. For the given model, P^Δ has no characteristic quantity, but it can be introduced artificially. Let us assume P^Δ P_o. As P_o, it is possible to take pressure in the system before the start of the development of a hydraulic shock. Substituting values ω^Δ, x^Δ, t^Δ, P^Δ into remaining complexes (5.28), we obtain

$$\pi_1 \qquad \frac{\rho \cdot \omega^\Delta \cdot x^\Delta}{t^\Delta \cdot P^\Delta} \qquad \frac{\rho \cdot \omega_o \cdot l \cdot c}{l \cdot P_o} \qquad \frac{\rho \cdot \omega_o \cdot c}{P_o};$$

$$\pi_2 \qquad \frac{\rho \cdot c^2 \cdot \omega^\Delta \cdot t^\Delta}{x^\Delta \cdot P^\Delta} \qquad \frac{\rho \cdot c^2 \cdot \omega_o \cdot l}{l \cdot P_o \cdot c} \qquad \frac{\rho \cdot \omega_o \cdot c}{P_o}. \qquad (5.27)$$

Comparison of complexes (5.29) shows their equality π_1 π_2 π. In nondimensionized model (5.26) and (5.27), only one nondimensionized complex remains – a similarity criterion.

At this stage, the process of model transformation usually ends. In the case under consideration, the characteristic value of the process t^Δ, absent in the boundary conditions, was successfully chosen for the time variable normalization. It is not always possible to do this. In such a situation, two similarity criteria, π_1 and π_2, would remain in the model under consideration.

Proposed method of nondimensionization

In accordance with (5.9), a tuple of dimensional quantities is constructed from model (5.23) and (5.24), in which elements 1-2 correspond to n_y, elements 3-4 $\Leftrightarrow n_s$, elements 5-6 $\Leftrightarrow n_\Omega$, elements 7-8 $\Leftrightarrow m$ of physical constants of the process. On the basis of (5.8), matrix A of the exponents for the corresponding variables is presented in the form for nondimensionized complexes (5.28):

$$
\begin{array}{c c c c c c c c c}
 & 1 & 2 & 3 & 4 & \vdots\ 5 & 6 & 7 & 8 \\
 & P & \omega & x & t & \vdots\ \omega_o & l & \rho & c \\
\pi_1 & -1 & 1 & 1 & -1 & \vdots\ 0 & 0 & 1 & 0 \\
\pi_2 & -1 & 1 & -1 & 1 & \vdots\ 0 & 0 & 1 & 2 \\
\pi_3 & 0 & -1 & 0 & 0 & \vdots\ 1 & 0 & 0 & 0 \\
\pi_4 & 0 & 0 & -1 & 0 & \vdots\ 0 & 1 & 0 & 0
\end{array} \qquad (5.28)
$$

For matrix (5.30) $rank[A]$ 4. Therefore, already at this stage, before implementing transformations, we can say about the possibility of achieving self-similarity by all criteria.

After applying the Gauss Jordan algorithm, the transformed matrix has the form:

$$
\begin{array}{c}
 \\
\pi_1 \\
\pi_2 \\
\pi_3 \\
\pi_4
\end{array}
\begin{array}{ccccccccc}
P & \omega & x & t & \vdots & \omega_o & l & \rho & c \\
\left[\begin{array}{cccc} 1 & 0 & 0 & 0 \end{array}\right. & & & & \vdots & -1 & 0 & -1 & \left.\begin{array}{c} -1 \end{array}\right] \\
0 & 1 & 0 & 0 & \vdots & -1 & 0 & 0 & 0 \\
0 & 0 & 1 & 0 & \vdots & 0 & -1 & 0 & 0 \\
0 & 0 & 0 & 1 & \vdots & 0 & -1 & 0 & 1
\end{array}
$$

$$
\begin{array}{ccc}
E & \vdots & B
\end{array}
$$

(5.29)

Normalizing quantities using matrix (5.31) are formed as follows:

 - we select any line from (5.31), for example, π_1. In part [E], in this line, "1" is located in the column corresponding to value P. For it, normalization $P^\triangle$ is determined;
 - using values from this line located in part [B] of matrix (5.21), the normalization form $P^\triangle$ is formed. It is constructed in accordance with (5.12). These quantities act as exponentials with the opposite sign for the values of the boundary conditions and physical constants denoting corresponding columns in this part of the matrix.

In accordance with this algorithm:

$$
\begin{array}{llll}
\text{From line } \pi_1 \rightarrow & P^\triangle & \omega_o^1 \cdot \rho^1 \cdot c^1 & \omega_o \cdot \rho \cdot c; \\
\text{From line } \pi_2 \rightarrow & \omega^\triangle & \omega_o^1 & \omega_o; \\
\text{From line } \pi_3 \rightarrow & x^\triangle & l^1 & l; \\
\text{From line } \pi_4 \rightarrow & t^\triangle & \frac{l^1}{c^1} & \frac{l}{c}.
\end{array}
$$

(5.30)

Substitution of the normalizing quantities (5.32) into (5.28) transforms all nondimensionized complexes into equal values $\pi = 1$. In other words, self-similarity is achieved for all similarity criteria. Results (5.32) were obtained without heuristic searches on the basis of a formal procedure that can be performed by a researcher of any qualification.

Nondimensionization of a MM written with allowance for a dissipative term

A more difficult variant of a MM of the process of a hydraulic shock can be presented by models (5.23) and (5.24) with the addition of a linearized dissipative term.

$$
\begin{cases}
-\dfrac{\partial P}{\partial x} & \rho \cdot \left(\dfrac{\partial \omega}{\partial t} + 2a \cdot \omega\right); \\[4mm]
-\dfrac{\partial P}{\partial t} & \rho \cdot c^2 \cdot \dfrac{\partial \omega}{\partial x}.
\end{cases}
\tag{5.33}
$$

boundary conditions:

$$
\text{initial} \begin{cases} t & 0 \end{cases} \quad \begin{matrix} \omega & \omega_o; \\ P & 0; \end{matrix} \quad \text{border} \begin{cases} x & 0 \\ x & l \end{cases} \quad \begin{matrix} P & 0; \\ \omega & 0. \end{matrix}
\tag{5.34}
$$

Here a – coefficient of resistance.

Conventional method of nondimensionization

Using transformations and normalizations $P^\triangle$, $\omega^\triangle$, $t^\triangle$, $x^\triangle$ similar to the procedure described in Section 2.1.6.1. we obtain a model in a nondimensionized form:

$$
\begin{cases}
-\dfrac{\partial \bar{P}}{\partial \bar{x}} & \pi_1 \cdot \dfrac{\partial \bar{\omega}}{\partial \bar{t}} + \pi_3 \cdot \bar{\omega}; \\[4mm]
-\dfrac{\partial \bar{P}}{\partial \bar{t}} & \pi_2 \cdot \dfrac{\partial \bar{\omega}}{\partial \bar{x}}.
\end{cases}
$$

boundary conditions:

$$
\text{initial} \begin{cases} \bar{t} & 0 \end{cases} \quad \begin{matrix} \bar{\omega} & 1; \\ \bar{P} & 0; \end{matrix} \quad \text{border} \begin{cases} \bar{x} & 0 \\ \bar{x} & 1 \end{cases} \quad \begin{matrix} \bar{P} & 0; \\ \bar{\omega} & 0. \end{matrix}
$$

$$
\text{where } \pi_1 \quad \pi_2 \quad \frac{\rho \cdot c \cdot \omega_o}{P_o}; \; \pi_3 \quad \frac{\rho \cdot 2a \cdot l}{P_o}; \; \pi_4 \quad \pi_5 \quad 1. \tag{5.35}
$$

The appearance of a new term in the equation in comparison with model (5.23) and (5.24) leads to the appearance of yet another nondimensionized complex – a similarity criterion π_3.

Proposed method of nondimensionization

Nondimensionized complexes for model (5.33) and (5.34) are written in a general form similar to (5.28):

$$\pi_1 \quad \frac{\rho \cdot \omega^\Delta \cdot x^\Delta}{t^\Delta \cdot P^\Delta}; \quad \pi_2 \quad \frac{\rho \cdot c^2 \cdot \omega^\Delta \cdot t^\Delta}{x^\Delta \cdot P^\Delta};$$

$$\pi_3 \quad \frac{\rho \cdot a \cdot \omega^\Delta \cdot x^\Delta}{P_o}; \quad \pi_4 \quad \frac{\omega_o}{\omega^\Delta}; \quad \pi_5 \quad \frac{l}{x^\Delta}. \qquad (5.36)$$

A tuple of dimensional quantities is constructed from model (5.33) and (5.34) and matrix A similar to (5.30) is formed:

$$
\begin{array}{cccccccccc}
 & 1 & 2 & 3 & 4 & 5 & \vdots & 6 & 7 & 8 & 9 \\
 & P & \omega & x & t & a & \vdots & \omega_o & l & \rho & c
\end{array}
$$

$$
\begin{array}{c}
\pi_1 \\
\pi_2 \\
\pi_3 \\
\pi_4 \\
\pi_5
\end{array}
\left[
\begin{array}{ccccccccc}
-1 & 1 & 1 & -1 & 0 & \vdots & 0 & 0 & 1 & 0 \\
-1 & 1 & -1 & 1 & 0 & \vdots & 0 & 0 & 1 & 2 \\
-1 & 1 & 1 & 0 & 1 & \vdots & 0 & 0 & 1 & 0 \\
0 & -1 & 0 & 0 & 0 & \vdots & 1 & 0 & 0 & 0 \\
0 & 0 & -1 & 0 & 0 & \vdots & 0 & 1 & 0 & 0
\end{array}
\right] \qquad (5.37)
$$

From (5.39), for all linearly independent lines $rank[A] \quad 5$ follows. To solve this system, five variables are needed. With available $4^{\underline{X}}$ (P, ω, x, t) for modeling, it is necessary to allocate one more quantity that will act as such. For this, in the case under consideration, we single out, for example, a – coefficient of resistance. In accordance with (5.25), it can be determined from the relation $\bar{a} \quad a/a^\Delta$, where $\bar{a}$ is the nondimensionized value of a coefficient of resistance. As a result, in complex π_3 in (5.36), instead of a dimensional value "a", a normalizing quantity a^Δ appears. After applying the Gauss Jordan algorithm, the transformed matrix has the form:

$$
\begin{array}{c}
\pi_1 \\
\pi_2 \\
\pi_3 \\
\pi_4 \\
\pi_5
\end{array}
\left[
\begin{array}{ccccccccc}
1 & 0 & 0 & 0 & 0 & \vdots & -1 & 0 & -1 & -1 \\
0 & 1 & 0 & 0 & 0 & \vdots & -1 & 0 & 0 & 0 \\
0 & 0 & 1 & 0 & 0 & \vdots & 0 & -1 & 0 & 0 \\
0 & 0 & 0 & 1 & 0 & \vdots & 0 & -1 & 0 & 1 \\
0 & 0 & 0 & 0 & 1 & \vdots & 0 & 1 & 0 & -1
\end{array}
\right] \qquad (5.38)
$$

$$\quad E \qquad\qquad\qquad B$$

By analogy with (5.32), we got normalizing quantities from (5.38):

$$
\begin{aligned}
&\text{From line } \pi_1 \;\rightarrow\; P^\Delta \quad \omega_o^1 \cdot \rho^1 \cdot c^1 \quad \omega_o \cdot \rho \cdot c; \\
&\text{From line } \pi_2 \;\rightarrow\; \omega^\Delta \quad \omega_o^1 \quad \omega_o; \\
&\text{From line } \pi_3 \;\rightarrow\; x^\Delta \quad l^1 \quad l; \\
&\text{From line } \pi_4 \;\rightarrow\; t^\Delta \quad \tfrac{l^1}{c^1} \quad \tfrac{l}{c}; \\
&\text{From line } \pi_5 \;\rightarrow\; a^\Delta \quad \tfrac{c}{l}.
\end{aligned}
\tag{5.39}
$$

With their use, original MM (5.33) and (5.34) in a nondimensionized form will be written as:

$$
\left\{
\begin{aligned}
-\frac{\partial \bar{P}}{\partial \bar{x}} \quad & \frac{\partial \bar{\omega}}{\partial \bar{t}} + 2\bar{a} \cdot \bar{\omega}; \\[2mm]
-\frac{\partial \bar{P}}{\partial \bar{t}} \quad & \frac{\partial \bar{\omega}}{\partial \bar{x}}.
\end{aligned}
\right.
\tag{5.40}
$$

boundary conditions:

$$
\text{initial} \left\{
\begin{aligned}
\bar{t} \quad & 0 \\
\end{aligned}
\right.
\quad
\begin{aligned}
\bar{\omega} \quad & 1; \\
\bar{P} \quad & 0;
\end{aligned}
\quad
\text{border} \left\{
\begin{aligned}
\bar{x} \quad & 0 \\
\bar{x} \quad & 1
\end{aligned}
\right.
\quad
\begin{aligned}
\bar{P} \quad & 0; \\
\bar{\omega} \quad & 0.
\end{aligned}
$$

As in the previous case, all the nondimensionized similarity criteria are π 1. As a result, self-similarity is achieved by all criteria. But in the transformed model (5.40), in addition to nondimensionized variables $\bar{P}$, $\bar{\omega}, \bar{x}, \bar{t}$ s, another quantity $\bar{a}$ appears. On the one hand, in (5.37), it is introduced as a variable. On the other hand, in the process of solving a particular problem it remains a constant value, similar to a similarity criterion. This is its peculiarity. In the final analysis, it is noteworthy that using the proposed method, unlike a conventional method of nondimensionization, it was possible to reduce the number of variables that determine the transformed model. So, with a conventional method, the model includes six quantities: $\bar{P}$, $\bar{\omega}, \bar{x}, \bar{t}$ and also π_1, π_3 from (5.35). In the case of applying the proposed method, there remain 5: $\bar{P}$, $\bar{\omega}$, $\bar{x}$, $\bar{t}$ and $\bar{a}$ as well.

Discussion of the results of the solution on the basis of the developed method

Applying different methods of nondimensionization to identical MMs should potentially give a unified result. Therefore, their effectiveness should be estimated by the number of nondimensionized quantities that make up models after a transformation procedure.

Let us consider model (5.23) and (5.24). It includes $N = 8$ dimensional variables (5.30) for $n = 3$ basic units of measurement [M], [L] and [T]. Based on the pi theorem, a nondimensionized model should include $\pi = N-n = 8-3 = 5$ nondimensionized quantities. Such a result was obtained after applying a generally accepted procedure: $\bar{P}$, $\bar{\omega}$, $\bar{t}$, $\bar{x}$ and π_1 π_2 π from (5.29). The application of the proposed method of nondimensionization made it possible to obtain the form of normalizing quantities (5.32) on the basis of a formalized procedure, leading to a further reduction in the number of nondimensionized variables. As a result, for the values from (5.29) π_1 π_2 1 is obtained, which corresponds to the achievement of self-similarity by similarity criteria. Such a result is the maximum possible in such procedures. This was due to taking into account MM structure in the process of nondimensionization.

A hydraulic shock MM allows to show a possibility of using other methods to achieve self-similarity by criteria. This is due to its simplicity. It follows from (5.29) that product $\rho \cdot w_o \cdot c$ has the dimension of pressure. The use of this expression as a normalizing quantity P^Δ leads, as in the previous case, to π_1 π_2 1. This result is trivial and is possible in this case due to the presence of only one criterion (5.29). In the presence of more criteria in models and complex connections between them, limiting reduction in the number of nondimensionized quantities is theoretically possible, but practically very difficult. The proposed method of nondimensionization is deprived of this shortcoming due to formalization of a nondimensionization procedure.

In MM (5.33) and (5.34), taking into account dissipative forces, in comparison with (5.23) and (5.24), an additional term appears and one more dimensional quantity (a coefficient of resistance). The total is $N = 9$. At the same time, the number of basic units of measurement did not change and $n = 3$ remained. As a result, on the basis of the pi theorem, a nondimensionized model should include $\pi = N-n = 9-3 = 6$ nondimensionized quantities. Such a result was obtained after applying a conventional procedure: $\bar{P}$, $\bar{\omega}$, $\bar{t}$, $\bar{x}$, and π_1 π_2 and π_3 as well from (5.35). The application of the proposed method of nondimensionization made it possible, as in the previous case, to obtain, on the basis of a formalized procedure, the form of normalizing quantities (5.39) leading to self-similarity by all the similarity criteria.

5.2.2 Examples of Using the Results of Reducing Models to a Nondimensionized Form

The purpose of the example in the previous section based on the model of a hydraulic shock in pipes was a demonstration of the mechanism of nondimensionization. On the other hand, the problem of considering a possibility of its application in solving problems and mapping the obtained results was not posed. The following examples illustrate some of the emerging opportunities in this direction.

5.2.2.1 Compiling data when displaying solution results
Non-stationary heat transfer

The problem of heating (cooling) of bodies that have the form of geometric primitives (an infinite plate, cylinder, sphere) is classic. The results of its analytical solution are given in monographs and in many textbooks on heat transfer. For this reason, it is a convenient example for demonstrating the possibilities of displaying the results of calculations on the basis of the proposed method of nondimensionization.

In known sources, the results of solving the problem are given in a nondimensionized form for generality. The corresponding procedure is performed in a standard way. As a result, functional dependence of the temperature of bodies on time, for example, for the central points of the bodies is represented in a two-criteria form: depending on the Fourier number of nondimensionized time and the Biot criterion. In a graphic form, this corresponds to a set of curves for each point under consideration. For each of the bodies under consideration, a solution is constructed and, accordingly, an individual set of curves.

An approximate model of heating (cooling) of geometric primitives is proposed in [10]. Generalization of measuring the results of its solution on the basis of the proposed method made it possible to obtain a single simple functional dependence of temperature on only one complex quantity – a modified number of homochronicity (greek. *homos* – equal, *chronos* – time), for example, for the central point of all the geometric primitives under consideration:

$$\bar{\theta}_c \quad 1 - \exp(-\widehat{H}\,o). \tag{5.41}$$

Temperature nondimensionization was done in a standard way:

$$\bar{\theta}_c \quad \frac{t_c - t_o}{t_s - t_o},$$

where t_c – current temperature in the center of the body in question; t_o – initial body and environment temperature; t_s – ambient temperature during body heating (cooling).

The modified number of homochronicity has the form of a nondimensionized complex:

$$\hat{H}o \quad Fo \cdot K_g \cdot \frac{Bi}{1 + k \cdot Bi}.$$

Here *Fo* and *Bi* – the Fourier number and the Biot criterion, respectively, K_g – the shape factor of the geometric primitive in question, and k – an integral coefficient due to the approximate nature of the original model.

Coefficient K_g is calculated. Coefficient k in [10] is also estimated analytically. But for more precise calculations, it is determined from a comparison with the available solution, in this case analytic. As a result, the following values were obtained for all types of the bodies considered (Table 5.1):

Table 5.1 Geometric characteristics and shape factors

Body	K_g	k
plate	1	0.42
cylinder	2	0.39
sphere	3	0.36

In the case under consideration, using the proposed method of nondimensionization, the results of calculating temperature in the center of the geometric primitives under consideration can be generalized using a single expression (5.41). Evaluation of the adequacy of such a generalization is accomplished by calculating the end time of heating of the bodies. The end time was fixed when the calculated temperature deviated from the ambient temperature by no more than 5%. The range of Bio variation ϵ [0.005 ... 1000] was considered. Throughout the entire range, an error in the calculations is (5.41) does not exceed 4% with respect to the results of the exact analytic solution. It should be taken into account that an error in determining a heat transfer coefficient used in *Bi* can reach 15–25% and even, according to some sources, 50%.

Expression (5.41) demonstrates a possibility of deeper generalization of the results of calculations, even in the presence of an exact analytic solution. This possibility is even more important in case of its absence. In solving technical problems of heat exchange, heat transfer through a flat wall plays a great role, especially in non-stationary processes. Expression (5.41) concerning a plate describes the case of its symmetrical heating: the heat exchange coefficients and the environment temperature are the same on its both sides. But in practice cases of different coefficients and temperatures are more interesting.

The first step in solving this problem is considering a case of asymmetrical heating of a plate: the temperatures on both sides are the same, but the heat exchange coefficients are different. In [10], while solving this problem and applying the proposed method of nondimensionization, we got an expression completely identical to (5.41). The expression of the modified number of homochronicity is different:

$$\hat{H}o \quad Fo \cdot 1 \cdot \frac{Bi_1(1 + \frac{Bi_2}{Bi_1} + 2 \cdot k \cdot Bi_2)^2}{(1 + k \cdot Bi_2) \cdot (1 + \frac{Bi_2}{Bi_1} + 2 \cdot k \cdot Bi_2 + k \cdot Bi_1 + k^2 \cdot Bi_1 \cdot Bi_2)}.$$

Here, Bi_1 and Bi_2 criteria define heat transfer on both sides of the plate depending on two different coefficients and k-integral coefficient due to the approximate nature of the initial model.

An exact (analytical) solution to this problem has not been found in the available literature. Therefore, the nature of the change in the temperature profile inside the plate, as well as the value of the coefficient k, were determined on the basis of numerical calculations. As a discrete counterpart, the control volume method was used. To ensure comparability of the results of calculations, it was also reduced to a nondimensionized form [11].

Due to the asymmetry of the heat exchange conditions on the sides of the plate, the temperature profile inside it is also asymmetric (Figure 5.1(a)):

In addition to the minimum temperature $\bar{\theta}_c$, the coordinate of its position inside the plate was determined (in a nondimensionized form) (Figure 5.1(b)). The thickness of the plate is assumed to be equal to $2l$:

$$\frac{x_{n1}}{2l} \quad \frac{1 + k \cdot Bi_2}{1 + \frac{Bi_2}{Bi_1} + 2 \cdot k \cdot Bi_2}.$$

As the next step in [10], the problem of non-stationary heat transfer through a flat wall at different temperatures and heat transfer coefficients on both sides is considered. With the help of numerical calculations [11], in a nondimensionized form, the trajectory of the displacement of a special point of the temperature profile inside the plate is determined. For any character of the heat transfer process, the trajectory consists of two sections of intersecting lines. Their location in space is determined by the position of the initial and final temperature profiles corresponding to the stationary states. On the basis of these data, an analytical solution in a nondimensionized form of a simplified model of nonstationary heat transfer through a flat wall is obtained.

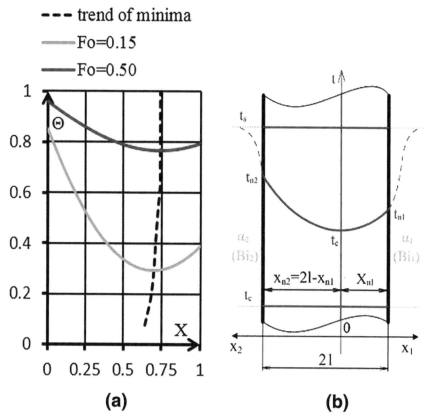

Figure 5.1 Temperature profile for asymmetric plate heating: a – examples of the results of numerical calculation of the relative temperature Θ as a function of the relative coordinate X for different values of Fo; b – accepted scheme for analytical calculations.

Although the form of solution in this case differs from (5.41), it has the same character:

$$\theta_c \quad A_1 \cdot x_{n1} \cdot Bi_1 \cdot \left[1 - \exp(-\widehat{H}o_1)\right], \quad\quad (5.42)$$

where

$$\widehat{H}o_1 \quad Fo \cdot \frac{1}{x_{n1}^2} \cdot \frac{1}{A_1 \cdot [1 + x_{n1} \cdot Bi_1 \cdot (1 - k)]};$$

$$A_1 \quad \frac{[1 + (1 - x_{n1}) \cdot Bi_2]}{x_{n1} \cdot (Bi_1 + Bi_2 + Bi_1 \cdot Bi_2)}.$$

Here x_{n1} – coordinate of a singular point. In its quality, we consider a point located at maximum distance from the temperature profile in a finite stationary state. In [10], an algorithm for determining its position is described.

Comparison of the obtained results shows the effectiveness of the application of the proposed method of nondimensionization for their display:

- the results are displayed using brief, easy-to-analyze expressions (5.41) and (5.42);
- the solution of problems, although of a general nature, but still different, has either an identical (5.41) or similar (5.42) form;
- the number of values involved in the display of information is reduced. This is equivalent to compressing information while displaying without losing it.

Natural oscillation frequency of a liquid with free surface in containers of various shapes

Problems connected with determining natural oscillation frequencies of liquids with free surfaces in containers of various shapes have remained relevant for a long time. The solution of a number of problems is related to the theory of ship- and hydro-construction. Problems connected with motion of a liquid with free surface were also covered in the solution of tasks of transporting liquid cargoes, when seismic loads were acting on the vessel. Particular attention was paid to the solution of similar problems in the design of aviation and rocket technology. Liquid motion in tanks in the form of rectangular parallelepipeds, cylindrical, conical, toroidal tanks, in the presence of central bodies, in rotational tanks of arbitrary shape, etc., was studied. Numeric-analytic, numeric, and experimental methods of solution were used. The interest in these issues has not been lost nowadays. Close attention to such tasks can be explained, on the one hand, by their prevalence and, on the other hand, by the complexity of the solution.

Having carried out a series of experiments using transparent containers of various shapes and visualization of currents inside liquid, we found out the features of its motion . On the basis of these data and discarding some minor details in [12], an approximate integral model of the oscillation of a liquid with a free surface is constructed. Such a model made it possible to obtain a solution in the form of a simple algebraic expression up to certain integral coefficients. These coefficients compensate for the impact of unaccounted (rejected) effects on the process under study.

To bring the obtained results to numerical values, it is necessary to determine the values of integral coefficients. As in the case of solving the problem of non-stationary heat transfer, their magnitude can be estimated analytically. But more accurate values are determined from experiments. Here it is useful to represent the model in a nondimensionized form. Reducing the dimensionality of the modeling space significantly reduces the number of necessary experimental studies.

When solving the problem, the following calculation scheme was used (Figure 5.2):

Σ, Σ' – traces of a free surface when moving and unperturbed position of the liquid;

M – trail of the meridian plane;

$\eta^{\Delta}, v^{\Delta}$ – scales of the amplitude and velocity of fluid oscillations;

V_{ef}, h_{ef} – effective volume and depth of fluid oscillations;

l_2 – characteristic free surface dimension;

h – total depth of liquid in a tank;

v_2^{Δ} – diagram of the liquid velocity in the transverse direction in the effective volume of the liquid.

The movement of the liquid occurs mainly in the effective region. It is bounded by a part of the spherical surface h_{ef} l_2. Its radius is equal to the characteristic dimension l_2 of the free surface. The characteristic size is half the transverse dimension of the container. At $h > h_{ef}$, the lower part of the liquid remains practically at rest.

Based on the above-mentioned assumptions, a solution was obtained with an accuracy of up to two integral coefficients: k_{ω} is a coefficient that depends on the shape of the container; k_{ef} is a coefficient that depends on the filling depth of the container.

$$\bar{\omega} \quad \frac{\omega}{\sqrt{\frac{g}{(l_2+r)}}} \quad \bar{h}_{ef} \cdot k_{\omega};$$

$$\bar{h}_{ef} \quad \frac{h}{l_2+r} \quad \begin{cases} 1 - \exp(-k_{ef} \cdot \bar{h}), & \text{provided } \bar{h} < 1 \\ 1, & \text{provided } \bar{h} \geq 1 \end{cases}, \quad (5.43)$$

where $\omega, \bar{\omega}$ are dimensional and nondimensionized, respectively, natural oscillation frequency of the liquid in the tank;

g – acceleration of gravity;

r – radius of an inner body in the container.

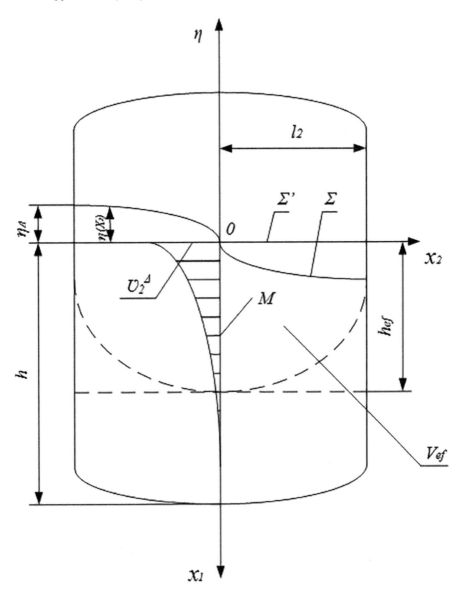

Figure 5.2   Calculation scheme of motion of a fluid with a free surface.

The value of r appears when there is an internal body in the container. An example is a container formed by two coaxial cylinders. In the absence of an inner body, r 0.

Consider expression (5.43). At a large filling depth ($h \geq h_{ef}$) we have $\bar{h}_{ef} = 1$ and, respectively:

$$\frac{\omega}{\sqrt{\frac{g}{(l_2+r)}}} \quad \bar{\omega} \quad k_\omega \tag{5.44}$$

In other words, the nondimensionized natural frequency of oscillations of the free surface of a liquid depends only on the shape of the container and remains constant for any size. At a shallow fill depth of the container ($h < h_{ef}$), we have $\bar{h}_{ef}$ $1 - \exp(-k_{ef} \cdot \bar{h})$ and, respectively:

$$\frac{1}{k_\omega} \cdot \frac{\omega}{\sqrt{\frac{g}{(l_2+r)}}} \quad \frac{1}{k_\omega} \cdot \bar{\omega} \quad 1 - \exp(-k_{ef} \cdot \bar{h}). \tag{5.45}$$

An assumption was made, confirmed experimentally later, that the coefficient of the form k_ω remains constant for large and small filling of the container.

Proceeding from the above, an algorithm for the experimental determination of k_ω and k_{ef} is as follows:

- for each of the considered tank shapes with a constant large filling ($h \geq h_{ef}$), a series of experiments is performed to determine the natural frequency of oscillations of the free surface. Starting from (5.44), the coefficient k_ω is defined as the average value of the experimental values reduced to a nondimensionized form;
- for each of the container shapes under consideration with a variable value of small fillings ($h < h_{ef}$), a series of experiments is carried out to determine the natural frequency of oscillations of the free surface. With the known k_ω, the experimental data are approximated using expression (5.45). The optimized value is k_{ef}.

The values k_ω and k_{ef} determined in this way for certain containers and methods of their location in space are given in Table. 5.2.

Considering the results obtained, it is interesting to note the coincidence of expressions (5.41) and (5.45) used to describe completely heterogeneous phenomena: non-stationary heat transfer and fluid motion with free surface.

The reliability of expression (5.43) can be estimated by comparing the results obtained with its help and calculations based, for example, on the variational method [13]. Consider the calculation results for a vertical cylinder.

Table 5.2   The value of coefficients for calculating the natural frequencies of fluid oscillations in tanks

№	The Shape of the Container, the Position and Direction of the Impacts	k_{ef}	k_ω	The Law of Distribution of the Normalized Deviation of the Calculated Results from the Experimental Ones
1	Rectangular straight parallelepiped, lateral faces are vertical, $\bar{h} \geq 0.3$	3,67	1,26	(0.011; 0.00042)
2	The circular cylinder, the axis is vertical, across the axis, $\bar{h} \geq 0.4$	4,29	1,36	(0; 0.00064)
3	The circular cylinder, the axis is horizontal, along the axis, $0.2 \leq \bar{h} \geq 3$	3,5	1,6	(0; 0.0028)
4	The circular cylinder, the axis is horizontal, across the axis, $0.2 \leq \bar{h} \geq 1.8$	2,43	1,29	(0; 0.00037)
5	The cone is truncated, the axis is vertical, the opening angle is 20^0, across the axis	5	1,21	(0; 0.000676)
6	The cone is truncated, the axis is vertical, the opening angle is 50^0, across the axis	7,57	0,905	(0; 0.000525)
7	Coaxial circular cylinders, the axis is vertical, across the axis, $\bar{h} \geq 0.4$	4,3	1,39	(0.02; 0.001)

It follows from [13] that:

$$\omega_n \quad \sqrt{\frac{ng}{R}} \cdot \sqrt{th(\varepsilon_i \cdot \frac{h}{R})} \cdot \sqrt{\varepsilon_i}. \qquad (5.46)$$

Here ε_i – roots of the differential equation with respect to the Bessel functions of the first and second kind of the first order; i – tone number of oscillations.

For the first four tones, the following values are given: ε_1 1.84, ε_2 5.33, ε_3 8.54, and ε_4 11.71.

Comparison of expressions (5.43) and (5.46) shows that their structure is the same. The first factor for r = 0 is the same for both expressions. The third factor is a form factor and for the first tone of the oscillation it is $\sqrt{\varepsilon_1}$ $\sqrt{1,84}$ $1,356$, which almost coincides with k_ω 1.36 (the value from the second point of Table 5.2). The second factor depends on the value of the relative filling and, as it increases from a certain value, becomes equal to 1. As can be seen, both methods give practically equal values for this value: $\bar{h}$ h/R 1. But if $\bar{h} < 1$, the results of the solution using the result of the variational method (5.46) are less than the results obtained with the integral coefficients method (5.43). Taking into account that (5.43) is an approximation of the experimental data, we can say that the solution of the variational problem for the case under consideration gives an underestimate. Thus, in the working range for expression (5.43), the maximum deviation at $\overline{h}$ 0.4 is 14%.

Let us consider a more complicated case of determining the natural oscillation frequency of a free fluid surface between coaxial cylinders (Figure 5.3). In [13], the graph of the oscillation frequency is represented as a function of the depth of filling (Figure 5.4). The results on the basis of which it was constructed use numerical calculations and correspond to only one combination of sizes: R = 1 m, r = 0.4 m. The filling depth h is indicated in meters.

Frequency of oscillation $\tilde{\omega}_1^2$ is displayed in a nondimensionized form. The nondimensionization is carried out by means of an expression similar to (5.44).

The graph is more of a qualitative nature. Nevertheless, some estimates can be made on its basis. Thus, for a given dimension ratio, it follows from (5.43) that the filling depth does not affect oscillation frequency at h 1.4 m.

In this case, starting from (5.44), nondimensionized frequency of oscillations is $\bar{\omega}$ k_ω 1.39. The values obtained agree well with the data of the graph (Figure 5.4).

It should be noted that the data on the basis of which the graph is built (Figure 5.4) are obtained by numerical calculations and correspond to only one size relationship. Expression (5.43) has a simple algebraic form, but it reflects all possible relationships between the dimensions of the container and the depths of its filling.

Let us estimate the internal correspondence of the coefficients for the capacities of some types (Table 5.2). When considering the container formed by coaxial cylinders (Figure 5.3), it is assumed that the shape factor k_ω retains its value for any size relationship, including when r $\rightarrow$ 0. But in this case the container of the form in Figure 5.3 degenerates into a circular cylinder with a vertical axis. Accordingly, the coefficients of the form k_ω for these tanks must be equal. Comparison of the corresponding positions in lines 2 and 7 of Table 5.2 (1.36) and (1.39) confirm this fact. Some discrepancy ($\sim$2%) can be attributed to the accuracy of experimental studies.

Comparison and analysis of the obtained results show the effectiveness of applying the proposed method of nondimensionization for generalization of experimental data:

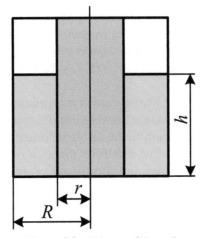

Figure 5.3   Scheme of the tank.

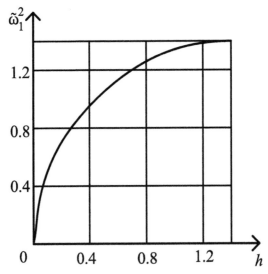

Figure 5.4   Influence of filling depth on frequency of natural oscillations of liquid between coaxial cylinders.

- the expressions obtained have a compact form;
- in the expressions received the internal connection of the processed data is displayed.

5.2.2.2 Approximate solution of nonlinear ordinary differential equations. Solution method

When studying the behavior of control systems, dynamic regimes are mainly considered. A mathematic tool for studying such systems is most often the apparatus of ordinary differential equations. General methods of solution have only their linear forms. For this reason, linear (linearized) approaches are used to construct object and controller models. But more accurate models are represented by nonlinear differential equations. This situation leads to the fact that the results obtained can be used only in the range of small changes in the parameters of the controlled processes near the linearization point. The search for new approaches to the analysis of nonlinear models is an important element in the solution of the problem of optimal control.

An effective way to simplify the study of complex models is their nondimensionization. Reducing the dimensionality of the modeling space allows us not only to reduce the number of experimental studies (natural and numerical) by orders of magnitude, but also to facilitate obtaining analytical dependencies. As another method of generalized analysis of differential equations, the theory of group methods for solving them can be considered. Initially, the theory of groups was created (the end of the XIX century) as a demonstration of the universality of the approach to the already known disparate solutions of ordinary differential equations, and in this sense did not come up with new solutions. However, the incorporated idea turned out to be productive in the study of models in the newly emerging fields of research, using new types of equations. As a rule, these are partial differential equations [14]. And on the basis of the theory of groups, both analytic and numeric solutions can be obtained [15]. In addition to exact solutions in the theory of groups, there is a direction of using approximate symmetries [16] and, accordingly, obtaining approximate solutions.

A special feature of the application of group methods when trying to solve an equation is the use of arbitrary, convenient or even random initial solutions. One of the most well-known mathematic applications of continuous groups is found in the control theory [17]. In the literature reviewed, it was not possible to find examples of the use, as the initial most developed solutions, of linearized models for investigating processes described by nonlinear equations.

The unidirectionality of the methods of nondimensionization and the theory of groups to simplify the solution of differential equations led to an attempt to unify them [18]. In a more narrow application in [19], the nondimensionization of the original differential equations was used to simplify the analysis of the derived from them, by means of the Laplace transform, algebraic equations. Such an operation is a tool for identifying a group of homogeneous stretches. But in the literature reviewed, it has not been possible to find examples of a combination of special methods of nondimensionization, which make it possible to achieve self-similarity in terms of parameters (similarity criteria) and ideas of group methods of solution.

The aim of the studies, the results of which are given below, was the development of a method for approximate solution of nonlinear differential equations using the results of nondimensionization and group methods of solution.

An approximate solution can be considered as an approximation of exact results. The quality of this procedure is determined by a number of factors:

a. The quality of approximation increases with decreasing the number of values considered in the approximation dependence;

b. the better the approximating dependence reflects the physical essence of the described process, the higher the quality of approximation is.

The way of nondimensionization of the mathematic model can influence the enhance of the effect of these factors.

The procedure of nondimensionization can only be applied to homogeneous equations. Nonlinear equations are not the ones. To homogenize the initial equations in the solution process, a linearization procedure is applied. The results obtained in this case are used for further solution, but they are not the ones. The proposed method of solution is based on the execution of a certain algorithm:

1. The original nonlinear differential equation is linearized and subjected to nondimensionization. The goal is the maximum possible reduction in the number of parameters in the model.

2. The resulting equation is solved. As the linearized equation is nondimensioned by the proposed method, both it and its solution contain the minimum possible number of parameters.

3. The solution obtained is used as the initial solution. The general solution of the original nonlinear equation, as is customary in group methods, is sought as a combination of the initial solution and the stretching coefficient k. At this stage, this coefficient is unknown.

4. The initial nonlinear equation is nondimensionized by means of the normalizing values obtained in paragraph 1 of the algorithm.

5. The expression for the general solution (paragraph 3 of the algorithm) is substituted into the determined nonlinear equation instead of the functions to be determined. All the differentiation operators are executed. By varying the value of the stretching factor k, a minimum (zero) residual

is ensured in the written equation. The value of k found is the desired value of the stretching factor.

6. In accordance with clause 3 of the algorithm, a general solution is sought in the form of a combination of the stretching factor defined in clause 5 and the initial solution defined in clause 2.

Example of solution

Analysis of transfer functions of control objects shows that in many cases they are represented in the form of an inertial link of the second order. As a consequence, the initial equations used to describe processes are ordinary linear second-order differential equations. Such equations describe oscillatory processes. Most often, the linearity of the equations is a forced measure and is determined by the methods used to solve them. Within the framework of the theory of automatic control, there is a need to solve such non-linear equations, let it be approximate, but nevertheless analytical. A model for the motion of a mathematical pendulum at large angles of deviation can serve as their analogue. Similar examples are considered below.

Model of motion of a mathematical pendulum at large deviation angles without allowance for dissipative forces

The solution is sought on the basis of the algorithm outlined in Section 2.2.2.1.

1. (algorithm). We consider an initial equation of the form:

$$\frac{d^2x}{dt^2} + g \cdot \sin(\alpha) \quad 0. \tag{5.47}$$

Here $\alpha = x/r$ – angle of deviation of the pendulum suspension from the equilibrium position, r – the length of a suspension thread.

After linearization and nondimensionization using the proposed method, the model with allowance for the initial conditions has the form:

$$\frac{d^2\bar{x}}{d\bar{t}^2} + \bar{x} \quad 0$$

$$npu \begin{bmatrix} \bar{t} & 0 & \bar{x} & 1; \\ \bar{t} & 0 & \dfrac{d\bar{x}}{dt} & 0. \end{bmatrix} , \tag{5.48}$$

where normalizing quantities are defined as follows:

$$x^\Delta \quad \delta; \quad t^\Delta \quad \sqrt{r/g}. \tag{5.49}$$

Here δ – the coordinate of a body with its initial deviation from the equilibrium position (amplitude of oscillations), g – the field strength of the mass forces (acceleration of gravity).

In the original equation (5.47), the value of the function depends on three quantities: $x = f(t, \alpha, g)$. After linearization, $- x = f(t, r, g)$. With the usual method of nondimensionization, the value of the function depends on two quantities: $\bar{\bar{x}}$ $f(\bar{\bar{t}}, \omega)$, where ω – natural oscillation frequency. The use of the proposed method of nondimensionization made it possible to obtain expression (5.48), in which the function depends only on one variable: $\bar{x}$ $f(\bar{t})$. In fact, this means the introduction of a new time unit: when solving problems, time will not be measured in seconds (minutes, hours), but in fractions of the period of oscillations.

2. (algorithm). Such a simplification of the model (5.48) makes it possible to obtain a solution in an extremely simple form:

$$\bar{x} \quad \cos(\bar{t}). \tag{5.50}$$

3. (algorithm). The general solution of the original nonlinear equation (5.47) is sought in the form of a combination of the initial solution and the stretching coefficient k:

$$\bar{x} \quad \cos(\bar{t}/k). \tag{5.51}$$

4. (algorithm). The original nonlinear equation (5.47) is nondimensionized by means of the normalizing quantities (5.49) obtained in paragraph 1 of the algorithm:

$$\alpha_\delta \frac{d^2\bar{x}}{d\bar{t}^2} + \sin(\alpha_\delta \cdot \bar{x}) \quad 0. \tag{5.52}$$

Here it is taken into account that $\alpha_\delta = \delta/r$ – angular amplitude of oscillations of a pendulum.

5. (algorithm). Equation (5.51) is substituted into equation (5.52). Differentiation is carried out for constant k. The value of k is chosen in such a way that the resulting equation is satisfied.

Let us discuss the value of k in more detail. We know the solution of the equation of the form (5.47). It can be found using the Jacobi integral. Within this decision, the period of oscillations is defined as:

$$T_0 \quad \sqrt{\frac{r}{g}} \cdot 2\pi \cdot k_p, \tag{5.53}$$

where k_p – elliptic integral of the first kind.

Table 5.3 Coefficients of stretching and error of analytical solutions for different initial angles of deviation (angular amplitudes) in the absence of energy dissipation

α_δ, deg	30°	60°	90°
k	1.0174	1.0732	1.1804
ε_{max}, %	0.12	0.48	1.19

For calculation it is practically convenient to expand the elliptic integral in series:

$$k_p \quad 1 + \left(\frac{1}{2}\right)^2 b^2 + \left(\frac{1\cdot 3}{2\cdot 4}\right)^2 b^4$$

$$+ \left(\frac{1\cdot 3\cdot 5}{2\cdot 4\cdot 6}\right)^2 b^6 \ldots \left[\frac{(2n-1)!!}{(2n)!!}\right]^2 b^{2n},$$

$$b \quad sin(\alpha_\delta/2). \tag{5.54}$$

The first part of expression (5.53) corresponds to the oscillation period of the pendulum, described by the linearized equation (5.50). In this case, k_p is corrective and can be considered similar to k from (5.51). The quantities k_p from (5.54) and k from solution (5.52) in clause 5 of the algorithm practically coincide. This indicates that the value of k found is the sum of the series (5.54) and, correspondingly, the solution of the elliptic integral for calculating k_p
6. *(algorithm)*. The found value of k in accordance with (5.51) determines the general solution of equation (5.47). As an example, the solution was obtained for three values of the angles of the initial deviation of a pendulum: 30°, 60°, 90°. The results are given in Table 5.3. For comparison, equation (5.47) was solved using the fourth-order Runge Kutta method. Calculations are performed in the range of the argument variation $\bar{t}$ 0 . . . 20 with step $h =$ 0.1. At each calculated point, the results obtained with the help of expression (5.51) and the numerical method are compared: the error ε_{max} referring to the doubled amplitude (the amplitude of the oscillations) is determined. Among the errors for all points, the maximum one was chosen.
Figure 5.5 displays:

- the results of the solution based on (5.50) ($k = 1$);
- the results of the solution based on (5.51) ($k = 1.1804$) for the initial angle of deviation of the pendulum $\alpha\delta = 90°$;
- results of the solution on the basis of the Runge Kutta method.

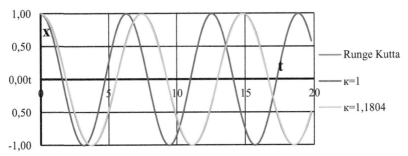

Figure 5.5   The nondimensionized coordinates x of the position of the pendulum as a function of the nondimensionized time t obtained on the basis of various methods for solving the nonlinear nondimensionized equation (5.21).

The visual absence of the third graph in Figure 5.5. demonstrates the practical coincidence of the results of calculations with the help of (5.51) and Runge Kutta method.

In practical calculations, the values k_p can be approximated as a function of α_δ and used in (5.51) to calculate different $\bar{x}$ values.

The results of the calculations based on the solutions obtained show:

- the proposed method of algebraic approximate solution of equation (5.47) is simpler than the method based on the use of the Jacobi integral;
- the error of calculations based on the proposed method compared with the numerical results for angles of the initial deviation of the pendulum to 90° does not exceed 1%.

Model of motion of a mathematical pendulum at large angles of deviation with allowance for dissipative forces

Like in the previous case, the solution is sought on the basis of the algorithm outlined in section *Approximate solution of nonlinear ordinary differential equations. Solution method.*

1. (algorithm). We consider the initial equation of the form (5.47) with the addition of a term taking into account the action of the forces of resistance of the environment:

$$m\frac{d^2x}{dt^2} + c\frac{dx}{dt} + mg\sin(\alpha) \quad 0. \tag{5.55}$$

Where c – a coefficient of resistance.

After linearization and nondimensionization with the help of the proposed method, the model with allowance for the initial conditions has the form:

$$\frac{d^2\bar{x}}{d\bar{t}^2} + \frac{1}{\bar{m}}\frac{d\bar{x}}{d\bar{t}} + \bar{x} \quad 0$$

with parameters
$$\begin{bmatrix} \bar{t} & 0 & \bar{x} & 1; \\ \bar{t} & 0 & \frac{d\bar{x}}{d\bar{t}} & 0, \end{bmatrix}$$ (5.56)

where the normalizing quantities are defined as follows:

$$x^{\Delta} \quad \delta; t^{\Delta} \quad \sqrt{r/g}; m^{\Delta} \quad c\sqrt{r/g}.$$ (5.57)

2. (*algorithm*). Depending on the size $\bar{m}$ solution (5.56) takes form:
1.1) when $\bar{m} > 0.5$

$$\bar{x} \quad e^{-\frac{\bar{t}}{2\bar{m}}}\left[\cos\left(\frac{p}{2\bar{m}}\bar{t}\right) + \frac{1}{p}\sin\left(\frac{p}{2\bar{m}}\bar{t}\right)\right],$$ (5.58)

where $p \quad \sqrt{4\bar{m}^2 - 1}$.
1.2) when $\bar{m} < 0.5$

$$\bar{x} \quad \frac{1}{2}\left(1 + \frac{1}{p}\right) \cdot e^{(a_1 \cdot \bar{t})} + \frac{1}{2}\left(1 - \frac{1}{p}\right) \cdot e^{(a_2 \cdot \bar{t})},$$ (5.59)

where $p \quad \sqrt{1 - 4\bar{m}^2}; a_1 \quad -\frac{(1-p)}{2\bar{m}}; a_2 \quad -\frac{(1+p)}{2\bar{m}}$.
3. (*algorithm*). In (5.58) and (5.59) we introduce a coefficient of stretching k:

$$\bar{x} \quad e^{-\frac{\bar{t}/k}{2\bar{m}}}\left[\cos\left(\frac{p}{2\bar{m}} \cdot \frac{\bar{t}}{k}\right) + \frac{1}{p}\sin\left(\frac{p}{2\bar{m}} \cdot \frac{\bar{t}}{k}\right)\right],$$ (5.60)

$$\bar{x} \quad \frac{1}{2}\left(1 + \frac{1}{p}\right) \cdot e^{(a_1 \cdot \frac{\bar{t}}{k})} + \frac{1}{2}\left(1 - \frac{1}{p}\right) \cdot e^{(a_2 \cdot \frac{\bar{t}}{k})}.$$ (5.61)

4. (*algorithm*). The original non-linear equation (5.55) is nondimensionized by means of normalizing quantities (5.57):

$$\alpha_\delta \frac{d^2\bar{x}}{d\bar{t}^2} + \frac{1}{\bar{m}}\frac{d\bar{x}}{d\bar{t}} + \sin(\alpha_\delta \cdot \bar{x}) \quad 0.$$ (5.62)

5. (*algorithm*). Equation (5.60) or (5.61) is substituted into equation (5.62). Differentiation is carried out for constant k. The value of k is chosen in such a way that the resulting equation is satisfied. The value of k found is the desired

Table 5.4   Coefficients of stretching and errors of analytic solutions for different initial angles of deviation in the presence of the environment resistance

α_δ, deg	30°		60°		90°	
–	0.4	20	0.4	20	0.4	20
K	1	1.0126	1.105	1.0515	1.25	1.126
ε_{max},%	1.25	0.66	1.00	2.73	2.18	5.50

value of the stretching factor.

6. *(algorithm).* The value of k found in accordance with (5.60) or (5.61) determines the general solution of equation (5.55) for the cases $\bar{m} > 0.5$ and $\bar{m} < 0.5$, respectively.

As an example, like in the previous case, the solution was obtained for three values of the angles of the initial deviation of a pendulum: 30°, 60°, and 90°. For each of the angles, two variants of values are considered, corresponding to the absence and presence of oscillations: $\bar{m}$ 0.4 and $\bar{m}$ 20. For comparison, equation (5.56) was solved using the fourth-order Runge Kutta method. Calculations are performed in the range of the argument variations $\bar{t}$ 0...20 with step h = 0.1. At each calculation point, the results obtained with the help of expressions (5.60), (5.61) and the numerical method are compared, thus an error ε_{max} is determined. Among the errors for all points, the maximum one is chosen. The results of the calculations are given in Table 5.4.

Figure 5.6 displays:

– the results of the solution on the basis of (5.59) and (5.58) ($k = 1$);
– the results of the solution on the basis of (5.61) ($k = 1.25$) and (5.60) ($k = 1.126$) for the initial angle of deviation of the pendulum $\alpha_\delta = 90°$;

results of the solution on the basis of the Runge Kutta method

In Figure 5.6, the calculation results for the initial deviation angle 90° are displayed. This angle is chosen because of the maximum error for it. But even in this case, the maximum error at $\bar{m}$ 0.4 does not exceed 2.2% and at $\bar{m}$ 20–5.5%.

As in the case considered above, in practical calculations, the values of k_p can be approximated as a function of α_δ and used in (5.60) and (5.61) to calculate different values of $\bar{x}$.

The results of calculations based on the solutions obtained show:

– the proposed method makes it possible to obtain an approximate analytical solution of a nonlinear differential equation;

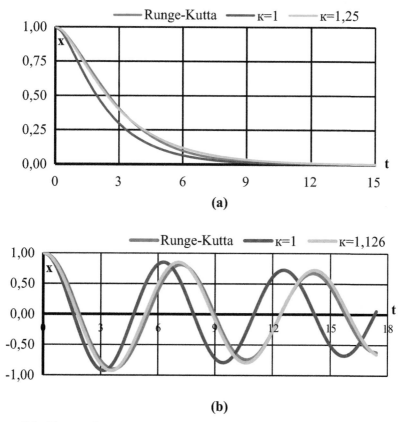

Figure 5.6 The nondimensionized coordinates x of the position of the pendulum as a function of the time t, obtained on the basis of various methods for solving nonlinear nondimentionized equation (5.29): $a = 0.4$; $b = 20$.

- the error in the considered range of parameter variation does not exceed the value permissible for engineering calculations ($\sim$5%).

5.3 Conclusion

As part of the research:

- an original algorithm is developed that formalizes the process of nondimensionisation of a mathematical model. In the course of the developed procedure, the number of nondimentionized variables and parameters decreases to smaller values than it is prescribed by the pi theorem;

– a procedure for determining the lower limit of the possible number of nondimensionized quantities in a mathematical model is established. Numerically, this value is defined as rank [A]. The number of lines in matrix [A] is determined by the number of allocated nondimensionized complexes. Matrix [A] is made up of the exponents of the quantities involved in these complexes. In a case when the number of complexes does not exceed the total number of functions and variables of the model, it is possible to achieve complete self-similarity over the complexes. In other words, it is possible to find expressions for the normalizing quantities, leading all complexes to values equal to 1;

– standard methods of nondimensionization always lead to the formation of a certain number of nondimensionized complexes – similarity criteria. The possibility of achieving self-similarity by all criteria indicates a deeper transformation of models and coverage of all previously known cases;

– the results of the given examples of the application of the proposed method of nondimensionization indicate the possibility of a deeper generalization of experimental data. As a result, it is possible to obtain simple approximate solutions of problems in various fields of engineering;

– the proposed method of nondimensionization can be used together with other methods of solving problems. As a result of a peculiar synergetic effect, it is possible to obtain simpler solutions in comparison with those already available or new ones that were previously non-existent.

References

[1] Atherton, M. A. Dimensional Analysis Using Toric Ideals: Primitive Invariants [Text] /M. A. Atherton, R. A. Bates, H. P. Wynn // PLoS One 9(12): e112827. 2014. doi: 10.1371/journal.pone.0112827.

[2] AA Sonin. A generalization of the Π-theorem and dimensional analysis [Text]/A. A. Sonin // Proceedings of the National Academy of Sciences of the United States of America. 2004, 101 (23): 8525–8526. doi: 10.1073/pas.040293110)1.(in Russian)

[3] Ekici, Ö. Lattice Boltzmann Simulation of Mixed Convection Heat Transfer in a Lid-Driven Square Cavity Filled With Nanofluid: A Revisit [Text] / . Ekici // J. Heat Transfer 140(7), 072501 (Mar 23, 2018) (9 pages) Paper No: HT-16-1785; doi: 10.1115/1.4039490.

[4] Brennan, S. Dimensionless Robust Control With Application to Vehicles [Text] / S. Brennan, A. Alleyne / IEEE Trans. on Control Systems Technology, Vvl. 13, no. 4, July 2005, pp. 624–630.

[5] Legenky, VI On the bundle of algebraic equations. [Text] / V. I. // Symmetries of differential equations. Sat. scientific works. Moscow: MIPT. 2009, pp. 118–128 (in Russian)

[6] Lehenky, V. I. Dimensionless variables: group-theoretic approach [Text] / V. I. Lehenky, G. N. Yakovenko // Symmetries of differential equations. Sat. scientific works. Moscow: MIPT. 2009. From 1 to 12 (in Russian).

[7] Azih, C. Similarity Criteria for Modeling Mixed-Convection Heat Transfer in Ducted Flows of Supercritical Fluids [Text] / C. Azih, M. I. Yaras // J. Heat Transfer 139(12), 122501 (Jun 27, 2017) (13 pages) Paper no: HT-16-1567; doi: 10.1115/1.4036689.

[8] Sheremet, M. A. Natural Convection in a Wavy Porous Cavity With Sinusoidal Temperature Distributions on Both Side Walls Filled With a Nanofluid: Buongiorno's Mathematical Model [Text] / M. A. Sheremet, I. Pop // J. Heat Transfer 137(7), 072601 (Jul 01, 2015) (8 pages) Paper No: HT-14-1436; doi: 10.1115/1.4029816.

[9] Seshadri, R. Group Invariance in Engineering Boundary Value Problems / R. Seshadri, T. Y. Na // Springer-Verlag, New York Inc., 224 p, 1985.

[10] Brunetkin, O. Development of the method of approximate solution to the nonstationary problem on heat transfer through a flat wall [Text] / O. Brunetkin, M. Maksymov, O. Maksymova, A. Zosymchuk // Eastern-European Journal of Enterprise Technologies, vol. 6(5), pp. 31–40, 2017. Doi: 10.15587/1729-4061.2017.118930.

[11] Brunetkin, O. A simplified method for the numerical calculation of nonstationary heat transfer through a flat wall [Text] / O. Brunetkin, M. Maksymov, O. Lysiuk // Eastern-European Journal of Enterprise Technologies, vol. 2(5), pp. 4–13, 2017. Doi: 10.15587/1729-4061.2017.96090.

[12] Integrated approach to solving the fluid dynamics and heat transfer problems [Text] / A. I. Brunetkin // Proceedings of the Odessa Polytechnic University, vol. 2(44), pp. 108–115, 2014.

[13] Kolesnikov K. S. Dynamics of rockets [Text] / K. C. Kolesnikov // 2 nd ed., Corrected. and additional. – M.: Mechanical Engineering, p. 520, 2003 (in Russian).

[14] Oliveri, Fr. Lie Symmetries of Differential Equations [Text] / Fr. Oliveri // Classical Results and Recent Contributions. Symmetry, vol. 2, pp. 658–706, 2010. Doi:10.3390/sym2020658. ISSN 2073-8994.

[15] Chhay, M. Lie Symmetry Preservation by Finite Difference Schemes for the Burgers Equation [Text] / M. Chhay, A. Hamdouni // Symmetry, no. 2, pp. 868–883, 2010. Doi:10.3390/sym2020868, ISSN 2073-8994.

[16] Gazizov, R. K. Integration of ordinary differential equation with a small parameter via approximate symmetries: Reduction of approximate symmetry algebra to a canonical form [Text] / R. K. Gazizov N. H. Ibragimov V. O. Lukashchuk // Lobachevskii Journal of Mathematics April 2010, vol. 31(2), pp. 141–151.

[17] Starrett, J. Solving Differential Equations by Symmetry Groups [Text] / J. Starrett // American Mathematical Monthly, vol. 114, no. 9, pp. 778–792(15), 2007.

[18] Legenky, VI On the bundle of algebraic equations. [Text] / V. I. // Symmetries of differential equations. Sat. scientific works. Moscow: MIPT, pp. 118–128, 2009 (in Russian).

[19] Brennan, S. Dimensionless Robust Control With Application to Vehicles [Text] / S. Brennan, A. Alleyne / IEEE Trans. on Control Systems Technology, vol. 13, no. 4, pp. 624–630, July 2005.

PART II

Control Systems Applications

6

Energy Efficiency of Smart Control Based on Situational Models

Volodymyr M. Dubovoi and Mariya S. Yukhymchuk

Department "Computer Control Systems" in Vinnytsia National Technical University, 95 Khmelnytske Shose, Vinnytsia, 21021, Ukraine
E-mail: v.m.dubovoy@gmail.com; umcmasha@gmail.com

The model for estimating energy losses during logical, fuzzy, and statistical control is developed. The research of this model with the use of simulation allowed make conclusions of the advantages of fuzzy and statistical situational control when considering the near and far horizon of events.

6.1 Introduction

The problem of saving and efficient use of energy affects all areas of activity. If in the areas in which energy is used as a direct element of the production process (heat engineering, energy, transport, etc.), this problem has been formulated and solved for many decades, the energy aspects of information processing, management of complex systems, and their impact on energy efficiency of the systems as a whole became relevant relatively recently.

The problems of control in complex systems that can be in a variety of states and in each state characterized by a large-dimensional vector of parameters are complicated with an increasing complexity of the controlled systems. The transition to automation of control processes makes their solution a necessary condition for increasing the systems' efficiency.

The most noticeable is complication and increasing of the systems' dimension in connection with the active development and dissemination of the concept of "Internet of Things". The growth of the number of objects connected with Internet is shown in Figure 6.1 [1].

145

Namber of Connected Objects Expected to Reach 50bn by 2020

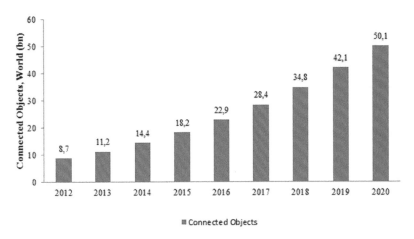

Penetration of connected objects in total 'things'
expected to reach 2,7% in 2020 from 0,6% in 2012

Figure 6.1 The growth of the number of objects connected with Internet.

Simplification of connection to computer networks of various devices led to an avalanche-like growth in the number of connected devices, an increase in the volume of incoming data and to a corresponding complication of control algorithms, and an increase in energy consumption.

A number of features of managing complex systems problems do not allow to apply classical approaches to the creation of energy-efficient control systems [2]:

- poor structuring of the task, due to the fact that a number of factors considered in the decision-making process are difficult or impossible to correctly formalize;
- incompleteness, inaccuracy, insufficient reliability of information, on the basis of which decisions are made;
- presence of both indistinctness and stochastic uncertainty in describing the state of the systems;
- multi-step nature of the management process.

In this regard, for the management of modern complex systems, artificial intelligence methods are increasingly being used.

These circumstances make it necessary to conduct research and develop the methods for energy efficient intellectual control.

In complex systems with poor structuring of control tasks, incomplete certainty of the initial information, large dimension of the source data vector and their representation in different scale systems [3], control is carried out mainly on the basis of a situational approach [4].

A lot of situations and transitions between them form a situational network. For dynamic systems managed on the basis of situational models, a situational-event approach was proposed [5].

In case of qualitative (linguistic) information or information from experts' presence, fuzzy logic uses. There are various methods that implement a fuzzy situational approach. Most of them are based on the representation a set of situations in the form of fuzzy graphs of system states [6–8]. Classification of the models of fuzzy situational-event networks [2] is shown in Figure 6.2.

Models in the form of fuzzy situational-event networks are used in the analysis of systems of different types and purposes. So in [9], the behavior of users is investigated when searching for information on the Internet. In [10], the processes of changing the state of the "smart house" are investigated.

A separate class of systems in which it is advisable to use the situational-event approach is hybrid discrete-continuous systems. The foundations of a formal describe of such systems are proposed in [11]. A typical example of the use of the situational-event approach for the investigation of a hybrid discrete-continuous system is the simulation of heating processes of the "smart house" [12].

An approach to modeling systems in conditions of combined influence of random and fuzzy influences on the basis of the operator method is proposed in [13]. In this chapter, continuous systems are mainly studied. Later, the method was extended to systems of other types. In particular, the models of systems with logical conditions are considered in [14], which allows extending the application of the operator method to situational-event systems.

Another feature of complex situational-event systems is the presence of a large number of interconnected parallel processes. Management of such processes is usually carried out on a hierarchical basis, which allows reducing the time of development of operational control influences and improving the efficiency of process prediction [15].

With research in the field of development of control systems based on artificial intelligence methods with the application of the situational-event approach, more and more attention is paid to the energy efficiency of control

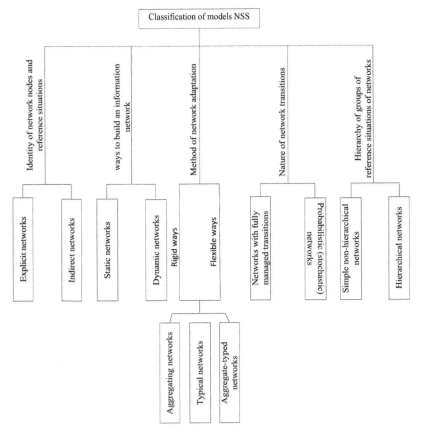

Figure 6.2　Classification of the models of fuzzy situational-event networks.

systems of various types and purposes [9, 16, 17]. The limit of energy efficiency of information processes was investigated by Rolf Landauer [23] and then refined repeatedly. The relict radiation temperature T (about 3 kelvin) sets the lower limit of the energy consumed to perform calculations with one switching of the logic element, approximately at $4k_BT$, where k_Bk is Boltzmann constant. If the device is cooled below this temperature during operation, the energy used for cooling will exceed the effect obtained from the lower operating temperature. Rolf Landauer formulated the principle that in any computer system, regardless of its physical realization, with the loss of 1 bit of information, heat is released in an amount of at least $W = k_BT\ln 2$ joules, where T is the absolute temperature of the computer system in kelvins.

The actual information-energy efficiency of the functioning of systems differs significantly from the limiting values [22, 24]: From a technological perspective, energy dissipation per logic operation in present-day silicon-based digital circuits is about a factor of 1,000 greater than ultimate Landauer limit, but is predicted to quickly attain it within the next couple of decades [25].

The analysis of literature sources leads us to the conclusion that, despite the effectiveness of the situational-event approach, the availability of methods for modeling of systems under conditions of combined uncertainty and the existing research in optimizing the energy efficiency of control, the task of applying energy criteria in the development of intelligent/smart control systems under conditions of combined uncertainty based on situational models is not solved.

The aim of the article is to improve and develop a methodology for researching systems with poorly structured control tasks, combined uncertainty of source information, and a large dimension of the source data vector. Thus, the main objectives can be listed as follows: (a) formalization of the energy efficiency criterion of intellectual control on the basis of situational models; (b) development of a method for simulation of the systems with intelligent control based on situational models in conditions of combined uncertainty of the initial information; (c) development of an algorithm for estimating the energy efficiency of situational models based intelligent control; (d) comparison of energy efficiency of logical, fuzzy and statistical control algorithms.

6.2 Situation-Event Graph Model

A complex control system in most generalized form can be represented in the form shown in Figure 6.3.

Situations in which control is executed forms a set $S = \{s_1, s_2, \ldots s_n\}$. In the considered model of the dynamic system, we will assume that the set of situations is discrete and countable, but in each situation, the state variables of system are represented by the time functions $G = \{g_{i1}(\tau), g_{i2}(\tau), \ldots g_{im}(\tau)\}$, where is τ – the time interval from the moment of the beginning of this situation. In addition to state variables, the situation is characterized by the values of elements of a set of external conditions (impacts) $\Theta = \{\theta_1(t), \theta_2(t), \ldots \theta_k(t)\}$.

The process of changing the situation is represented in the form of a situation-event graph, an example of which for $n = 8$ is shown in Figure 6.4.

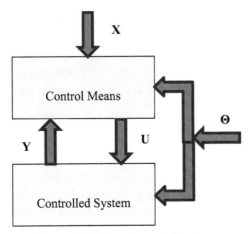

Figure 6.3   Complex control system.

The vertices of the graph correspond to situations, and the arcs correspond to events (changes in situations). The situation-event graph is described by a weight matrix, whose elements are the transition probabilities $P = \{p_{ij}; i, j = 1 \ldots n\}$.

As a result of the impact on the system by the controls $U = \{u_1, u_2, \ldots u_l\}$ both state variables of the current situation $Z_i(s_i, U, \Theta)$ and the probabilities of changes the situation $P_i(s_i, U, \Theta) = \sum_{\substack{j=1 \\ j \neq i}}^{n} p_{ij}$ are changing.

The process of controlling the state of the system is accompanied by the expenditure of energy $E(X, Y, \Theta, W)$, where X is a matrix of parameters of the given system states; Y is a vector of controlled system parameters, depending on the state parameters: $Y(Z)$; W is a law (algorithm) of control: $U = W(X, Y, \Theta)$. External influences Θ are random processes and characterized by probability distributions $F = \{f(\theta_i); i = 1 \ldots k\}$.

In a system with fuzzy logic-based intelligent/smart control, influences are characterized by membership functions $M = \{\mu(u_i); i = 1 \ldots l\}$.

The matrix of parameters of given states of the system is represented by deterministic functions of time.

Thus, in order to estimate energy consumption for control E and optimize control by criterion $\min_U [E]$, it is necessary to develop a model for changing the state of the system in a given situation, which depends on fuzzy, stochastic and deterministic influences, as well as a probabilistic model of the situation change. For the development of the first model, the method of modeling the

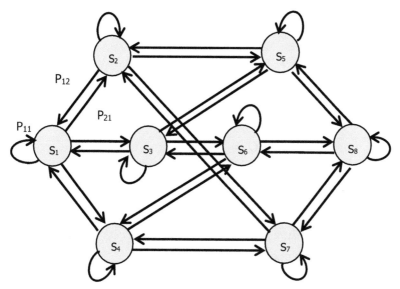

Figure 6.4 Situation-event graph.

system in conditions of combined uncertainty [13] and for second the Markov model [10] used.

6.3 Energy Loss Model

6.3.1 Formalizing of the Criterion of Energy Efficiency of Smart Control on the Basis of Situational Models

The energy of any system characterized by a given number of freedom degrees, in general, is the sum of potential energy and kinetic energy. In particular cases, the system's energy could be only potential or only kinetic. The system's potential energy which performs small oscillations at an equilibrium position is a homogeneous quadratic function of the generalized coordinates is

$$\Pi = \frac{1}{2}(c_{11}g_1^2 + c_{22}g_2^2 + \ldots)$$

$$+c_{12}g_1g_2 + c_{13}g_1g_3 + \ldots + c_{23}g_2g_3 + \ldots, \tag{6.1}$$

where $\{g_i, i = 1\ldots v\}$ are generalized coordinates (state parameters).

Let us bound the consideration with linear systems, for which model (6.1) extends over the entire coordinate range of.

The kinetic energy T can be represented as a quadratic function of generalized velocities is

$$T = \frac{1}{2}\left(m_{11}\dot{g}_1^2 + m_{22}\dot{g}_2^2 + \ldots\right)$$

$$+m_{12}\dot{g}_1\dot{g}_2 + m_{13}\dot{g}_1\dot{g}_3 + \ldots + m_{23}\dot{g}_2\dot{g}_3 + \ldots \tag{6.2}$$

The total system's energy is

$$Q = T + \Pi. \tag{6.3}$$

In dissipative systems, the loss of total energy (the scattering function)

$$E_R = \frac{1}{2}(R_{11}\dot{g}_1^2 + R_{22}\dot{g}_2^2 + \ldots)$$

$$+R_{12}\dot{g}_1\dot{g}_2 + R_{13}\dot{g}_1\dot{g}_3 + \ldots + R_{23}\dot{g}_2\dot{g}_3 + \ldots, \tag{6.4}$$

where R - is the Rayleigh coefficient.

Let us use the Lagrange equation of the second kind to find the generalized forces acting in the system (controls and external influences). We find the generalized force as a coefficient for a generalized displacement in the expression for the work when the total energy of the system changes. In a linear system, all impacts are independent and the generalized force is the sum of the control and external influences. Using the Lagrange equation, we can write the expression for the generalized force as

$$(u_i + \theta_i) = -\frac{\partial E_R}{\partial \dot{g}_i} = \frac{d}{dt}\left(\frac{\partial T}{\partial \dot{g}_i}\right) - \frac{\partial T}{\partial g_i} + \frac{\partial \Pi}{\partial g_i}. \tag{6.5}$$

The dependence of the state parameters g_i (generalized coordinates) on controls u_i and external influences θ_i (model of the object) is represented by a system of equations of dynamics

$$\begin{cases} \sum_{j=0}^{\delta g1}\left[a_{1j}\left(\frac{d}{dt}\right)^j g_1(t)\right] \\ \quad = \sum_{i=1}^{l}\sum_{j=0}^{\delta u1}\left[b_j\left(\frac{d}{dt}\right)^j u_i\right] + \sum_{i=1}^{k}\sum_{j=0}^{\delta \theta 1}\left[d_j\left(\frac{d}{dt}\right)^j \theta_i\right] \\ \ldots \\ \sum_{j=0}^{\delta gv}\left[a_{vj}\left(\frac{d}{dt}\right)^j g_v(t)\right] \\ \quad = \sum_{i=1}^{l}\sum_{j=0}^{\delta uv}\left[b_j\left(\frac{d}{dt}\right)^j u_i\right] + \sum_{i=1}^{k}\sum_{j=0}^{\delta \theta v}\left[d_j\left(\frac{d}{dt}\right)^j \theta_i\right] \end{cases} \tag{6.6}$$

The dependence of the observed signals y_i on the state parameters g_i and external influences θ_i (measurement control model) will be described by a similar system of equations of dynamics

$$
\begin{cases}
\sum_{j=0}^{\delta_{y1}} \left[e_{1j} \left(\frac{d}{dt} \right)^j y_1(t) \right] \\
\quad = \sum_{i=1}^{l} \sum_{j=0}^{\delta_{g1}} \left[q_j \left(\frac{d}{dt} \right)^j g_i \right] + \sum_{i=1}^{k} \sum_{j=0}^{\delta_{\theta1}} \left[r_j \left(\frac{d}{dt} \right)^j \theta_i \right] \\
\cdots \\
\sum_{j=0}^{\delta_{yv}} \left[e_{vj} \left(\frac{d}{dt} \right)^j y_v(t) \right] \\
\quad = \sum_{i=1}^{l} \sum_{j=0}^{\delta_{gv}} \left[q_j \left(\frac{d}{dt} \right)^j g_i \right] + \sum_{i=1}^{k} \sum_{j=0}^{\delta_{\theta v}} \left[r_j \left(\frac{d}{dt} \right)^j \theta_i \right]
\end{cases}
\tag{6.7}
$$

Dependence of control influences u_i on the observed signals y_i, external effects θ_i of the final conditions $G_n[S(t)]$ for a given situation, taking into account its time variation of $s_i = S(t)$ is

$$
\begin{cases}
U = H \left[D^{(\delta_Y)} (Y) - X \right] \\
X = G_n [S(t)]
\end{cases}
\tag{6.8}
$$

Thus, the linear system is described by a system of equations of the formal model for estimating the energy consumption:

$$
\begin{cases}
\Pi = \dfrac{1}{2} \left(c_{11} g_1^2 + c_{22} g_2^2 + \ldots \right) + c_{12} g_1 g_2 + c_{13} g_1 g_3 \\
\quad + \ldots + c_{23} g_2 g_3 + \ldots \\[4pt]
T = \dfrac{1}{2} \left(m_{11} \dot{g}_1^2 + m_{22} \dot{g}_2^2 + \ldots \right) + m_{12} \dot{g}_1 \dot{g}_2 \\
\quad + m_{13} \dot{g}_1 \dot{g}_3 + \ldots + m_{23} \dot{g}_2 \dot{g}_3 + \ldots \\[4pt]
Q = T + \Pi \\[4pt]
E_R = \Delta Q = \dfrac{1}{2} \left(R_{11} \dot{g}_1^2 + R_{22} \dot{g}_2^2 + \ldots \right) \\
\quad + R_{12} \dot{g}_1 \dot{g}_2 + R_{13} \dot{g}_1 \dot{g}_3 + \ldots + R_{23} \dot{g}_2 \dot{g}_3 + \ldots \\[4pt]
(u_i + \theta_i) = -\dfrac{\partial E_R}{\partial \dot{g}_i} = \dfrac{d}{dt} \left(\dfrac{\partial T}{\partial \dot{g}_i} \right) - \dfrac{\partial T}{\partial g_i} + \dfrac{\partial \Pi}{\partial g_i} \\
\vdots
\end{cases}
$$

$$
\left\{
\begin{array}{l}
\vdots \\[2pt]
\sum_{j=0}^{\delta_{g1}} \left[a_{1j} \left(\frac{d}{dt} \right)^j g_1(t) \right] \\[6pt]
\qquad = \sum_{i=1}^{l} \sum_{j=0}^{\delta_{u1}} \left[b_j \left(\frac{d}{dt} \right)^j u_i \right] + \sum_{i=1}^{k} \sum_{j=0}^{\delta_{\theta 1}} \left[d_j \left(\frac{d}{dt} \right)^j \theta_i \right] \\[6pt]
\cdots \\[4pt]
\sum_{j=0}^{\delta_{gv}} \left[a_{vj} \left(\frac{d}{dt} \right)^j g_v(t) \right] \\[6pt]
\qquad = \sum_{i=1}^{l} \sum_{j=0}^{\delta_{uv}} \left[b_j \left(\frac{d}{dt} \right)^j u_i \right] + \sum_{i=1}^{k} \sum_{j=0}^{\delta_{\theta v}} \left[d_j \left(\frac{d}{dt} \right)^j \theta_i \right] \\[6pt]
\sum_{j=0}^{\delta_{y1}} \left[e_{1j} \left(\frac{d}{dt} \right)^j y_1(t) \right] \\[6pt]
\qquad = \sum_{i=1}^{l} \sum_{j=0}^{\delta_{g1}} \left[q_j \left(\frac{d}{dt} \right)^j g_i \right] + \sum_{i=1}^{k} \sum_{j=0}^{\delta_{\theta 1}} \left[r_j \left(\frac{d}{dt} \right)^j \theta_i \right] \\[6pt]
\cdots \\[4pt]
\sum_{j=0}^{\delta_{yv}} \left[e_{vj} \left(\frac{d}{dt} \right)^j y_v(t) \right] \\[6pt]
\qquad = \sum_{i=1}^{l} \sum_{j=0}^{\delta_{gv}} \left[q_j \left(\frac{d}{dt} \right)^j g_i \right] + \sum_{i=1}^{k} \sum_{j=0}^{\delta_{\theta v}} \left[r_j \left(\frac{d}{dt} \right)^j \theta_i \right]
\end{array}
\right. \tag{6.9}
$$

$$
\left\{
\begin{array}{l}
U = H \left[D^{(\delta_Y)} (Y) - X \right] \\[4pt]
X = G_n \left[S(t) \right]
\end{array}
\right.
$$

6.3.2 Development of the Method for Simulation of Systems with Smart Control Based on Situational Models in Conditions of Combined Uncertainty of the Initial Information

Let us consider the effect of transition probabilities and the reliability of the identification of situations on the system's energy efficiency.

Incorrect identification of the situation leads to an additional energy loss for transferring the system to an adequate state. The process in this case is

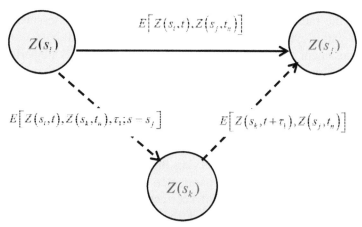

Figure 6.5 The case of incorrect identification of situation.

shown in the form of a graph in Figure 6.5. In the figure, the initial situation is s_i, a new situation is s_j, but as a result of an incorrect identification of the situation, the control system began to change the state parameters according to the situation s_k. During time τ_1 the situation assessment is improved, and the control process is transferred to the parameter's change mode in accordance with the situation s_j.

Since the energy consumption criterion is additive, then by the triangle theorem, we have

$$E_{ik}\left[Z\left(s_i, t\right), Z\left(s_k, t_n\right), \tau_1; s = s_j\right]$$
$$+ E_{kj}\left[Z\left(s_k, t + \tau_1\right), Z\left(s_j, t_n\right)\right] > E_{ij}\left[Z\left(s_i, t\right), Z\left(s_j, t_n\right)\right],$$

(6.10)

and losses

$$\Delta E = E_{ik} + E_{kj} - E_{ij}.$$

Let us estimate the time interval τ_1.

Incorrect identification of a situation usually connected with uncertainty of estimations of situation parameters. During the transition process $(0, \tau_1)$ the assessment interval is refined. RMSE decreases to

$$\sigma_{at} = \frac{\sigma_s(t = 0)}{\tau_1/T},$$

where σ_{at} is the radius of the set of points in the space of assessments of the situation within which the situation is correctly identified; $\sigma_s(t = 0)$ is RMSE of assessment of the situation at the initial moment; T is the period of the situation evaluation (with discrete estimation) or the correlation interval (with continuous estimation). From here, we obtain

$$\tau_1 = T \frac{\sigma_s(t = 0)}{\sigma_{at}} \tag{6.11}$$

We will compose a matrix for classifying the situation $\Phi = \{\varphi_{ij}; i, j = 1...n\}$ on the basis of measuring the parameters of the situation. Each row of the matrix consists of the probability of a correct solution $[\varphi_{ij} : i = j]$ and the probabilities of errors of the second kind (the adoption of a false hypothesis) $[\varphi_{ij} : i \neq j]$. The meaning of the probabilities of errors is shown in Figure 6.6 by the example of three situations $\{s_i, s_j, s_k\}$ in the space of two parameters $\{z_1, z_2\}$. The set of coordinates of the points of two regions of parameters belonging to different situations intersection satisfies the condition

$$D\{z_1, z_2\} : \left[(z_1 - z_{1i})^2 + (z_2 - z_{2i})^2 \leq r_i^2\right] \&$$
$$\left[(z_1 - z_{1j})^2 + (z_2 - z_{2j})^2 \leq r_j^2\right], \tag{6.12}$$

where (z_{1i}, z_{2i}) and (z_{1j}, z_{2j}) are coordinates of the centers of the regions of the parameters of situations s_i and s_j respectively; r_i and r_j are confidence intervals for probability distributions of the situation parameters.

Assuming the probability distribution $\varphi_i(Z)$ of the parameters' vector of the situation s_i to be normal, we find the probabilities of incorrect identification of the situation

$$\Phi_{ij} = \int_{Z \in D_{ij}} \varphi_i(Z) \, dZ, \quad j = 1...n, \quad j \neq i \tag{6.13}$$

Confidence interval with confidence level $1 - \Phi$ is

$$\frac{ns^2}{\chi_{1-\Phi/2,n-1}^2} < \sigma_{at}^2 < \frac{ns^2}{\chi_{\Phi/2,n-1}^2}, \tag{6.14}$$

where $\chi_{1-\varepsilon/2,n-1}^2, \chi_{\varepsilon/2,n-1}^2$ are quantiles χ^2 of Student's distributions with $n - 1$ freedom degrees of levels $(1 - \frac{\varepsilon}{2})$ and $(\frac{\varepsilon}{2})$ respectively; $n = \frac{T}{\tau_1}$.

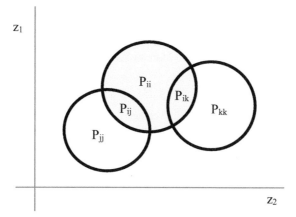

Figure 6.6 Probabilities of errors and correct solution.

Table 6.1 (not complete) of the logical identification of the situation

1	2	3	T	S
0	0	0	$7^{00} - 8^{00}$	1
0	0	1	$8^{00} - 9^{00}$	X
...	...	...	...	...
1	0	1	$14^{00} - 15^{00}$	7
...	...	...	...	...
0	1	1	$19^{00} - 20^{00}$	5
...	...	...	...	...
1	0	0	$1^{00} - 2^{00}$	X
...	...	...	...	...

Note: X is an unplanned situation.

The identification of the situation in the transient modes and the generation of control influences in the stationary mode in the situational control system are carried out with the help of the controller. There are three main variants of controllers: logical automaton (LA), statistical decision-making subsystem (SDM), and fuzzy (smart) control (FC).

The logic controller performs truth table-based identification of situation. For the graph in Figure 6.4 and four parameters [10], the truth table has the form as the Table 6.1.

Further, in accordance with the situation, the system settings are calculated on the basis of the model (6.9) by the criterion of minimum loss ΔE under constraints $G \in G_S$.

The use of an intelligent decision-making system makes it possible to increase the authenticity of identification by taking into account the possible

Table 6.2   (incomplete) of fuzzy identification of the situation

1	2	3	T	W	S
0	0	0	Morning	w_1	1
0	0	1	Morning	w_2	X
...	...	...	...	...	...
1	0	1	Day	...	7
...	...	...	...	...	...
0	1	1	Evening	...	5
...	...	...	...	...	...
1	0	0	Night	...	X
...	...	...	...	...	...

vagueness of the results of assessing the parameters of the situation. So, in the above example, with four situation parameters, of which three binary (0; 1) and one linguistic are the time of day, both in determining the time of day and sensors' switching, errors are possible, the base of fuzzy rules shown in Table 6.2 is similar to Table 6.1; however, the set of values of some parameters is shortened to reduce the number of rules.

In the process of fuzzification and fuzzy inference, the uncertainty of the estimates of the situation parameters is taken into account, and the decrease in the number of terms is compensated by fuzzy approximation using membership functions. During the training of the FC, it is necessary to configure the parameters of membership functions and weights of the rules to increase the reliability of the identification of the situation and minimize losses.

After identification, the system parameters are controlled in accordance with the base of fuzzy rules for the given situation. The advantage of a fuzzy control algorithm is its learning ability. For training, data on the actual energy consumption are used from the data of measuring instruments and subjective assessment of the quality of control. The quality of control is determined by the compliance of the system parameters with the limitations $G \in G_S$ and duration of the transient processes to achieve these constraints.

The disadvantage of a fuzzy algorithm is the difficulty of predicting a change in the situation. The insertion of the probability of a change in the base of fuzzy rules multiplies the size of the base.

Statistical SDM during the formation of control influences in a stationary mode takes into account the probability of a situation change.

Suppose that the process of changing the situation is Markov with the transition matrix $P(t, \Delta t) = \{p_{ij}(t, \ \Delta t), i = 1..n, \ j = 1..n\}$. If at the initial moment the probabilities of situations is $P_S(t_0)$, then in a period Δt, the probabilities of situations will be

$$\boldsymbol{P}'_S(t + \Delta t) = P_S^T(t) \cdot P(t, \Delta t), \tag{6.15}$$

where T is the sign of transposition.

The transition matrix is formed gradually as a result of the accumulation of statistics of situation changes. This makes it possible to increase the reliability of identification using forecasting changes to reduce losses. If the probabilities of identifying a situation on the basis of parameters' monitoring are $\boldsymbol{P}_0(t) = \{p_{0i}(t), \ i = 1..n\}$, then the refined probability taking into account the prediction based on the model (6.15) are

$$\boldsymbol{P}'(t) = \alpha \cdot \boldsymbol{P}_0(t) + \beta \cdot \left[P_S^T(t - \Delta t) \cdot P(t, \Delta t) \right], \tag{6.16}$$

where α and β are confidence coefficients, $\alpha + \beta = 1$.

The average loss is used as a criterion of statistical optimization

$$\overline{\Delta E} = \sum_S \boldsymbol{P}'(s, t) \cdot \Delta E_S \tag{6.17}$$

with restrictions on the parameters of the system $G \in G_S$.

In transient modes, changes in state lead to additional energy losses. This is due to the asymmetry of the process of increasing and decreasing the energy of the system – the increase in energy is provided by consumption from an external source, and the decrease occurs due to a dissipative process. Taking into account that the number of states of the system is finite, and the process of transition between states is infinite, the number of transitions associated with the increase and decrease of energy over a sufficiently large time interval can be considered equal.

The probability of η transitions associated with an increase in energy over a time interval T can be found from the matrix of transition probabilities:

$$P_\eta(\tau) = \frac{\displaystyle\sum_{\substack{i,j=1 \\ i \neq j}}^{n} p_{ij}(\tau)}{2 \cdot \tau \cdot (n^2 - n)} \tag{6.18}$$

If the range of changes in the system energy when the situation changes D_E, then on average, each transition is associated with a change in energy by $\delta_E = \pm D_E / \frac{m}{2}$. If the situation is incorrectly identified, the number of transitions increases. Thus, the average additional loss for transients with incorrect identification is

$$\Delta E_{id} = \frac{2D_E}{m} \cdot \frac{\displaystyle\sum_{\substack{i,j=1 \\ i \neq j}}^{n} \Phi_{ij}(\tau)}{(n^2 - n)} \tag{6.19}$$

6.3.3 Development of an Algorithm for Estimating the Energy Efficiency of Situational Model-Based Smart Control

The algorithm in Figure 6.7 provides for several scenarios.

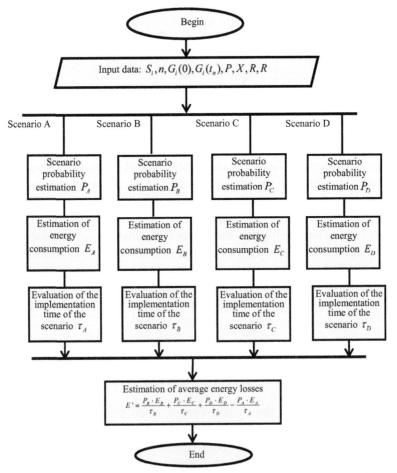

Figure 6.7 Algorithm for estimating the energy efficiency of intelligent control based on situational models.

Scenario A: The situation does not change. Probability of scenario implementation is P_A;

Scenario B: The situation changes and is identified correctly. The transition to a new situation is completely implemented. After that, the system functions for a while in a new situation. Probability of scenario implementation is P_B;

Scenario C: The situation changes and is identified correctly. The transition to a new situation is not completely implemented. After that, the situation changes again. Probability of scenario implementation is P_C;

Scenario D: The situation changes and is not identified correctly. The transition takes place to parameters of a wrong situation. This process continues until identification becomes correct or the situation changes. Probability of the implementation of the scenario is P_D.

6.3.4 Experimental Investigations of Developed Model

As an object for experimental verification and study of the proposed model, consider the management of the heating system of the multi-room house. In connection with the diversity of the systems under consideration, we use simulation in Scilab/Xcos environment as a research method.

Structure of the system for the case of two rooms is shown in Figure 6.8. In the scheme: N – heater; (r_1, r_2) – rooms; (m1, m2) – signals of motion sensors; (t_{10}, t_{20}) – the set of preset temperatures.

The estimation of energy losses for such system is easier, since in the model (6.9) we can neglect the scattering of kinetic energy.

In the simplest case, the logic controller controls the switch on/off the heat source. To do this, the situation is identified in accordance with the rules in Table 6.1. Optimal and actual values of the on/off switching criterions for the given situation are calculated as

$$T_R = \sum_{i=1}^{N} m_i t_{0i} + t_0' \sum_{i=1}^{N} (1 - m_i) \quad \text{and} \quad T_F = \sum_{i=1}^{N} m_i t_i + \sum_{i=1}^{N} (1 - m_i) t_i',$$

and a decision is made to turn on/off the heater

$$D = sign(T_R - T_F).$$

With a fuzzy control algorithm, the fuzzy controller FC performs the function of calculating the optimum average temperature, the deviation and determining of the moment of switch on/off the heater. To do this, the situation is

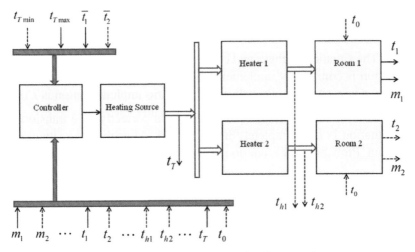

Figure 6.8   Structure of a multi-room heating control system.

identified in accordance with the rules of Table 6.2. Optimal and actual values of the on/off switching criteria for this situation are calculated and decision is made to switch on/off the heater with a fuzzy derivation of Mamdani-Zadeh. The model provides training for a fuzzy controller.

With a statistically optimal control, the controller performs an estimate of the average energy loss when switched on and when the heat source is switched off and it makes a decision that minimizes losses. In the case of statistically optimal control, in addition to the identification and decision-making processes, a set of statistics and a corresponding correction of the matrix of transition probabilities is performed.

The process of training a fuzzy controller or a set of statistics for adopting a statistically optimal solution allows you to increase efficiency (i.e., reduce energy losses). A change in the time of normalized energy losses for three values of the dimension of the situational graph of the system is shown in Figure 6.9.

An increase in the dimension of the system (the dimensions of the situational graph and the state of the state parameters vector) reduces the average loss for all three control types; however, if for the statistical control normalized average losses tend to zero, for logical (determined logic and fuzzy logic) control the limited value is bounded by the limitations on the number of logical conditions, i.e., the system's complexity. Comparison of

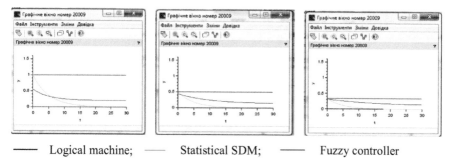

——— Logical machine; ——— Statistical SDM; ——— Fuzzy controller

Figure 6.9 Changing the time of the normalized value of the average energy loss in the control of the logical automaton, the fuzzy controller and the statistical SDM with the dimensions of the situational graph 4, 8, and 16.

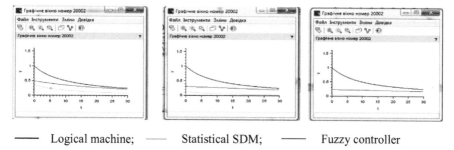

——— Logical machine; ——— Statistical SDM; ——— Fuzzy controller

Figure 6.10 Changing the normalized value of average energy losses with increasing system dimension and control of a logical automaton, fuzzy controller, and statistical SDM with horizons 1, 30, and 120 events.

energy losses when changing the system dimension for different training intervals is shown in Figure 6.10.

6.4 Analysis of Simulation Results

Comparison of energy consumptions when changing the dimension of the vector of situation parameters shows that the increase in dimension leads to a decrease in energy losses. This is due to the decrease in the amplitude of the oscillations of the system-state parameters. The influence of the amplitude of oscillations is due to the asymmetry of the active heating process and the dissipative cooling process.

Transient processes are asymmetrical as vibrational modes. An increase in the dimension of a situational graph also leads to a decrease in energy losses. This is due to a decrease in the variance of optimizing criterion for system parameters in the space of situations.

Comparison of the energy efficiency of systems with a fuzzy controller and a deterministic logical controller shows that in a system with a fuzzy controller, a more flexible control of the system with the use of forecasting and training is possible. This allows to reduce energy consumptions in transitional and oscillation processes to a level that is determined by the constraints on the complexity of the system (the size of the base rules). At the same time, the system for making statistical decisions provides minimum average losses, but to achieve such result, a large and reliable statistics are required. At an initial stage, until a sufficient set of statistics is generated, fuzzy control provides less energy loss.

6.5 Conclusion

The studies of the model of energy efficiency using simulation allow us to make conclusions on the advantages of fuzzy and statistical situational control when considering the near and far event horizons.

In further researches, it is expedient to investigate systems with the possibility of controlling the situation and algorithms of an energy-efficient control.

References

[1] N. Brisbourne. Growth in the internet of things (2013) –Available at http://www.theequitykicker.com/2013/07/31/growth-in-the-internet-of-things/.

[2] V. V. Borisov, M. M. Zernov. Realizacija situacionnogo podhoda na osnove nechetkoj ierarhicheskoj situacionno-sobytijnoj seti (2009) Artificial Intelligence and Decision-Making, vol. 1, pp. 17–30.

[3] Nominal, ordinal, interval, and ratio typologies are misleading. The American Statistician (American Statistical Association), no. 47, pp. 65–72, 1993. Doi: 10.2307/2684788.

[4] V. Seyranian. Contingency Theories of Leadership. (2009) Encyclopedia of Group Processes & Intergroup Relations. Edited by John M. Levine and Michael A. Hogg. Thousand Oaks, California: SAGE, pp. 152–156.

[5] N. Ashish, D. Kalashnikov, S. Mehrotra, N. Venkatasubramanian. An Event Based Approach To Situational Representation (2009) arXiv:0906.4096 [cs.DB]

[6] J. N. Mordeson, P. S. Nair. Fuzzy Graphs and Fuzzy Hypergraphs. Studies in Fuzziness and Soft Computing, vol. 46, Springer-Verlag, 2000.

[7] W. B. Vasantha Kandasamy, F. Smarandache. Fuzzy relational maps and neutrosophic relational maps, 2014. Doi: 10.6084/M9.FIGSHARE.101 5555. Available at https://arxiv.org/ftp/math/papers/0406/0406622.pdf.

[8] V. V. Borisov, M. M. Zernov. Vyvod na osnove nechetkoj situacionnoj seti(2008) Proceedings of the 11th National Conference on Artificial Intelligence with International Participation, vol. 1. Dubna, pp. 320–327.

[9] V. Dubovoi, O. Moskvin. Impact of the Internet Resources Structure on Energy Consumption While Searching for Information. "Green IT Engineering: Concepts, Models, Complex Systems Architectures" [Vyacheslav Kharchenko, Yuriy Kondratenko, Janusz Kacprzyk – Editors], pp.125–146, 2016. Doi: 10.1007/978-3-319-44162-7.

[10] R. H. Rovira, V. M. Dubovoi, M. S. Yukhimchuk, M. M. Bayas, W. D. Torres. A Model of Self-oscillations in Relay Outputs Control Systems with Elements of Artificial Intelligence. Proceedings of the International Conference on Information Technology & Systems. Advances in Intelligent Systems and Computing, [Rocha Á, Guarda T. (eds)], vol. 721, Springer, Cham, 2018. Doi: https://doi.org/10.1007/978-3-319-73450-7_33.

[11] Shpakov V. M. A Situation-Event Approach to Hybrid Processes Specifications. (2007) SPIIRAS Proceedings. Issue 4.

[12] V. M. Dubovoi, O. D. Nikitenko, M. S. Yukhymchuk. Modeling of the automated control system of heating in the "smart house". (2017) Automatics – 2017 XXIV International Conference on Automated Control, Kiev, Ukraine proceedings. Riev.. p. 68.

[13] O. Glon, V. Dubovoi. Generalization of Analytical Dependencies on a Case of Simultaneous Use of the Statistical and Fuzzy Data. (2001) Proceedings of International Conference on Modeling and Simulation. Lviv, pp. 176, 177.

[14] V. Dubovoi, M. Yukhimchuk. Evaluation of uncertainty of control by measurement with logical conditions. Proceedings of the Photonics Applications in Astronomy, Communications, Industry, and

High-Energy Physics Experiments 2016 conference.) SPIE Digital Library, 2016. Doi: http://dx.doi.org/10.1117/12.2248871.

[15] M. M. Bayas, V. M. Dubovoi, J. Shegebaeva K. Gromaszek. Optimization of hierarchical management of technological processes. Proc. SPIE 9816, Optical Fibers and Their Applications, 981622, 2015. Doi: 10.1117/12.2229201; http://dx.doi.org/10.1117/12.2229201.

[16] X. Xiaohua, Z. Jiangfeng. Energy Efficiency and Control Systems – from a POET Perspective. (2010) IFAC Proceedings Volumes, vol. 43(1), pp. 255–260. https://doi.org/10.3182/20100329-3-PT-3006. 00047.

[17] O. A. Voronina, V. A. Lobanova. Sovershenstvovanie sistemy upravlenija kak osnova reshenija problemy jekonomii jenergii na malyh neftepererabatyvajushhih zavodah. (2007) Proceedings of the V International Scientific and Practical Internet Conference "Energy and Resource Saving - XXI Century". Orel, p. 57.

[18] L. I. Minchala-Avila, K. Palacio-Baus, J. P. Ortiz, J. D. Valladolid, J. Ortega. Comparison of the performance and energy consumption index of model-based controllers. (2016) Ecuador Technical Chapters Meeting (ETCM), IEEE. pp. 1–6. Doi: 10.1109/ETCM.2016. 7750825.

[19] X. Bai, W. T. Tsai, R. Paul, K. Feng, L. Yu. Scenario-based modeling and its applications (2002). Object-Oriented Real-Time Dependable Systems, 2002. Proceedings of the Seventh International Workshop. ISSN: 1530-1443, San Diego, CA, USA. Doi: 10.1109/WORDS.2002.1000060.

[20] M. S. Kalavathi, C. S. R. Reddy. Performance evaluation of classical and fuzzy logic control techniques for brushless dc motor drive. Power Modulator and High Voltage Conference IEEE International, pp. 488–491, 2012.

[21] Z. Jiasheng, Y. Dingwen, Q. Shiqing. Structural research of fuzzy PID controllers. Control and Automation, ICCA '05. International Conference 26–29, June 2005, ISBN: 0-7803-9137-3 Budapest, Hungary. Doi: 10.1109/ICCA.2005.1528312.

[22] V. Dubovoi. Information Characteristics of Optical Sensors (2001) Optoelectronic Information Technologies. Proceedings of SPIE, vol. 4425, p. 478–484.

[23] R. Landauer. Irreversibility and heat generation in the computing process. IBM Journal of Research and Development, vol. 5, pp. 183–191, 1961.

[24] A. Bérut et al. Experimental verification of Landauer's principle linking information and thermodynamics. Nature 483.7388, pp. 187–189, 2012.

[25] C. F Cerofolini, D. Mascolo. Hybrid Route From CMOS to Nano and Molecular Electronics. Nanotechnology for electronic materials and devices, ISBN 978-0387-23349-9, pp. 16–18, 2006.

7

Ellipsoidal Pose Estimation of an Uncooperative Spacecraft from Video Image Data

Vyacheslav F. Gubarev[*], Nikolay N. Salnikov and Serhii V. Melnychuk

Space Research Institute, National Academy of Sciences of Ukraine and State Space Agency of Ukraine, Glushkov, Ukraine
*Corresponding Author: v.f.gubarev@gmail.com

This chapter presents a pose estimation method for an autonomous approach and docking with an uncooperative known spacecraft. The method includes evaluating the parameters of the relative position and attitude using a computer vision and applying dynamic filtering. Pose determination is based on informative features of images with the use of machine learning, which reduces computational complexity of algorithm implementation. This method allows obtaining a solution with a known maximum error. For filtering a guaranteed estimation algorithm, using ellipsoids is applied. Due to inclusion of the dynamic equations of motion, high-precision estimates of the parameters of mutual motion are achieved.

7.1 Introduction

An automatic rendezvous and docking in space is an important problem of modern space technology. The active spacecraft (the Chaser) is maneuvering to approach the passive spacecraft (the Target) for docking. Safe rapprochement requires a current tracking of the relative pose (position and attitude) of Target. The method of solving the problem depends on the Target type (is it cooperative or uncooperative, of known or unknown form) and the sensors used (laser range finders, LIDAR scanning systems, and visual systems).

Here we consider the problem of vision-based pose estimation for unco-operative spacecraft with known geometric configuration [1–5]. The usual approach [6, 7] is based on the capability of extracting and comparing some features of the Target image and its known three-dimensional model such as point descriptors [8], lines [9, 10], corners [11], contours [12], and others. The pose determination in this case consists of two tasks:

- finding the correspondence between three-dimensional elements and two-dimensional images of these elements (correspondence problem);
- calculation of the position and attitude parameters (matching problem).

The main drawback of visual navigation is the computational complexity of algorithms. It is connected with a large amount of video information, high requirements for accuracy of pose evaluation, and limitations of on-board computational resources. In such conditions, the use of machine learning methods seem to be promising, because the system training for a particular target can be performed in advance. In this way, the runtime computations can be significantly reduced.

The method of pose estimation proposed here refers to a geometric feature learning technique supplemented by the use of a decision tree. The main runtime computations are limited only to image processing and extraction of certain representative features of the geometric form. These features are compared with the values obtained in the training mode. Due to discreteness of data and limited computing resources, the accuracy of direct measurements cannot be arbitrarily increased. Consequently, the resulting values of the pose parameters inevitably contain errors. The proposed method allows obtaining a solution with an unknown but bounded error.

The accuracy of pose estimation can be improved by a filtering algorithm. Using the dynamic model of the Target, it is possible to determine the speed of approach and the angular velocity of its rotation, which is necessary for maneuvering control.

The dynamics of the orbital motion and rotation around the center of mass can be considered separately. When docking, the accuracy of the rotation parameters estimating is more crucial. Hence, in this chapter, we consider dynamic filter application to estimate these parameters.

The most common approach to filtering consists in the assumption of the stochastic nature of errors. It allows to use the Kalman filter and its numerous modifications [13, 14]. However, a reasonable application of such filters requires a sufficiently accurate knowledge of the distribution laws for random variables or their basic stochastic moments: mathematical expectation and variance. Obtaining such information involves a large number of experiments, which require considerable time and resources. In addition, the properties of random variables are often not stationary as practice shows [15]. Then there are difficulties in determining time variation of error stochastic characteristics.

That is why, one of the modifications [16, 17] of guaranteed estimation algorithms using ellipsoids [18–24] is used for filtering in this chapter. The main advantages of the modification used are a high rate of convergence, ease of implementation, applicability to nonlinear systems, and robustness with respect to possible violations of a priori hypotheses about the properties of uncertain quantities. The only assumption is that unknown quantities belong to given bounded sets.

7.2 Relative Pose Determination

Two space vehicles located at a distance of up to 100 m are considered. The Chaser is equipped with a computer vision system (CVS) with an optical digital camera. The camera is fixed to the body of the Chaser in a known position. The Target is an uncooperative spacecraft (i.e., is not equipped with known markers) with a non-deformable rigid body of known geometry (given in the form of a three-dimensional model).

The task of the CVS is to evaluate the position and attitude of the Target relative to the Chaser using the captured image and 3D model.

To determine the relative pose, we introduce two coordinate systems. The reference frame $O_1 x_1 y_1 z_1$ is associated with the camera. It has an origin in a projection center, vector x_1 coincides with the optical axis of the camera, vectors y_1 and z_1 are parallel to the image plane and correspond to the "up" and "right" direction on the image. The reference frame $O_2 x_2 y_2 z_2$ is associated with the Target, its origin is placed in the Target center of mass and axis coincide with its main inertia axis (Figure 7.1).

Consider the position and attitude of the Target relative to the Chaser. Position of $O_2 x_2 y_2 z_2$ is defined by vector r_{21} $O_2 - O_1$. In the reference frame $O_1 x_1 y_1 z_1$, we denote its value as r_{21} $(x, y, z)^{\mathrm{T}}$. Attitude of $O_2 x_2 y_2 z_2$ is defined by coordinates of unit vectors x_2, y_2, z_2 in the reference frame $O_1 x_1 y_1 z_1$. They form columns of the orthogonal rotation matrix $T_{21} \in R^{3 \times 3}$, which can be uniquely determined by means of different parameters sets.

We will use the minimal parameter set: the Euler angles, that define transformation T_{21} as a composition of elemental rotations. Rotations around

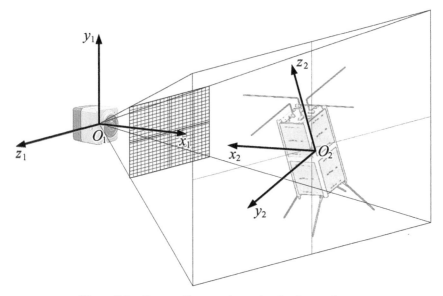

Figure 7.1   Camera, Target and associated reference frames.

three coordinate axes are used: by the roll angle γ around the x axis, by the yaw angle ψ around the y axis, and by the pitch angle ϑ around the z axis. The matrices of elementary rotations by an angle θ are, respectively,

$$
T_X \quad \begin{pmatrix} 1 & 0 & 0 \\ 0 & \cos\theta & \sin\theta \\ 0 & -\sin\theta & \cos\theta \end{pmatrix}, \quad
T_Y \quad \begin{pmatrix} \cos\theta & 0 & -\sin\theta \\ 0 & 1 & 0 \\ \sin\theta & 0 & \cos\theta \end{pmatrix},
$$

$$
T_Z \quad \begin{pmatrix} \cos\theta & \sin\theta & 0 \\ -\sin\theta & \cos\theta & 0 \\ 0 & 0 & 1 \end{pmatrix}.
$$

For the sequence "pitch-yaw-roll", the rotation matrix T_{21} $T_Z^{\mathrm{T}}(\vartheta)T_Y^{\mathrm{T}}(\psi)$ $T_X^{\mathrm{T}}(\gamma)$ is expressed as

$$
T_{21} \quad \begin{pmatrix}
\cos\vartheta\cos\psi & \begin{array}{c}\cos\vartheta\sin\psi\sin\gamma \\ -\sin\vartheta\cos\gamma\end{array} & \begin{array}{c}\sin\vartheta\sin\gamma \\ +\cos\vartheta\sin\psi\cos\gamma\end{array} \\[2ex]
\sin\vartheta\cos\psi & \begin{array}{c}\cos\vartheta\cos\gamma \\ +\sin\vartheta\sin\psi\sin\gamma\end{array} & \begin{array}{c}\sin\vartheta\sin\psi\cos\gamma \\ -\cos\vartheta\sin\gamma\end{array} \\[2ex]
-\sin\psi & \cos\psi\sin\gamma & \cos\psi\cos\gamma
\end{pmatrix}.
$$

$$(7.1)$$

To determine the dynamics of rotation movement, it is more convenient to deal with another set of parameters, namely coordinates of normalized quaternion q $(q_0, q_1, q_2, q_3)^{\mathrm{T}}$. In this case, the rotation matrix is

$$T_{21}(q) \quad \begin{pmatrix} 1 - 2\left(q_2^2 + q_3^2\right) & 2\left(q_1 q_2 - q_0 q_3\right) & 2\left(q_1 q_3 + q_0 q_2\right) \\ 2\left(q_1 q_2 + q_0 q_3\right) & 1 - 2\left(q_1^2 + q_3^2\right) & 2\left(q_2 q_3 - q_0 q_1\right) \\ 2\left(q_1 q_3 - q_0 q_2\right) & 2\left(q_2 q_3 + q_0 q_1\right) & 1 - 2\left(q_1^2 + q_2^2\right) \end{pmatrix}. \quad (7.2)$$

It is convenient to collect the coordinates of r_{21} and Euler angles into a pose vector

$$p \quad (x, y, z, \vartheta, \psi, \gamma)^{\mathrm{T}}, \quad (7.3)$$

which determines the relative position and attitude of the Target with respect to the Chaser. The task of the CVS is to find appropriate p on some bounded set P

$$p \in P \subset R^6, \quad (7.4)$$

defined by the limits of practical application.

7.3 Position and Attitude Determination by Learning

The input data of CVS is a captured image and 3D model of the Target. The captured digital image is a rectangle pixel array of size $W \times H$, where W is the number of columns and H is the number of rows. This image can be represented by a vector a.

Suppose that Target is in the field of view. Then the captured image a is completely determined by the pose vector, p, characteristics of external lighting l and noise of different nature ε; hence, there is a dependence

$$a \quad f(p, l, \varepsilon). \quad (7.5)$$

Let us assume the CVS operates in a normal mode if the following conditions hold

$$l \in L, \quad \varepsilon \in E, \quad (7.6)$$

where L is the set of admissible illuminations and E is the set of permissible noises. Using memberships (7.4) and (7.6), we denote the set A of the possible captured images by

$$a \in A \subset R^{WH}. \quad (7.7)$$

The presence of unknown illumination l and noise ε in (7.5) does not allow to obtain a direct functional dependence of the image only on the pose vector p.

Each p corresponds to a set of images $[a]_p \subset A$ generated by possible realizations of l and ε.

Suppose there exists such a vector-function g of dimension N mapping the set of images into a vector space R^N, that g is sensitive to p and is not sensitive to l and ε

$$g : R^{WH} \to R^N,$$

$$\forall p \in P, \ \forall l_1, l_2 \in L, \ \forall \varepsilon_1, \varepsilon_2 \in \mathrm{E} : \begin{array}{cc} a_1 & f(p, l_1, \varepsilon_1) \\ a_2 & f(p, l_2, \varepsilon_2) \end{array} \Rightarrow g(a_1) \quad g(a_2). \tag{7.8}$$

The coordinate functions g_i define certain numerical characteristics of image. We assume that constructive algorithm for computing g is known. Each admissible image is translated into an informative vector

$$\tilde{a} \quad g(a) \tag{7.9}$$

We need an informative vector to be an identifier of the Target position and attitude; hence, $\dim(P) \quad 6 < N << WH$. To obtain a direct relationship between the image and the pose vector p, we apply g to (7.5) and denote $h(p) \quad (g \circ f)(p, l, \varepsilon)$. Then

$$\tilde{a} \quad h(p). \tag{7.10}$$

A function h is a mapping P into a set $\tilde{A} \subset R^N$. Suppose h is injective, i.e., different $p_1 \ / \ p_2$ correspond to different $\tilde{a}_1 \ / \ \tilde{a}_2$. Consider some point $\hat{p} \in P$ and choose all coordinate functions $h_k(p), \ k \quad \overline{k_1, k_M}$, which are continuously differentiable at $\hat{p}$. Suppose $M \geq 6$. We will compose a vector-function

$$h^{reduced}(p) \quad (h_{k_1}(p), \ldots, h_{k_M}(p))^{\mathrm{T}} \tag{7.11}$$

and represent it in the form of Taylor series expansion in neighborhood of $\hat{p}$

$$h^{reduced}(p) \quad h^{reduced}(\hat{p}) + H_{\hat{p}} \cdot \Delta p + o||\Delta p||, \tag{7.12}$$

where $\Delta p \quad p - \hat{p}$ and $H_{\hat{p}}$ is a Jacobian matrix $\left[\frac{\partial h_k}{\partial p_j}(\hat{p}) \right], k \quad \overline{k_1, k_M}$, $j \quad \overline{1, 6}$. If $H_{\hat{p}}$ has full rank, it is possible to obtain an linear approximation for p in a neighborhood of $\hat{p}$ using least squares

$$p \approx \hat{p} + \left(H_{\hat{p}}^{\mathrm{T}} H_{\hat{p}} \right)^{-1} H_{\hat{p}}^{\mathrm{T}} \left(\Delta \tilde{a}^{reduced} \right), \tag{7.13}$$

where $\Delta \tilde{a}^{reduced} \quad h^{reduced}(p) - h^{reduced}(\hat{p})$. Thus, having the values of $k_1, \ldots, k_M, \tilde{a}(\hat{p})$, and $H_{\hat{p}}$, we can approximately calculate the vector p from the image a in a neighborhood of $\hat{p}$.

Since the 3D model of Target is specified, it is possible to obtain image renders for arbitrary values of p, i.e., to synthesize an image $a^{synt} f(p.l^{synt}, 0)$ for some selected realization of illumination l^{synt} in the absence of noise. Application $g(a^{synt}) \quad \tilde{a}$ allows us to get pairs of values $\langle p, \tilde{a} \rangle$ for further machine learning.

We cover set P by a discrete grid with sufficient small step Δp. Denote all nodes

$$\hat{p}_s \quad (x_s, y_s, z_s, \vartheta_s, \psi_s, \gamma_s), \quad s \quad \overline{1, S}, \tag{7.14}$$

by indexed elements of space R^6, where S is the total number of nodes. In each node $\hat{p}_s$ we compute the vector $\hat{\tilde{a}}_s \quad h(\hat{p}_s)$. Now, if we implement some procedures for finding the nearest grid node with respect to $\tilde{a}$, then the problem of finding p will be solved with an error up to the value Δp

$$p(\tilde{a}) \quad \left\{ \hat{p}_{s^*} : s^* \quad \min_{s \in [1,S]} \|\hat{\tilde{a}}_s - \tilde{a}\| \right\}. \tag{7.15}$$

To obtain an acceptable accuracy, a very large number of nodes may be required. The most efficient procedure of finding the nearest node (7.15) is a decision tree [25]. This approach allows to perform the most necessary calculations (i.e., building a tree) in advance on the training stage.

If the Target is asymmetric, each $\hat{p}_s$ corresponds to a unique $\hat{\tilde{a}}_s$. We divide all pairs $(\hat{p}_s, \hat{\tilde{a}}_s)$ into two equal groups according to a certain criterion $F_1(\tilde{a})$. In turn, these groups are divided into subgroups by some other criterions. The process will continue until one node is left in each group. The received hierarchical structure permits to define the nearest node by $\tilde{a}$ using a sequential criteria checking. Partitioning into groups can be done in different ways. Constructing an optimal tree is a nontrivial problem.

7.4 Informative Features of Image

In the proposed approach, several assumptions were used. Consider how it is possible to fulfill condition (7.8). Each component of the vector function g can be considered as a scalar function of computing a certain image attribute. A vector of these attributes is called an image descriptor $\tilde{a}$.

Image descriptors are used in many areas related to problems of content-based image retrieval [26], object recognition [27], and image registration [28]. Requirements to the descriptors properties are different and depend on task. For example, in the image recognition problem descriptors should be invariant to displacement, flat rotations, and scaling. Such characteristics are

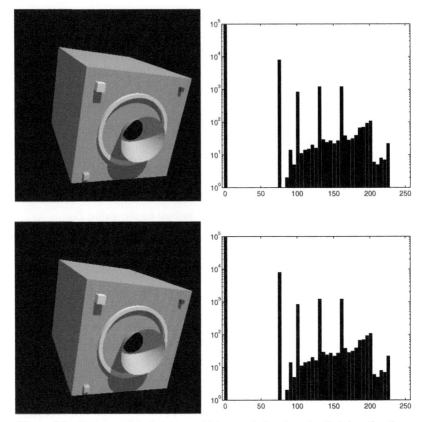

Figure 7.2   Variance of the brightness histogram to the changing lighting direction.

obviously not suitable for solving the pose determination problem. Consider separately three main groups of image features: of color, texture, and shape.

Color features describe the overall distribution of colors throughout the image. Such features are histograms [29] and statistical moments [30]. These characteristics do not satisfy condition (7.8), since the pixels brightness level are substantially sensitive to the direction of Target illumination, as shown in Figure 7.2.

Texture features are composed of numerical characteristics computed from local areas covering the entire image. Review and classification of specific textural features can be found in the literature, see for example [31]. To calculate some characteristics, a transition to the frequency domain is used [32]. As applied to the problem under consideration, these features

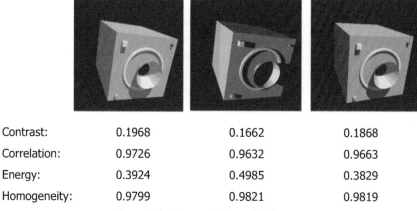

Contrast:	0.1968	0.1662	0.1868
Correlation:	0.9726	0.9632	0.9663
Energy:	0.3924	0.4985	0.3829
Homogeneity:	0.9799	0.9821	0.9819

Figure 7.3 Texture informative features.

are insensitive to the Target position and attitude. Figure 7.3 shows how the values of the texture features [33] change at

- change of external lighting and
- change of target attitude.

The features of shape are numerical characteristics of the object shape. For extracting these features, contours of objects are required. By construction, these features are more insensitive to variations in lighting, as shown in Figure 7.4.

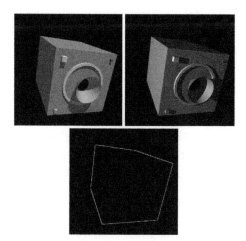

Figure 7.4 Outer contour invariance to ambient light variation.

The features of shape are divided into two groups. The first one includes descriptors that characterize the outer contour of object, and the second one - those that characterize the inner region. In turn, these groups of characteristics are divided into two subgroups: global (continuous) and structural (discrete).

Structural contour descriptors are different finite-dimensional contour approximation such as chain codes or polygonal decomposition. They are inconvenient to use because of the large number of coefficients.

Structural descriptors of regions include characteristics based on the region decomposition [34]. As well as structural contour descriptors, they are not convenient to direct use.

Global contour descriptors are different smooth continuous contour approximations, such as the function of turns [35], wavelets, or splines. They can be constructed in polar coordinates or in the complex plane. Such descriptors approximately describe the shape of bodies, which does not meet the requirement for high-precision pose evaluation.

Global descriptors of regions are divided into two groups. The first group is the geometric features, such as image square, center of the image mass, and others. The second group is represented by the statistical moments of the two-dimensional distribution of pixel coordinates on the region. It includes such features as invariant moments of Zernike [36] and Fourier descriptors [37].

To solve the problem under consideration, the global descriptors of regions are the most suitable as coordinate functions g_i. To obtain a greater number of characteristics, other informative features may be additionally used. A common property in the feature selection is their strong sensitivity to the movement and rotation of the subject and a weak sensitivity to changing color tones. The total number of features must ensure that all nodes of the grid (7.14) are distinguishable.

7.5 Estimation Problem Statement

The equations of the Target angular motion will be considered in a coordinate system (TCS) rigidly connected to the Target. Its origin is placed in the center of mass, and the axes coincide with the Target main inertia axes. In this coordinate system, the motion equations have the following form

$$J\dot{\omega} + \omega \times J\omega \quad M, \tag{7.16}$$

where $\omega \quad (\omega_1, \omega_2, \omega_3)^{\mathrm{T}}$, is an angular velocity vector determined by its components in the TCS, $M \quad (M_1, M_2, M_3)^{\mathrm{T}}$ is a vector of moments acting on the Target, matrix of inertia moments

$$J \quad \text{diag}(J_1, J_2, J_3)$$

is assumed to be known. Here J_i is inertia moment with respect to the i–th coordinate axis, $i \quad \overline{1, 3}$.

The navigation of Target and Chaser is considered in some inertial coordinate system (ICS) which is given. Target attitude change with respect to the ICS is described by the following equation:

$$\dot{q} \quad \frac{1}{2}\Omega(q) \cdot \omega, \tag{7.17}$$

where $q \quad (q_0, q_1, q_2, q_3)^{\mathrm{T}}$ is the Target attitude quaternion with respect to the ICS [38–40], matrix

$$\Omega(q) \quad \Omega(q_0, q_1, q_2, q_3) \quad \begin{pmatrix} -q_1 & -q_2 & -q_3 \\ q_0 & -q_3 & q_2 \\ q_3 & q_0 & -q_1 \\ -q_2 & q_1 & q_0 \end{pmatrix}. \tag{7.18}$$

The vector q should be normalized

$$\|q\| \quad q_0^2 + q_1^2 + q_2^2 + q_3^2 \quad 1. \tag{7.19}$$

The initial conditions for equations (7.16) and (7.17), vectors $q_0 \quad q(t_0)$, and $\omega_0 \quad \omega(t_0)$ are unknown. Hence, we cannot use straightforward these equations to calculate the vectors $q(t)$ and $\omega(t)$.

Estimations of TCS attitude angles with respect to the coordinate system rigidly connected with the Chaser (CCS) are effected at discrete time instants $t_k t_0 + k \cdot \Delta t, k \quad 1, 2, \ldots$, where Δt is a measurement period. Note that there are no measurements when the Chaser camera does not observe Target. As it follows from CVS algorithm description, the attitude angles $(\vartheta, \psi, \gamma)$ are measured with some additive error

$$\tilde{\vartheta}_k \quad \vartheta_k + \xi_k^\vartheta, \quad \tilde{\psi}_k \quad \psi_k + \xi_k^\psi, \quad \tilde{\gamma}_k \quad \gamma_k + \xi_k^\gamma, \quad k \quad 1, 2, \ldots. \tag{7.20}$$

Here errors $\xi_k^\vartheta, \xi_k^\psi, \xi_k^\gamma$ of the angle $(\vartheta_k, \psi_k, \gamma_k)$ determination are supposed to be bounded

$$|\xi_k^\vartheta| \le c^\vartheta, |\xi_k^\psi| \le c^\psi, |\xi_k^\gamma| \le c^\gamma. \tag{7.21}$$

Knowing the TCS attitude angles $(\vartheta_k, \psi_k, \gamma_k)$ with respect to CCS, one can uniquely calculate the corresponding quaternion q_k^T.

It is assumed that the Chaser attitude quaternion q_k^C with respect to the ICS is measured by the Chaser navigation system with a high accuracy. The Target attitude quaternion $q(t_k)$ q_k (from the equation (7.17)), is connected with q_k^C and q_k^T by the following exact relation

$$q_k \qquad q_k^C * q_k^T, \tag{7.22}$$

where $*$ means quaternion multiplication, which for two quaternions q^1 $(q_0^1, q_1^1, q_2^1, q_3^1)^{\mathrm{T}}$ and q^2 $(q_0^2, q_1^2, q_2^2, q_3^2)^{\mathrm{T}}$ is determined [38–40] by the following expression

$$q^1 * q^2 \qquad \begin{pmatrix} q_0^1 q_0^2 - \sum\limits_{i=1}^{3} q_i^1 q_i^2 \\ q_0^1 q_1^2 + q_1^1 q_0^2 + q_2^1 q_3^2 + q_3^1 q_2^2 \\ q_0^1 q_2^2 + q_2^1 q_0^2 - q_1^1 q_3^2 + q_3^1 q_2^2 \\ q_0^1 q_3^2 + q_3^1 q_0^2 + q_1^1 q_2^2 + q_2^1 q_1^2 \end{pmatrix} \qquad C(q^1) q^2 \tag{7.23}$$

Here

$$C(q^1) \qquad \begin{pmatrix} q_0^1 & -q_1^1 & -q_2^1 & -q_3^1 \\ q_1^1 & q_0^1 & -q_3^1 & q_2^1 \\ q_2^1 & q_3^1 & q_0^1 & -q_1^1 \\ q_3^1 & -q_2^1 & q_1^1 & q_0^1 \end{pmatrix}.$$

Each set of angles $(\vartheta_k, \psi_k, \gamma_k)$ defining attitude of TCS with respect to CCS is related to some quaternion [38]. Using conversion formulas and taking into account bounds (7.20) and (7.21), one can obtain for unknown quaternion q_k^T $(q_{0,k}^T, q_{1,k}^T, q_{2,k}^T, q_{3,k}^T)^{\mathrm{T}}$ interval estimates in the form

$$\underline{q}_{i,k}^T \leq q_{i,k}^T \leq \bar{q}_{i,k}^T, \quad i \quad \overline{0,3}. \tag{7.24}$$

Then for quaternion q_k^T, we have

$$q_k^T \qquad \tilde{q}_k^T + \xi_k^q, \tag{7.25}$$

where estimate $\tilde{q}_k^T$ $(\tilde{q}_{0,k}^T, \tilde{q}_{1,k}^T, \tilde{q}_{2,k}^T, \tilde{q}_{3,k}^T)^{\mathrm{T}}$ is determined by

$$\tilde{q}_{i,k}^T \qquad 0.5 \cdot (\underline{q}_{i,k}^T + \bar{q}_{i,k}^T), \quad i \quad \overline{0,3}, \tag{7.26}$$

and the components $\xi_{i,k}^q$ of the error vector ξ_k^q $(\xi_{0,k}^q, \xi_{1,k}^q, \xi_{2,k}^q, \xi_{3,k}^q)^{\mathrm{T}}$ will satisfy to the inequalities

$$|\xi_{i,k}^q| \leq c_{i,k}^q. \tag{7.27}$$

Here

$$c_{i,k}^q \quad 0.5 \cdot (\bar{q}_{i,k}^T - \underline{q}_{i,k}^T), \quad i \quad \overline{0,3}. \tag{7.28}$$

Taking into account expressions (7.23–7.26), equation (7.22) can be written in the form

$$q_k \quad C(q_k^C)(\tilde{q}_k^T + \xi_k^T). \tag{7.29}$$

Then, we conclude

$$G_k q_k - \tilde{q}_k^T \quad \xi_k^q, \quad G_k \quad C^{-1}(q_k^c) \tag{7.30}$$

From (7.27) and (7.30), it follows that the unknown quaternion q_k must satisfy the inequalities

$$|G_{i,k} q_k - \tilde{q}_{i,k}^T| \le c_{i,k}^T, \quad i \quad \overline{0.3}, \quad k \quad 1, 2, \ldots \tag{7.31}$$

Here $G_{i,k}$ is the i-th row of the matrix G_k.

Differential equations (7.16) and (7.17) represent nonlinear continuous time dynamic system with state vector

$$x \quad (\omega^T, q^T)^T. \tag{7.32}$$

At each instant of time, state vector should satisfy inequalities (7.31) associated with the measurements. In addition, for the component q, the normalization (7.19) must be satisfied.

Estimate $\hat{x}_k$ of the state vector $x_k \quad x(t_k)$ for discrete time t_k, $k \quad 1, 2, \ldots$ in limit should satisfy a condition

$$\lim_{k \to \infty} \|\hat{x}_k - x_k\| \quad 0. \tag{7.33}$$

Due to dependence of limit on properties of measurement errors, it is difficult to verify and to provide (7.33) in practice. Therefore, we consider the solution under weakened conditions. It consists in estimating $\hat{x}_k$, calculated according to equations (7.16) and (7.17), which ensure the fulfillment of inequalities (7.31) starting from some finite time K.

Measurement errors written in the form (7.21) or (7.27) allow using a guaranteed approach [18–24] to solve the problem. Since information on the initial state vector is absent and system is nonlinear, we use the modified algorithm of ellipsoidal estimation [16, 17].

7.6 Method of Nonlinear Ellipsoidal Filtration

Equations (7.16) and (7.17) can be written as

$$\dot{x} \quad f(x(t), u(t), \zeta(t)), \quad t \geq t_0, \tag{7.34}$$

where $x(t)$ is state vector at continuous time t, $u(t)$ is vector of known or measured input variables, and $\zeta(t)$ is a vector of uncontrolled disturbances. In considered case, $\zeta(t)$ is a vector of moments acting on the Target,

$$\zeta(t) \in Z, \tag{7.35}$$

where Z is some bounded closed set, $Z \subset R^3$.

The constraints on the state vector given by (7.31) can be written in the following form:

$$|f_j(x_k) - y_{jk}| \leq c_j, \quad j \quad \overline{1, 4}. \tag{7.36}$$

The initial vector $x(t_0)$ x_0 for equation (7.34), as well as its perturbation values, are unknown according to the problem statement. Therefore, it is impossible to obtain vector $x(t)$ for an arbitrary time t by integrating Equation (7.34).

Let us describe the state vector estimation procedure according to the method of ellipsoidal estimation [18–22]. Suppose that at a time t_k it is known that

$$x_k \quad x(t_k) \in E_k, \tag{7.37}$$

where set E_k $\{x : (x - \hat{x}_k)^T H_k^{-1}(x - \hat{x}_k) \leq 1\}$ is an ellipsoid, characterized by its center vector $\hat{x}_k$ and a positive definite symmetric matrix H_k $H_k^T > 0$. The ellipsoid E_k in equation (7.37) is usually called ellipsoidal estimate of vector x_k. The center of the ellipsoid is taken as a point estimate of vector $\hat{x}_k$. Considering equation (7.34) on the time interval $[t_k, t_{k+1}]$ for all possible initial conditions satisfying (7.37) and realizations of perturbations $\zeta(\cdot)$ satisfying condition (7.35) on this interval, we can obtain the set

$$X_{k+1|k} \quad \{x \quad x(t_{k+1}, x_k, \zeta_u(\cdot)), \forall x_k \in E_k \forall \zeta(\tau) \in Z \, \forall \tau \in [t_k, t_{k+1}]\}$$

of all possible values of a vector x_{k+1} $x(t_{k+1})$ at a discrete time $k + 1$. In general case, this set is not an ellipsoid. This set can be obtained, for example, by integrating the system (7.34) on the interval $[t_k, t_{k+1}]$ under different initial conditions satisfying (7.35) and for all possible perturbation realizations $\zeta(\tau) \in Z \, \forall \tau \in [t_k, t_{k+1}]$. Obviously, this is a very laborious process, requiring a large number of calculations. On the other hand, at the time t_{k+1} the state vector x_{k+1} must satisfy conditions (7.36), which can be written in the form

$$x_{k+1} \in \bar{X}_{k+1} \quad \{x : |f_j(x) - y_{j,k+1}| \le c_j, \ j \quad \overline{1,4}\}.$$

The set $\bar{X}_{k+1}$ contains state vectors that are compatible with measurements for given a priori bounds on measurement errors. As a result, we can conclude that

$$x_{k+1} \in X_{k+1} \quad \overline{X}_{k+1} \cap X_{k+1|k}.$$

The set X_{k+1} can be approximated by ellipsoid E_{k+1}, which contains it, $X_{k+1} \subset E_{k+1}$. Thus, for time $k+1$, we obtain an inclusion (7.37) and hence recurrent procedure for constructing estimates. Despite the obvious simplicity and logical rigor of the approach, its practical implementation in general case encounters insuperable computational difficulties connected with the construction of sets $X_{k+1|k}, \bar{X}_{k+1}, X_{k+1}$ and E_{k+1}. Hence, it is proposed to use somewhat different more effective approach considered in [16, 17]. In accordance with it, at the time t_{k+1}, an ellipsoid

$$E_{k+1|k} \quad \{x : (x - \hat{x}_{k+1|k})^{\mathrm{T}} H_{k+1|k}^{-1}(x - \hat{x}_{k+1|k}) \le 1\},$$

is constructed. Its center vector $\hat{x}_{k+1|k}$ is found by integrating the following equation

$$d\tilde{x}/dt \quad f(\tilde{x}(t), u(t)), \quad \tilde{x}(t_k) \quad \hat{x}_k, \quad t \in [t_k, t_{k+1}], \tag{7.38}$$

taking $\hat{x}_{k+1|k} \quad \tilde{x}(t_{k+1})$. The ellipsoid matrix $E_{k+1|k}$ is taken in the following form

$$H_{k+1|k} \quad A_k H_k A_k^T, \tag{7.39}$$

where

$$A_k \quad \exp(\partial \ f(0,5 \cdot (\hat{x}_{k+1|k} + \hat{x}_k), U_{k+1/2} \Delta T) \tag{7.40}$$

is $n \times n$–matrix (n is a dimension of vector x). Here $\partial \ f(\cdot, \cdot)$ is a Jacobi matrix of the functions $f(\cdot, \cdot)$ on variable x,

$$U_{k+1/2} \quad \frac{1}{\Delta T} \int_{t_k}^{t_{k+1}} u_m(\tau) d\tau. \tag{7.41}$$

As it follows from equations (7.39–7.41) the linearization and averaging procedures were used for constructing an ellipsoid matrix $E_{k+1|k}$. In general case, the ellipsoid $E_{k+1|k}$ is some approximation of $X_{k+1|k}$, but $E_{k+1|k} \neq X_{k+1|k}$.

Let us consider the ellipsoid

$$\tilde{E}_{k+1|k} \quad \{x : (x - \hat{x}_{k+1|k})^{\mathrm{T}} \tilde{H}_{k+1|k}^{-1}(x - \hat{x}_{k+1|k}) \leq 1\}, \tag{7.42}$$

whose center coincides with the center of the ellipsoid $E_{k+1|k}$, and matrix

$$\tilde{H}_{k+1|k} \quad \alpha^2 H_{k+1|k}, \tag{7.43}$$

where parameter $\alpha \geq 1$. Geometrically ellipsoid $\tilde{E}_{k+1|k}$ is obtained by uniform α–expansion of $E_{k+1|k}$ with respect to its center.

Statement 1. Let the function $f(\cdot, \cdot)$ in equation (7.31) is such that matrix A_k defined by (7.40) is nonsingular and bounded, perturbations $\zeta(t)$ are limited. Then there exists a finite number $\alpha \geq 1$, such that

$$X_{k+1|k} \subseteq \tilde{E}_{k+1|k}. \tag{7.44}$$

Algorithm for ellipsoidal estimation of the state vector x_k consists in constructing a sequence of ellipsoids $\{E_k\}_{k=0}^{\infty}$ with centers $\hat{x}_k \in \overline{X}_k$. It is proposed to construct this sequence as follows. Suppose, at time k, the ellipsoid E_k is already built. Then the parameters of ellipsoid $E_{k+1|k}$ can be determined using formulas (7.38–7.41). If

$$E_{k+1|k} \cap \overline{X}_{k+1} \neq \emptyset, \tag{7.45}$$

then E_{k+1} may be taken as the ellipsoid containing set $E_{k+1|k} \cap \bar{X}_{k+1}$ with a multidimensional volume $V(E_{k+1})$ less than the $E_{k+1|k}$ volume, that is $V(E_{k+1}) < V(E_{k+1|k})$. In particular, such construction can be performed using successive optimal procedures [19, 20] covering by an ellipsoid an intersection of a strip with ellipsoid, which was obtained early. Ellipsoid $E_{k+1|k}$ is taken as the initial ellipsoid and when the iterative process stops E_{k+1} will be the resulting ellipsoid.

If it turned out that condition (7.45) is not satisfied, then ellipsoid $\tilde{E}_{k+1|k}$ should be taken instead of $E_{k+1|k}$. Besides parameter $\alpha \geq 1$ is chosen in such way that the condition (7.45) to be satisfied [16]. After that an ellipsoid E_{k+1} is constructed as described above. The center of resulting ellipsoid E_{k+1} always belongs to $\overline{X}_{k+1}, \hat{x}_{k+1} \in \overline{X}_{k+1} \forall k$.

7.7 Estimation of Target Attitude and Angular Velocity

Let us consider the implementation of the above approach using equations (7.16) and (7.17) and taking into account inequality (7.31). The right-hand side of equation (7.16) includes the moment vector M. The moments acting on the Target can be caused by solar radiation action, particle flows, or by gravitational moment [41]. However, all these moments are rather weak and their effects are manifested over a large time interval, which is substantially longer than time required for rapprochement and docking. The vision system starts working at distance of 30 m, and the time required for docking takes no more than an hour. During this time, the action of these moments can be neglected and assume $M \approx 0$ in equation (7.16). We write it in a coordinate form

$$\begin{cases} \dot{\omega}_1 = J_1^{-1}(J_2 - J_3)\omega_2\omega_3, \\ \dot{\omega}_2 = J_2^{-1}(J_3 - J_1)\omega_1\omega_3, \\ \dot{\omega}_3 = J_3^{-1}(J_1 - J_2)\omega_1\omega_2. \end{cases} \tag{7.46}$$

Suppose that at time t_k vector $x_k \in E_k$, where $E_k = \{x : (x - \hat{x}_k)^T H_k^{-1}(x - \hat{x}_k) \leq 1\}$, is known. It can be written that

$$x_k = \hat{x}_k + \Delta x_k, \tag{7.47}$$

where an unknown vector $\Delta x_k \in E_k(0) = \{x : x^T H_k^{-1} x \leq 1\}$, $E_k(0)$ is ellipsoid with center at the origin. We denote by $\tilde{x}(t) = (\tilde{\omega}^T, \tilde{q}^T)^T$ the solution of equations (7.46) and (7.17) on the interval $[t_k, t_{k+1}]$ under the initial condition $\tilde{x}(t_k) = \hat{x}_k$. Substituting

$$x(t) = \tilde{x}(t) + \Delta x(t)$$

into equations (7.46) and (7.17), after discarding terms of order greater than the first, we obtain linear differential equations for $\Delta x(t) = (\Delta \omega^T, \Delta q^T)^T$. In particular, for equation (7.46), we have

$$\Delta\dot{\omega} = \begin{pmatrix} 0 & (J_2 - J_3)J_1^{-1}\omega_3 & (J_2 - J_3)J_1^{-1}\omega_2 \\ (J_3 - J_1)J_2^{-1}\omega_3 & 0 & (J_3 - J_1)J_2^{-1}\omega_1 \\ (J_1 - J_2)J_3^{-1}\omega_2 & (J_1 - J_2)J_3^{-1}\omega_1 & 0 \end{pmatrix} \Delta\omega$$

or

$$\Delta\dot{\omega} = A(\omega_k)\Delta\omega, \tag{7.48}$$

where

$$A(\omega_k) \quad \begin{pmatrix} 0 & (J_2 - J_3)J_1^{-1}\omega_3 & (J_2 - J_3)J_1^{-1}\omega_2 \\ (J_3 - J_1)J_2^{-1}\omega_3 & 0 & (J_3 - J_1)J_2^{-1}\omega_1 \\ (J_1 - J_2)J_3^{-1}\omega_2 & (J_1 - J_2)J_3^{-1}\omega_1 & 0 \end{pmatrix}.$$

(7.49)

After linearization of the equation (7.17), we obtain

$$\Delta\dot{q} \quad \frac{1}{2}\Omega(q) \cdot \Delta\omega + \frac{1}{2}\Omega(\Delta q) \cdot \omega \quad \frac{1}{2}\Omega(q) \cdot \Delta\omega + \frac{1}{2}A_\Omega(\omega)\Delta q, \quad (7.50)$$

where

$$A_\Omega(\omega) \quad \begin{pmatrix} 0 & -\omega_1 & -\omega_2 & -\omega_3 \\ \omega_1 & 0 & \omega_3 & -\omega_2 \\ \omega_2 & -\omega_3 & 0 & \omega_1 \\ \omega_3 & \omega_2 & -\omega_1 & 0 \end{pmatrix}.$$

(7.51)

Combining equations (7.48) and (7.50), we obtain

$$\begin{pmatrix} \Delta\dot{\omega} \\ \Delta\dot{q} \end{pmatrix} \quad \begin{pmatrix} A(\omega_k) & \Theta_4 \\ 0.5\Omega(q_k) & 0.5A_\Omega(\omega_k) \end{pmatrix} \begin{pmatrix} \Delta\omega \\ \Delta q \end{pmatrix}$$

In accordance with the general approach to ellipsoid determination, the following approximate difference analog of the equation can be used

$$\begin{pmatrix} \Delta\omega_{k+1}, \\ \Delta q_{k+1} \end{pmatrix} \quad A_{k+1} \begin{pmatrix} \Delta\omega_k, \\ \Delta q_k \end{pmatrix},$$

or

$$\Delta x_{k+1} \quad A_{k+1}\Delta x_k, \quad (7.52)$$

where

$$A_{k+1} \quad \exp\left[\begin{pmatrix} A(\omega_{k+1/2}) & \Theta_4 \\ 0.5\Omega(q_{k+1/2}) & 0.5A_\Omega(\omega_{k+1/2}) \end{pmatrix} \Delta t\right].$$

Here $\omega_{k+1/2}$ $0.5(\tilde{\omega}_{k+1} + \hat{\omega}_k)$, and $q_{k+1/2}$ is equal to the normalized quaternion $0.5(\tilde{q}_{k+1} + \hat{q}_k)$. Since Δx_k takes all possible values belonging to ellipsoid $E_k(0)$, $\Delta x_k \in E_k(0)$, vector Δx_{k+1} satisfies

$$\Delta x_{k+1} \in E_{k+1|k}(0) \quad \{x : x^\mathrm{T} H_{k+1|k}^{-1} x \leq 1\},$$

where

$$H_{k+1|k} \quad A_{k+1} H_k A_{k+1}^\mathrm{T}.$$

Finally, we assume

$$E_{k+1|k} \quad \{\tilde{x}_{k+1}+E_{k+1|k}(0)\} \quad \{x : (x-\hat{x}_{k+1|k})^{\mathrm{T}}H_{k+1|k}^{-1}(x-\hat{x}_{k+1|k}) \le 1\}. \tag{7.53}$$

Here $\hat{x}_{k+1|k}$ $\tilde{x}_{k+1}$ and summation is understood as vector sum of sets.

Let us consider the method applied to constructing an ellipsoid E_{k+1}. Set $\overline{X}_{k+1}$ connected with measurements can be represented in the following form

$$\overline{X}_{k+1} \quad \bigcap_{j=1}^{4}\overline{X}_{j,k+1}, \tag{7.54}$$

where each of the sets

$$\overline{X}_{j,k+1} \quad \{x \in R^n : |C_{i,k+1}x - \tilde{q}_{j-1,k+1}^n| \le c_{j-1,k+1}^q\}, \quad j \quad \overline{1,4}. \tag{7.55}$$

is connected with one of the inequalities (7.31) considered for the moment $k+1$. In equation (7.55), $C_{i,k+1}$ is i-th row of matrix C_{k+1} $(\Theta_{4\times3} : G_{k+1})$.

Each of the sets $\overline{X}_{j,k+1}$, j $\overline{1,4}$, as it follows from the equation (7.55), is a multidimensional layer in R^n. The recurrent procedure [17] is used to find an ellipsoid E_{k+1} containing the intersection of $E_{k+1|k}$ with $\overline{X}_{k+1}$. According to its sequence of ellipsoids E_s, s $\overline{0, S}$ is calculated. Each ellipsoid E_{s+1} contains intersection of E_s with layer $\overline{X}_{j,k+1}$ for some j and is built with the use of algorithm [19] and its modification [16]. The procedure stops after finite S steps [16]. The ellipsoid $E_{k+1|k}$ is taken as initial $E_{s=0}$ $E_{k+1|k}$ and resulting ellipsoid E_S is taken as E_{k+1}, E_{k+1} E_S. According to algorithm [19] the center of the resulting ellipsoid, E_{k+1} will satisfy the inequalities (7.55).

The above procedure is supplemented by normalization of the quaternion estimate, which is performed using compression operation of space R^n on part of variables. Matrix H_s of the ellipsoid E_s can be uniquely represented in the form H_s $G_s^{\mathrm{T}}G_s$, where G_s is some matrix. For any vector $x \in E_s$, the following representation holds

$$x \quad x_s + G_s\varepsilon, \quad \|\varepsilon\| \le 1. \tag{7.56}$$

Hence, for part q of state vector, we can write

$$q \quad q_s + G_s(4 : 7, 1 : 7)\varepsilon, \quad \|\varepsilon\| \le 1. \tag{7.57}$$

Here 4×7–matrix $G_s(4 : 7, 1 : 7)$ is formed of matrix rows G_s from 4-th to 7-th. Let λ_s $|q_s|^{-1}$ $(q_{s,0}^2 + q_{s,1}^2 + q_{s,2}^2 + q_{s,3}^2)^{-1/2}$ / 1. Every vector q

of ellipsoid E_s (determined by equation (7.57)) is divided by the number λ_s. As a result, we obtain ellipsoid

$$E_{s+1} \quad \{x \quad x_s(\lambda_s) + G_s(\lambda_s)\varepsilon, \; \|\varepsilon\| \leq 1\}, \qquad (7.58)$$

where

$$x_s(\lambda_s) \quad (\omega_s^{\mathrm{T}}, \lambda_s^{-1} q_s^{\mathrm{T}})^{\mathrm{T}}, \; G_s(\lambda_s) \quad \begin{pmatrix} G_s(1:3,1:n) \\ \lambda_s^{-1} G_s(4:7,1:n) \end{pmatrix}. \qquad (7.59)$$

Obviously, after such transformation $\|q_{s+1}\| \equiv 1$ and the values range of other ellipsoid E_s variables do not change.

7.8 Numerical Simulation

The characteristics of the TOPEX/Poseidon satellite were taken as a model example for the Target. This satellite has mass about 2200 kg; the dimensions of the main satellite body are 5.5 m × 6.6 m × 2.8 m; the dimensions of the solar cells are 8.9 m × 3.3 m. The satellite moments of inertia are J_1 8098 kgm^2, J_2 7618 kgm^2, and J_3 3616 kgm^2. The initial angular velocity $\omega(t_0)$ $(0.005, 0.010, 0.020)^{\mathrm{T}}$ and quaternion of TCS orientation with respect to ICS $q(t_0)$ $(0.1005, 0.5025, 0.3015, 0.8040)^{\mathrm{T}}$ were chosen arbitrary, but considered as unknown. These values were used in integrating equations (7.16) and (7.17) to calculate so-called true values of these parameters in time. Plots of time-varying components of the angular velocity vector $\omega(t)$ and quaternion $q(t)$ are shown in Figures 7.5 and 7.6, respectively.

It was assumed that orientation of the Chaser relative to the ICS did not change. Without loss of generality, it is assumed that the CCS and ICS axes are coincided. In this case, the quaternion of the Chaser remains unchanged $q_k^s \equiv (1, 0, 0, 0)^{\mathrm{T}}$, matrix $C(q_k^s) \equiv I$ and, hence, $G_k \equiv I$ too.

The measured TCS quaternion $\tilde{q}_k^T$ was prepared according to the algorithm described in Section 6. Simulation of the Target attitude determination was carried out as follows. According to the quaternion $q(t_k)$ calculated by integration of the equations (7.46) and (7.17), the satellite image was created as a matrix of pixels with the use of satellite graphic computer model and a mathematical model of image construction on the camera matrix. Then this image was analyzed using algorithm described in the first part of this article and values of angles $(\tilde{\vartheta}_k, \tilde{\psi}_k, \tilde{\gamma}_k)$ were calculated. Using these angles and their accuracy, values $\tilde{q}_k^n$ and $c_{i,k}^q, i$ $\overline{0,3}$ were calculated according

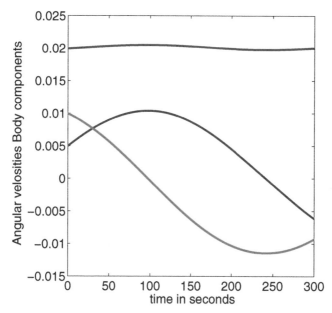

Figure 7.5 Change of body components of angular velocity in time

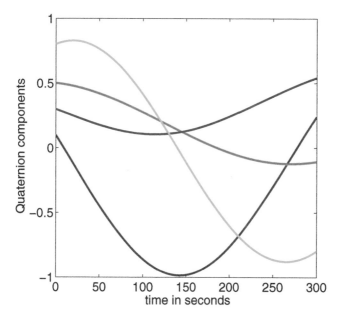

Figure 7.6 Change of quaternion components in time (red stands for q_0, blue for $-q_1$, green $-q_2$, turquoise $-q_3$)

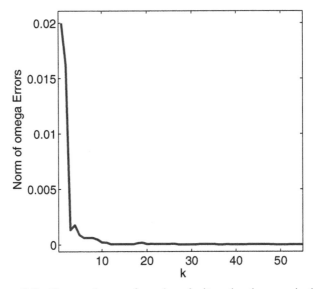

Figure 7.7   Change of norm of angular velocity estimation error in time

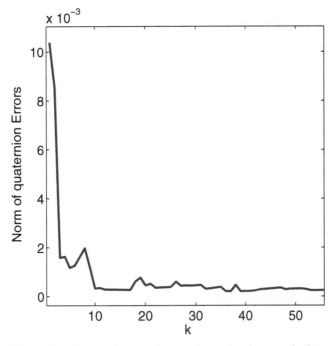

Figure 7.8   Change of norm of quaternion estimation error in time

to the algorithm of Section 6. These values were used in the above-described estimation algorithm. Maximum error values $\xi_{i,k}^q$ did not exceed level of 0.02, $c_{i,k}^q \leq 0.02$ in the simulation process.

The simulation period was 300 seconds, which corresponds to 5 minutes. Estimation errors of angular velocity vector e_k^ω $\quad \|\omega(t_k) - \hat{\omega}_k\|_\infty$ and the Target quaternion e_k^q $\quad \|q(t_k) - \hat{q}_k\|_\infty$ are shown in Figures 7.7 and 7.8, respectively.

As can be seen from these plots, the accuracy of the angular velocity estimation reached 0.001 rad/sec quickly enough, which is no more than 5% of the nominal value. The accuracy of the quaternion estimation for the same time reached 0.001, which is 20 times smaller than maximum value of $\xi_{i,k}^q$. The most significant accuracy improvement occurred in the first four steps (4 s). This is the minimum necessary time for estimating the attitude angles and their derivatives (angular velocities).

7.9 Conclusion

The results of numerical simulation showed the ellipsoidal filter operability and efficiency for solving estimation problem of relative rotational motion of two spacecrafts, when the system of computer vision is used as an angular coordinate sensor. The Target quaternion relative to the inertial coordinate system and its angular velocity vector were restored based on the Target video images obtained by Chaser camera. These parameters are necessary for Chaser docking with Target. Due to effective filtering provided by ellipsoidal estimation significantly improve of the motion parameters accuracy compared to direct measurements were obtained.

The problem was solved with some simplifying assumptions, which do not significantly affect the research results. In particular, one of such simplifications was assumption that the camera was installed in the Chaser center of mass. In real situation, this is not the case. Taking into account the real situation, some refinement should be introduced in the equations connected with the measurements, but it is not an obstacle to implementation of the proposed algorithms. The perturbing moments acting on the Target were also not taken into account. Account for these and other details are necessary for implementation of the proposed algorithms in practice. It is direction where authors intend to develop this work further.

It should be mentioned important application of the proposed algorithms. In the authors' opinion, these algorithms are absolutely indispensable in automatic docking of Chaser with Target when it is in rotational motion. Only high-precision knowledge of the attitude and angular velocity will allow to provide maneuver on approaching and docking is such situation.

Acknowledgments

The authors are grateful to their colleagues from Scientific Engineering Company KURS, the developers and creators of automatic docking systems for Soyuz spacecraft, Dobrovolsky V. Yu. and Godunok L. A. for help in setting the task, consulting and supporting this work.

References

[1] R. Opromolla et al., 'A review of cooperative and uncooperative spacecraft pose determination techniques for close proximity operations', Progress in Aerospace Sciences, vol. 93, pp.53–72, 2017.

[2] J.-F. Shi et al., 'Uncooperative Spacecraft Pose Estimation Using an Infrared Camera During Proximity Operations', AIAA Space 2015 Conference and Exposition. Issue AIAA 2015–4429.

[3] J. M. Kelsey et al., 'Vision-Based Relative Pose Estimation for Autonomous Rendezvous and Docking', IEEE Aerospace Conference, 2006.

[4] Shijie et al., 'Monocular Vision-based Two-stage Iterative Algorithm for Relative Position and Attitude Estimation of Docking Spacecraft', Chinese Journal of Aeronautics, vol. 23(2), pp. 204–210, 2010.

[5] S. D'Amico, M. Benn, J. L. Jrgensen, 'Pose estimation of an uncooperative spacecraft from actual space imagery', Proceedings of 5th International Conference on Spacecraft Formation Flying Missions and Technologies. Munich, Germany, 2013.

[6] D. F. Dementhon, L. Davis, 'Model-Based Object Pose in 25 Lines of Code', International Journal of Computer Vision, vol. 15, pp. 123–141, 1995.

[7] P. David et al., 'SoftPOSIT: Simultaneous Pose and Correspondence Determination', International Journal of Computer Vision, vol. 59(3), pp. 259–284, 2004.

[8] D. G. Lowe, 'Object recognition from local scale invariant features', In Proceedings of the 7th International Conference on Computer Vision, Kerkyra, Greece, pp. 1150–1157, 1999.

[9] R. O. Duda, P. E. Hart, 'Use of the Hough Transformation to Detect Lines and Curves in Pictures', Communication of the ACM, 15, 1, January, 1972.

[10] P. David et al., 'Simultaneous Pose and Correspondence Determination using Line Features', *Proc. CVPR*, vol. 2, pp. 424–431, 2003.

[11] L. Trujillo, G. Olague, 'Automated design of image operators that detect interest points', Journal Evaluationary Computation, vol. 16, pp. 483–507, 2008.

[12] C. Harris, M. Stephens, 'A combined corner and edge detector', Proceedings of the 4th Alvey Vision Conference, pp. 147–151, 1988.

[13] S. Haykin, 'Kalman Filtering and Neural Networks', New York, Toronto: John Wiley&Sons, 2001.

[14] M. S. Arulampalam et al., 'A tutorial on particle filters for online nonlinear/non-Gaussian bayesian tracking', IEEE Trans. on Signal processing, vol. 50, no. 2, pp. 174–188, 2002.

[15] I. I. Gorban, 'The Statistical Stability Phenomenon', Springer, 2016.

[16] N. N. Salnikov, 'On One Modification of Linear Regression Estimation Algorithm Using Ellipsoids', Journal of Automation and Information Sciences, vol. 44, no. 3, pp. 15–32, 2012.

[17] N. N. Salnikov, 'Estimation of State and Parameters of Dynamic System with the Use of Ellipsoids at the Lack of a Priori Information on Estimated Quantities', Journal of Automation and Information Sciences, vol. 46, no. 4, pp. 60–75, 2014.

[18] F. C. Schweppe, 'Uncertain dynamic systems', Englewood Cliffs, N. J., Prentice-Hall, 1973.

[19] V. V. Volosov, 'Method of constructing ellipsoidal estimates in problems of non-stochastic filtering and identification of the parameters of control systems', Journal of Automation and Information Sciences, vol. 24, no. 3, pp. 22–30, 1991.

[20] F. L. Chernousko, 'State estimation for dynamic systems', Boca Raton: CRC Press, 1994.

[21] A. B. Kurzhanski, I. Valyi, 'Ellipsoidal Calculus for Estimation and Control', Boston: Birkhauser, 1997.

[22] S. B. Chabane et al., 'A New Approach for Guaranteed Ellipsoidal State Estimation', Preprints of the 19th World Congress, The International Federation of Automatic Control, Cape Town, South Africa, August 24–29, pp. 6533–6538, 2014.

[23] F. Blanchini, S. Miani, 'Set-Theoretic Methods in Control', Springer International Publishing Switzerland, 2015.

[24] A. Poznyak, A. Polyakov, V. Azhmyakov, 'Attractive Ellipsoids in Robust Control', Springer International Publishing Switzerland, 2014.

[25] J. L. Bentley, 'Multidimensional binary search trees used for associative searching', Communications of the ACM, vol. 18, pp. 509–517, 1975.

[26] N. S. Vassilieva, 'Content-based Image Retrieval Methods', Programming and Computer Software, vol. 35, no. 3, pp. 158–180, 2009.

[27] F. Van der Heijden, 'Image based measurement systems: object recognition and parameter estimation', John Wiley & Sons, 1994.

[28] B. Zitová, J. Flusser, 'Image registration methods: a survey', Image Vision Comput., 21(11), pp. 977–1000, 2003.

[29] V. Chitkara, 'Color-Based Image Retrieval Using Compact Binary Signatures', Tech. Report TR 01-08, University of Alberta Edmonton, Alberta, Canada, 2001.

[30] M. Stricker, M. Orengo, 'Similarity of Color Images', Proc. of the SPIE Conf., vol. 2420, pp. 381–39, 1995.

[31] R. M. Haralick, K. Shanmugam, I. Dienstein, 'Textural Features for Image Classification', IEEE Trans. Systems, Man Cybernetics, vol. 3, no. 6, pp. 610–621, 1973.

[32] K. Shanmugam, I. Dienstein, 'Texture Features for Browsing and Retrieval of Image Data', IEEE Trans. Pattern Analysis Machine Intelligence, vol. 18, no. 8, pp. 837–842, 1996.

[33] H. Tamura, S. Mori, T. Yamawaki, 'Textural Features Corresponding to Visual Perception', IEEE Trans. Systems, Man Cybernetics, vol. 8, pp. 460–472, 1978.

[34] M.-K. Hu, 'Visual Pattern Recognition by Moment Invariants', IEEE Trans. Information Theory, vol. 8(2), pp. 179–187, 1962.

[35] E. M. Arkin et al., 'An Efficiently Computable Metric for Comparing Polygonal Shapes', IEEE Trans. Pattern Analysis Machine Intelligence, vol. 13, no. 3, pp. 209–216, 1991.

[36] L. Wang, G. Healey, 'Using Zernike moments for the illumination and geometry invariant classification of multispectral texture', IEEE Trans. on Image Processing, vol. 7, pp. 196–203, 1998.

[37] Zhang, D. S. Generic Fourier Descriptor for Shape-based Image Retrieval / D. S. Zhang, G. Lu // Proc. of IEEE Int. Conf. on Multimedia and Expo (ICME2002), Lausanne, Switzerland, vol. 1, pp. 425–428, 2002.

[38] V. N. Branets., I. P. Shmyglevskiy, 'Primenenie kvaternionov v zadachah orientatsii tverdogo tela' (Application of quaternion in rigid body orientation problems)(in Russian), Moscow: Nauka Publ., 1971.

[39] V. F. Zhuravlev, 'Osnovy teoreticheskoy mekhaniki' (Basis of theoretical mechanics) (in Russian), Moscow: Fizmatlit Publ., 2001.

[40] C. Jekeli, 'Inertial navigation systems with geodetic applications', Berlin, New York: Walter de Gruyter, 2001.

[41] V. A. Sarychev, P. Paglione, A. D. Guerman, 'Stability of Equilibria for a Satellite Subject to Gravitational and Constant Torques', Journal of Guidance Control and Dynamics, vol. 31, no. 2, pp. 386–394, 2008.

8

Fuzzy Controllers for Increasing Efficiency of the Floating Dock's Operations: Design and Optimization

Yuriy P. Kondratenko[1], Oleksiy V. Kozlov[2] and Andriy M. Topalov[2,*]

[1]Petro Mohyla Black Sea National University, Mykolaiv, Ukraine
[2]Admiral Makarov National University of Shipbuilding, Mykolaiv, Ukraine
*Corresponding Author: topalov_ua@ukr.net

This chapter presents the developed by the authors combined step-by-step approach to design of fuzzy controllers (FC) for the automatic control systems (ACS) of the floating docks main operations. The proposed approach allows performing the synthesis of the structure and parameters of the Mamdani type FCs for the ACSs of the main control parameters of the floating docks operations using expert knowledge in conjunction with certain optimization procedures on the basis of mathematical programming methods. In particular, the structural-parametric optimization of the FCs is carried out on the basis of the ACS desired transients and methods of gradient descent of numerical optimization, that allows to provide high-quality indicators of control of the docking operations. In turn, the reduction of the FC rule base (RB) is performed on the basis of calculating the influence of each rule on the control process and allows reducing the total number of rules of the RB without deterioration of the quality indicators of the docking operations control. In order to study and validate the effectiveness of the developed approach, the design of the Mamdani type FC for the draft ACS of the floating dock for low-tonnage vessels is performed in this work. The designed FC has a relatively simple hardware and software implementation as well as allows to achieve high quality indicators of the draft control, high economic and operational indicators, low energy consumption, as well as high quality of the docking operations conducting, which confirms the high efficiency of the proposed approach.

8.1 Introduction

In order to meet the growing demand for ship docking, ship repair plants in different countries of the world are replenished with new means of vessels docking [1]. Among them, the floating docks are rather widespread, despite their complexity and higher operating cost, and in our time are the main means of lifting and repairing of the ships hulls [2]. This is especially true in recent times, as new advanced technologies of the composite floating docks allow lowering the original price and construction time. Also, the benefits of floating docks include their mobility as well as a shorter duration of docking operations, which plays a significant role in minimizing the costs of ship repair.

Conduction of docking operations is a complex technical task, because they require real-time control of all operating parameters with high accuracy and timely control of the executive mechanisms of the floating dock [1]. Any "human" errors can lead to increased ship docking time and, accordingly, to reduction of the cost-effectiveness of using the dock as well as, possibly, to emergency situations for both floating dock and vessel itself. Moreover, in practice of the floating docks, exploitation there are serious crash and emergency situations: docks flood, towers fracture, the appearance of corrugations on towers and dents on the floating docks stack-deck, emergency lists of vessels when immersing a dock for their withdrawal, etc. [1, 2].

The complexity of modern floating docks as technological plants necessitates the development of specialized control systems for their docking operations, whose effectiveness is significantly increased when operating in automatic modes [3]. Control processes automation of the floating docks docking operations allows to significantly improve the control accuracy of their main operating parameters as well as energy, economic, and performance indicators [4].

There is a number of publications in the scientific and technical literature concerning automation and mathematical modeling of marine objects [3–8]. The mathematical description of various types of marine floating objects and vessels is presented and detailed in the works [4–6]. In particular, in [7], the mathematical modeling of the motion of marine objects is given, the static and dynamics problems are considered. The work [8–11], in turn, describes the application of imitation simulations directly in the study of control systems for docking operations of floating docks.

Currently, a well-developed mathematical apparatus of automatic control of various technical objects is known [12–19]. A large number of methodologies for calculating and adjusting of classical PID controllers, the synthesis of closed-loop dynamic systems of automatic control that meets the quality control process indicators are also developed. However, for a complex dynamic object with certain nonlinear characteristics, building of an efficient control system to date is a complex technical task requiring knowledge in the field of theories of intellectual control. This problem is particularly difficult for dynamic plants with variable parameters and uncontrollable disturbances, for which the adequate mathematical description of the processes dynamics is very complicated, for example, for marine movable objects [6]. In some cases, it is very difficult to solve the problem of the synthesis of the control system for such plants. Therefore, the possibility of using in such situations various heuristics and expert methods as well as intellectual algorithms is a prerequisite for the widespread introduction of fuzzy control.

Fuzzy control systems ensure the quality of the control and regulation processes to be better than in systems synthesized by classical traditional methods for nonlinear and non-stationary objects. In addition, nowadays, there are widespread complex combined control systems for technological processes that include PI, PD, or PID controllers, functioning on a traditional basis and intelligent blocks, built on the basis of fuzzy logic [20–22].

Methods of fuzzy logic allow to formalize control object, using the membership functions to each of the logical terms and systematizing the information of the expert (operator) in the form of rule base [20] Thus, the fuzzy controller, using the strategy presented in the form of a RB, transforms measured and calculated values of deviations of the controlled parameters to the certain control action [21].

Analysis of algorithmic and software solutions for the design and implementation of automatic control systems for complex nonlinear and non-stationary plants with variable parameters, for which the mathematical model is very complicated, including the floating docks, shows the expediency of using the principles of intellectual control based on fuzzy logic [20–25].

So, fuzzy controllers of Mamdani and Sugeno types are used for control of complex technical plants with uncertain or non-linear parameters in various technical fields [24–38]. An important feature of the automatic control systems based on fuzzy controllers is that they are developed mainly on the basis of expert assessments, and their performance indicators significantly depend on the qualifications and experience of the developers (operators or experts) as well as on the number of subjective factors [39–41]. In particular, FCs of Mamdani type, developed on the basis of expert knowledge, have low efficiency at controlling of objects, for which there is no significant manual control experience gained by their operators [42–44]. Thus, the problem is the development of effective approaches, methods and algorithms for designing fuzzy control devices, which will reduce the influence of subjective factors on the design process, increase in general the quality indicators of control as well as energy and economic efficiency of the fuzzy ACSs of complex little-studied control plants [45].

In order to reduce the negative influence of subjective factors on the FC design process and to increase the efficiency of their operation, it is expedient to use specialized mathematically formulated optimization procedures [45–47] in the methods and algorithms of the FC synthesis. Thus, the methods and algorithms for FCs designing are presented in papers [48–50], which include optimization procedures of the parameters of the linguistic terms membership functions (LTMF) of their input and output variables on the basis of the ACSs desired transients. In turn, papers [51–53] present the method of FCs designing with the application of the structural optimization and reduction of their rule bases, which allows to significantly reduce the number of rules in the FC RB without quality indicators deterioration of the ACS. These methods and algorithms allow FCs designing for ACSs of complex plants that provide substantial increasing of their energy, economic and operational indicators. The applied in them optimization procedures in conjunction with expert knowledge and assessments can be successfully used at creating of a combined, mathematically formulated design approach of fuzzy controllers for the ACSs of the floating docks docking operations.

The aim of this chapter is development and efficiency research of the combined approach to design of fuzzy controllers for the floating docks main operations for providing high economic and operational indicators, low energy consumption as well as conducting high-quality docking operations.

8.2 Design Features of the Automatic Control Systems of the Floating Docks Main Operations

At performing docking operations of the floating docks, it is important to control the angles of inclination and draft, which is characterized by the general orientation of the marine object relative to the surface of water [54]. To eliminate unwanted inclines and possible critical deformation of the floating dock, it is expedient to carry out specific calculations of the ballast tanks

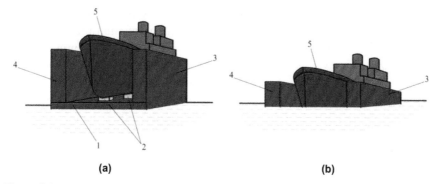

(a) **(b)**

Figure 8.1 Two towers floating dock with the vessel: *a* – submerged; *b* – after surfacing; 1 – pontoon; 2 – dock supports; 3 – right tower; 4 – left tower; 5 – vessel.

filling before placing the vessel in the dock. The purpose of calculations is to obtain such a ballasting, at which the list and the trim angles of the dock will be minimal, and inflection moment will be less than allowed for this floating dock. The inflection moments of the dock can be completely eliminated by leveling the load on each pontoon with the help of the ballast [8].

The calculation of ballast system needs to be carried out for set time on the natural flooding of the floating dock [3, 8]. The pressure head of the outboard water coming into the dock is constantly changing. It is impossible to obtain the same resistance of the pipeline from any outlet valve or damper to any ballast tank without excessive complication of the valves. Accordingly, levels of water in ballast tanks at different time intervals may differ from the given values. So, the calculation is carried out at the time of filling the most distant tank from the receiving hole.

This circumstance in case of simplification of the diagram and valves complicates operation of computer control system software and the Dockmaster-operator. For uniform filling of the dock, it is necessary to manipulate gate valves, accelerating, decelerating water inflows in this or that tank depending on water level indices in tanks.

Docking of the vessel is usually provided in the following order. The floating dock takes the ballast water into the ballast tanks of the pontoon and towers, and then it submerges to a depth sufficient for the vessel setting on its stock-deck with supporting devices (Figure 8.1(a)). After that the ballast water is pumped out from the ballast tanks and the vessel is set on the floating dock supports that are fixed of its stock-deck. The docking process is completed at the floating dock final surfacing (Figure 8.1(b)).

Thus, the ballast system is the main floating dock system for performing docking operations, since the processes of filling and emptying of the ballast tanks lead to changes in the draft of the floating dock [54]. Another important function of the ballast system is elimination of the critical deformation of the floating dock and unwanted inclines due to the distribution of liquid ballast among ballast tanks. The ballast system, in turn, includes ballast pumps, filling and emptying pipes, and gate valves (flow regulators).

In practice, two circuits of the ballast system are used: linear and circular [8]. In the linear scheme, each ballast pump is located on one pontoon board of the floating dock and connected to the distribution box, which distributes the appendages with the appropriate gate valves to the ballast compartments. The distribution boxes themselves are connected by linear pipelines with valves. Linear circuit assumes emergency discharge of any of the pumps pumping water by the pump of a neighboring distribution box. The circular scheme consists of two on-board highways connected in at least two places. Such a system allows in case of failure of the part of pumps on the ring system to pump water from any compartment by other pumps.

The main operational requirement for ballast systems of different types is to ensure the pumping of water by all the available ballast pumps or part of them in case of failure of one or more pumps. The possibility of simple and quick enabling and disabling of individual pumps and, as a consequence, the possibility of pumping out the water from any compartment allows the submerging and surfacing of the floating dock with a minimum inflection, that provides safe input or output of the ship. The diameters of the pipelines of the system should allow the dock to surface with the vessel for 50–80 minutes.

Regardless of the ballast system, the general view of the location of ballast compartments on a number of floating docks is shown in the Figure 8.2. From the drawings, one can see that the ballast tanks of large capacity BT13–BT18 are located in the central part of the dock, and ballast tanks of small capacity BT1–BT12 – in the bow and stern.

The ballast system of the floating dock is serviced by several pumps [50]. The centrifugal and propeller pumps are the most widely used types of pumps for ballast systems. Productivity and quantity of pumps are chosen for reasons of ensuring the speed of water in the ballast pipeline 2–2.5 m/s. It is also desirable to select the pumps in such a way that at reduced pressure, the productivity ("flow") is increased at constant values of revolutions and power consumption. In these conditions, ballasting is the fastest.

So, the automation of the floating dock involves controlling the ballast system in automatic modes to achieve the required values of draft, list, and

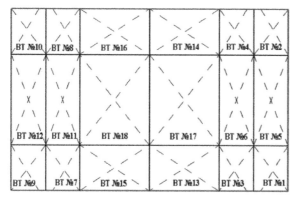

Figure 8.2 The layout of the ballast tanks.

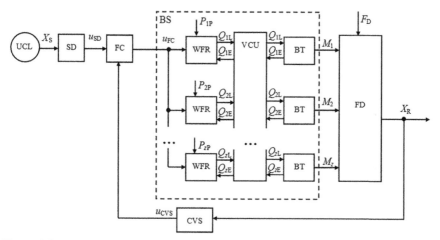

Figure 8.3 Functional structure of the generalized system of automatic control of the main parameters of the floating dock.

trim, while performing docking operations. The functional structure of the generalized system of an automatic control of the main parameters of the docking operations of the floating dock is shown in Figure 8.3 [54], where the following notations are adopted: UCL is the upper control level; SD is the setting device; FC is the fuzzy controller; BS is the ballast system; X_S, X_R are set and real controlled value; WFR is the water flow regulator; VCU is the valve control unit; BT is the ballast tank; CVS is the control value sensor; u_{SD}, u_{FC}, and u_{CVS} are the output signals of SD, FC, and CVS, respectively; Q_{iL} and Q_{iE} are the values of water flow in the g-th filling and emptying line

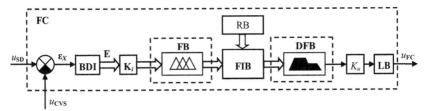

Figure 8.4 Structure of the FC.

$(g = 1, \ldots, z)$; P_{gp} is the overpressure created by the g-th pump $(g = 1, \ldots, z)$; M_g is the ballast mass in the g-th tank; F_D are the disturbances that effect on the hull of the floating dock.

The purpose of the control is to maintain a preset value of the controlled X_R value, in conditions of action on the object of controlled and uncontrolled disturbing influences. Among the main controlled parameters of the docking operations X the most important for the floating dock is draft H, list γ and trim θ [54]. The control of the above parameters is carried out in a closed-loop control system using an FC. Moreover, depending on the docking operation and the control parameter X, while controlling the floating dock, certain ballast compartments are involved, the configuration of which is defined in the VCU.

FC includes a fuzzy inference system of the Mamdani type and certain auxiliary blocks. The structure of FC is shown in Figure 8.4, where the following notations are accepted: BDI is the block of signals differentiation and integration; FB is the fuzzification block; FIB is the fuzzy inference block; DFB is the defuzzification block; LB is the limiting block; ε_X is the control error; **E** is the vector of input variables of the FC; $\mathbf{K}_i$ is the vector of coefficients of proportionality for the input variables of the FC; K_u is the coefficient of proportionality for the output variable u_{FC}.

The FC provides control of the corresponding docking operations parameter of the floating dock on the principle of feedback.

The BDI receives an error signal ε_X and, with the help of the embedded integrals and differentials of different orders, outputs a vector signal **E**, that may consist of error signal ε_X, its first $\dot{\varepsilon}_X$ and second $\ddot{\varepsilon}_X$ derivatives, integral $\int \varepsilon_X dt$, etc. The output variable of the FC, in turn, is the signal u_{FC}. In order to normalize the input signals of the FC and to convert them into relative units, according to the maximum values, the vector of coefficients of proportionality $\mathbf{K}_i$ is used, which may include coefficients K_P, K_D, K_{2D}, K_I, etc. For the implementation of the control signal $u_{FC,}$ additional adjustment and limiting the proportional coefficient K_u and the LB are applied, respectively.

At developing of the FC of the Mamdani type of the ACSs of the floating docks docking operations, it is advisable to use the proposed by the authors design approach.

8.3 Combined Design Approach of the FCs for the ACSs of the Floating Docks Main Operations

At designing of the effective FCs for controlling of the complex technical plants, it is reasonable to use expert knowledge and assessments in conjunction with certain optimization procedures. In particular, at the initial stage of design, it is advisable to synthesize FC structure and parameters on the basis of expert assessments and recommendations of operators. Then, for further improvement of the control quality, energy and economic efficiency of the ASC, it is necessary to perform specialized procedures of optimization of the FC parameters and structure on the basis of the corresponding methods of mathematical programming [55–66]. Herewith, it is advisable to carry out structural and parametric optimization of the FC based on the desired transients of the ACS, until the optimal value of the goal function is reached [67, 68]. After that, it is expedient to conduct reduction of the FC RB on the basis of calculating the influence of each rule on the control process to significantly simplify its further software and hardware implementation [51, 52, 56]. Application of such an approach allows to provide rather high-quality indicators of control as well as high operational efficiency of ACSs of the complex technical plants at a relatively simple implementation of their FCs.

The combined approach proposed by the authors is universal and can be used for the FCs designing of the docking operations ACSs for the floating docks of different types and sizes. It takes into account the above recommendations and consists of the following steps.

Step 1. Selection of the ACS goal function type and optimal value. The type of the ACS goal function and its optimal value is selected at this stage on the basis of desired transients of the main control parameters of the docking operations. In turn, the desired transients of the ACS of the control parameters of the floating dock docking operations should have the minimum duration and aperiodic character, which corresponds to a high-quality indicators control, low energy consumption and, consequently, high economic efficiency. In turn, the desired transient characteristic $X_D(t)$ can be obtained on the basis of the reference model (RM) of the ACS of the floating dock docking operations, which has the transfer function $W_{RM}(s)$

$$W_{RM}(s) \quad \frac{X_D(s)}{X_S(s)} \quad \frac{1}{(T_{RM}s + 1)^v}, \tag{8.1}$$

where T_{RM} and v are the time constant and order of the transfer function of the RM; s is the Laplace operator.

At the designing of the FC on the basis of desired transients of the control parameters of the floating dock docking operations in many cases, it is advisable to choose the mean integral quadratic deviation $I(t)$ of the real transient characteristic of the ACS $X_R(t)$ from the desired $X_D(t)$ [54] as the goal function

$$I(t) \quad \frac{1}{t_{max}} \int_0^{t_{max}} (X_D(t) - X_R(t))^2 dt, \tag{8.2}$$

where t_{max} is the total time of the ACS transient of the docking operations main control parameters.

The values of the time constant T_{RM} and order v of the RM as well as the optimal value of the goal function I_{opt} are determined experimentally for each particular floating dock.

Step 2. Selection of the input and output variables of the FC. The vector of input variables and output variable of the FC is selected at this stage. In turn, the input variables vector **E** in general form can be represented as follows

$$\mathbf{E} \quad \{\mathbf{E}_i\}, \quad i \quad \{1, \ldots, n\}, \tag{8.3}$$

where i is the number of input variable of the FC; n is the total number of variables of the FC.

Depending on the optimal value of the goal function I_{opt} of the floating dock docking operations ACS, the vector of the input variables of the FC can be represented by the expressions (8.4–8.8)

$$\mathbf{E} \quad \{\varepsilon_X, \dot{\varepsilon}_X\}; \tag{8.4}$$

$$\mathbf{E} \quad \left\{\varepsilon_X, \int \varepsilon_X dt\right\}; \tag{8.5}$$

$$\mathbf{E} \quad \{\varepsilon_X, \dot{\varepsilon}_X, \ddot{\varepsilon}_X\}; \tag{8.6}$$

$$\mathbf{E} \quad \left\{\varepsilon_X, \dot{\varepsilon}_X, \int \varepsilon_X dt\right\}; \tag{8.7}$$

$$\mathbf{E} \quad \left\{\varepsilon_X, \dot{\varepsilon}_X, \ddot{\varepsilon}_X, \int \varepsilon_X dt\right\}. \tag{8.8}$$

The signal u_{FC} is usually selected as the output variable of the FC.

Initially, the vector **E** is selected in the form of expression (8.4) and further development of the FC is carried out according to all next steps of the approach. If the optimal value of the goal function is not achieved with such an FC configuration, and a return to this step is performed, then the vector **E** is selected as the expression (8.5) and further FC designing is carried. At the next returning to this step, the expression (8.6) is selected as the vector **E** and so on up to expression (8.8), until the optimal value of the goal function I_{opt} is reached.

Step 3. Preliminary determination of the vector of coefficients of proportionality and setting of the operating ranges for the input and output variables of the FC. The vector of coefficients of proportionality **K** for the input and output variables of the docking operations FC and its initial values are determined at this stage [54]. Also, the operating ranges, within which the input and output variables may change during its operation, are set. In general, the vector of coefficients **K** can be represented as follows

$$\mathbf{K} = \{\mathbf{K}_i, K_u\}, \quad i = \{1, \ldots, n\}, \tag{8.9}$$

where $\mathbf{K}_i$ is the coefficients vector for the input variables **E**; K_u is the coefficient for the output variable u_{FC}.

Depending on the vector of the FC input variables **E**, selected at the *Step 2* the vector **K** can be represented by the expressions (8.10–8.14)

$$\mathbf{K} = \{K_P, K_D, K_u\}; \tag{8.10}$$
$$\mathbf{K} = \{K_P, K_I, K_u\}; \tag{8.11}$$
$$\mathbf{K} = \{K_P, K_D, K_{2D}, K_u\}; \tag{8.12}$$
$$\mathbf{K} = \{K_P, K_D, K_I, K_u\}; \tag{8.13}$$
$$\mathbf{K} = \{K_P, K_D, K_{2D}, K_I, K_u\}; \tag{8.14}$$

The initial values of the proportionality coefficients $\mathbf{K}_{i0}$ and K_{u0} for the input variables vector $\mathbf{E}_i$ ($i = 1, \ldots, n$) and output variable u_{FC} are determined on the basis of their maximum allowable values $\mathbf{E}_{imax}$ and u_{FCmax}

$$\mathbf{K}_{i0} = \frac{1}{\mathbf{E}_{i\,max}}, \quad i = \{1, \ldots, n\}; \tag{8.15}$$
$$K_{u0} = u_{FCmax}. \tag{8.16}$$

Since the input variables $\mathbf{E}_i$ ($i = 1, \ldots, n$) with the previous multiplication by the corresponding coefficients $\mathbf{K}_i$ ($i = 1, \ldots, n$) arrive to the FC input in relative units from the maximum value and can have both positive and negative values, then their operating ranges in the FC should be set from − 1 to 1. In turn, it is also advisable to set the operating range of the output variable u_{FC} from −1 to 1, as it is further multiplied by a corresponding to its maximum value coefficient K_u.

Step 4. Selection of the number of linguistic terms for the input and output variables of the FC. The values of the number of linguistic terms m_i for each i-th input variable $(i = 1, ..., n)$ and w for the output variable u_{FC} of the FC are selected at this stage. It is expedient to select an odd number of linguistic terms m_i and w, which, depending on the regulated parameter q, can be calculated by the formula

$$m_i \quad w \quad q \cdot 2 + 1. \tag{8.17}$$

Initially, the parameter q is set equal to one $(q \quad 1)$ and further development of the FC is carried out according to all next steps of the approach. If the optimal value of the goal function is not achieved with such an FC configuration, and a return to this step is performed, then the parameter q is set equal to two $(q \quad 2)$ and further FC designing is carried out. At the next returning to this step, the parameter q is set equal to three $(q = 3)$ and so on up to $q = 4$ until the optimal value of the goal function is reached. If at $q = 4$, the optimal value of the goal function is not achieved, then the transition to *Step 2* is performed.

Step 5. Selection of the LTMF types and preliminary determination of their parameters vector for the input and output variables of the FC. The types of the LTMF for each i-th input and output variable of the FC are selected at this stage. The triangular, trapezoidal, Gaussian of 1st and 2nd types, π-like and others types of LTMF can be selected for the input **E** and output u_{FC} variables of the FC for the ACS of docking operations main control parameters [20–24]. The vector **P** of the parameters (vertices) of the LTMF of the input and output variables of the FC and its initial values $\mathbf{P}_0$ are also determined on the basis of knowledge of experts or operators at this stage. In turn, the vector of the LTMF parameters has the form

$$\mathbf{P} \quad \{\mathbf{P}_{j,k}^{E_i}, \mathbf{P}_{j,k}^{u_{FC}}\}, \quad i \quad \{1,\ldots,n\}, \quad j \quad \{1,\ldots,m\}, \quad k \quad \{1,\ldots,l\}, \tag{8.18}$$

where $\mathbf{P}_{j,k}^{E_i}$ is the parameters vector of the LTMF of the input variables **E**; $\mathbf{P}_{j,k}^{u_{FC}}$ is the parameters vector of the LTMF of the output variable u_{FC}; j is the term number of the certain input or output FC variable; k is the parameter (vertex) number of the j-th term; l is the total amount of parameters (vertices) of the j-th term.

At the absence of expert knowledge, the linguistic terms of the input and output variables of the FC can be evenly distributed in their operating ranges, selected at *Step 3*.

Step 6. Preliminary synthesis of the rule base of the FC. The preliminary synthesis of the RB of the FC is carried out on the basis of the input **E** and output u_{FC} variables, as well as their linguistic terms selected at *Steps 2* and *4*, respectively, at this stage. The total number of rules s is determined by the number of all possible combinations of the linguistic terms of the input variables **E** of the FC [24]

$$s \quad \prod_{i=1}^{n} m_i. \tag{8.19}$$

The consequents for each r-th rule ($r = 1, ..., s$) are selected by the expert or operator from a set of linguistic terms of the output variable u_{FC} of the FC, the number of which w is chosen at *Step 4*.

Step 7. Selection of computational procedures of fuzzy inference and defuzzification method of the FC. The selection of operations of aggregation, activation, and accumulation of fuzzy inference as well as defuzzification method of the FC is carried out at this stage. As an aggregation operation, for example, operations "min" or "prod" can be selected [23]. As an activation operation the operation "max" can be selected and as an accumulation operation – "max" or "sum" [21]. Also, one of the following methods can be selected as the defuzzification method of the FC of docking operations ACS: the gravity center method, the square center method, the left modal value method, the right modal value method, and the other [22].

Step 8. Calculation and estimation of the value of the ACS goal function with the FC current configuration. The calculation and estimation of the value of the ACS goal function I with the FC current configuration is carried out at this stage. If the optimal value of the goal function is reached ($I = I_{opt}$), then the transition to *Step 10* is performed. In the opposite case, the transition to *Step 9* is carried out.

Step 9. Parametric optimization of the FC. The optimization procedures of the vector of coefficients **K** and vector of the LTMF parameters **P** of the input and output variables of the FC are carried out at this stage on the basis of the ACS goal function I and methods of gradient descent of numerical optimization.

To optimize the vectors $\mathbf{K}$ and $\mathbf{P}$ of the LTMF parameters, the goal function I and its optimal value I_{opt}, the initial values (initial hypothesis $\mathbf{K}_0$ and $\mathbf{P}_0$) and the constraints of the given vectors $[\mathbf{K}_{min}, \mathbf{K}_{max}]$ and $[\mathbf{P}_{min}, \mathbf{P}_{max}]$ are selected. Also, the type of the procedures of the iterative optimization is chosen [48].

The goal function I is calculated by the equation (8.2) and its optimal value I_{opt}, is previously selected at *Step 1*. As the initial values (initial hypothesis) of the vectors $\mathbf{K}$ and $\mathbf{P}$, the values $\mathbf{K}_0$ and $\mathbf{P}_0$ are selected, which are previously determined at *Steps 3* and *5*, respectively. The constraints for the possible values of the vector $\mathbf{K}$ of the coefficients of proportionality are determined at this step. As the constraints for the possible values of the vector $\mathbf{P}$ of the LTMF parameters it is advisable to select the boundaries of the operating ranges of the input and output variables of the FC, which are set at *Step 3*.

It is expedient to conduct the iterative procedures of the vectors $\mathbf{K}$ and $\mathbf{P}$ optimization on the basis of the vector equations (8.20) and (8.21) according to the method of gradient descent [54]

$$\mathbf{P}[\tau + 1] \quad \mathbf{P}[\tau] - \gamma[\tau]\frac{\partial I(\mathbf{P})}{\partial \mathbf{P}}\bigg|\mathbf{P}[\tau]; \tag{8.20}$$

$$\mathbf{K}[\tau + 1] \quad \mathbf{K}[\tau] - \gamma[\tau]\frac{\partial I(\mathbf{K})}{\partial \mathbf{K}}\bigg|\mathbf{K}[\tau], \tag{8.21}$$

where γ is the vector of steps of the gradient descent and τ is the iteration number. In turn, the vector γ can be determined in different ways depending on the chosen method of gradient descent (with a fixed step, with a crushed step, the fastest descent, etc.) [54].

These iterative procedures are carried out before fulfilling the condition of completion of the optimization of the vector of coefficients $\mathbf{K}$ and vector of the LTMF parameters $\mathbf{P}$ of the input and output variables of the FC. The condition of completion of the optimization procedures can be considered to be fulfilled in the following cases: (a) if the optimal value of the goal function is reached ($I = I_{opt}$); (b) if the maximum number of iterations is performed ($\tau = \tau_{max}$); (c) in the case when for a certain number of iterations τ the value of the goal function I has not decreased. If the condition (a) is fulfilled, the transition to *Step 10* is performed. In turn, if the conditions (b) or (c) are fulfilled, the transition to *Step 4* is carried out.

Step 10. Reduction of the RB of the FC. The reduction of the RB of the FC is carried out at this stage on the basis of the evaluation of the influence of each

r-th rule (r = 1, ..., s) on the control process and further exclusion of those rules whose influence is negligible. The procedure of reduction of the FC RB consists of identifying the influence degree of the RB rules on the output signal u_{FC} of the FC, forming the corresponding ranking series of rules **R** by the reduction of this parameter, as well as exclusion of those rules from the RB, whose influence on the formation of the signal u_{FC} is insignificant [54].

The identification of the influence degree of the RB rules on the output signal u_{FC} of the FC should be based on the calculation of changing of the truth degree of rules $\mu^R(t)$ in the control process and on the evaluation functional $G[\mu^R(t)]$, which argument is $\mu^R(t)$ [48].

The changing of the truth degree $\mu^R(t)$ of the r-th rule in the control process is presented by the function from time t

$$\mu_r^R(t) \quad \bigcap_{i=1}^{n} \mu_i^r(\mathbf{E}_i(t)) \quad \inf_{i=1}^{n} \mu_i^r(\mathbf{E}_i(t)), \tag{8.22}$$

where μ_i^r is the fuzzification result of the i-th input variable of the FC $\mathbf{E}_i$ (i = 1, ..., n) by the corresponding linguistic term of the r-th rule.

The evaluation functional $G[\mu^R(t)]$ of the influence of the r-th rule on the control process, in turn, is calculated according to the equation

$$G_r[\mu_r^R(t)] \quad \frac{1}{t_{max}} \int_0^{t_{max}} \mu_r^R(t)dt$$

$$\frac{1}{t_{max}} \int_{i=1}^{n} \inf_{i=1}^{n} \mu_i^r(\mathbf{E}_i(t))dt. \tag{8.23}$$

At the same time (for an adequate evaluation of the influence of the RB rules of the FC on the control process), it is expedient to carry out the simulation of the docking operations ACS in calculating changing of the truth degree $\mu^R(t)$ and evaluation functional $G[\mu^R(t)]$ of the RB rules in the time interval from 0 to t_{max} in all possible functioning modes (at submerging and surfacing with and without the vessel).

The ranking series of rules **R** consists of all the rules of the FC RB, located in the order of decreasing of their values of the calculated evaluation functional $G[\mu^R(t)]$ [50].

The exclusion of the RB rules, whose influence on the formation of the signal u_{FC} is insignificant, should be carried out consistently one by one rule (beginning from the end of the formed ranking series **R**) to a certain optimal value of their number s_{opt}, at which the value of the goal function I does not exceed the optimal value I_{opt}, and the quality indicators of the ACS of the main control parameters of the docking operations remain in the acceptable limits.

Step 11. Implementation of the developed FC using the appropriate element base. The structure of the FC will be simplified by the reduction the RB conducted at *Step 10.*

Step 12. Usage and experimental research of the developed FC in the ACS of the floating docks docking operations. The using and researching of the developed and optimized FC in the ACS of the main parameters of the floating docks main operations is performed at this stage.

The effectiveness study of the given above combined approach is conducted in the design process of the Mamdani type FC for the draft ACS of the floating dock for low-tonnage vessels.

8.4 Design of the FC for the Draft ACS of the Floating Dock for Low-Tonnage Vessels

The floating dock for low-tonnage vessels (yachts, towing, etc.) has 10 ballast tanks [54] and the following geometric parameters: width $a = 49$ m; length $b = 60$ m; width of each tower $c = 5$ m; height of the pontoon $d = 3$ m; height of each tower $h = 22$ m. The floating dock maximum weight is 3000 tons, and maximum load capacity is 5000 tons. The control of the current draft value is accomplished by filling or emptying the ballast tanks of the dock. Ballast system of the floating dock for low-tonnage vessels includes 12 ballast pumps with capacity of 1800 m³/h each, as well as 20 automated water flow regulators [54].

Since in this chapter the WFR is considered to be a closed-loop automatic control system of the ballast water flow rate with an optimally adjusted flow controller, then it is advisable to use the following transfer function as it mathematical model [54]

$$W_{\text{WFR}}(s) \quad \frac{Q_{\text{WFR}}(s)}{u_{\text{WFR}}(s)} \quad \frac{k_{\text{WFR}}}{T_{\text{WFR}}^2 s^2 + 2\zeta_{\text{WFR}} T_{\text{WFR}} s + 1}, \quad (8.24)$$

where Q_{WFR} and u_{WFR} are the output water flow and the input control signal of the WFR; k_{WFR}, T_{WFR}, ζ_{WFR} are WFR transfer function gain, time constant, and damping factor, respectively, which are determined by the controlled valve, drive electric motor, reduction gear and power amplifier parameters, which are the part of the closed-loop automatic control system of the ballast water flow.

The gain coefficient k_{WFR} can be considered to be linear and constant due to the optimal adjustments of the flow controller that is included in the WFR [54].

The damping factor ζ_{WFR} mostly dependents on the parameters of the WFR flow controller and at its optimal adjustments can be equal to one ($\zeta_{WFR} = 1$).

The time constant T_{WFR} depends on the controlled valve, drive electric motor, reduction gear, and power amplifier parameters as well as may vary at changing of the output pressure values of the ballast pumps [54].

The mathematical models of the floating dock ballast tanks as the water level and mass control objects are generally used for the mathematical and computer modelling of the floating docks draft control systems. Water level value L_{WBTg} of the g-th ballast tank is calculated by the following equation [54]

$$L_{WBT_g} = \frac{V_{WBT_g}}{S_{BT_g}} = \frac{1}{S_{BT_g}} \int (Q_{WFRg1} - Q_{WFRg2})dt$$

$$= \frac{(Q_{WFRg1} - Q_{WFRg2})}{S_{BT_gs}}, \tag{8.25}$$

where V_{WBTg} is the ballast water volume value in the g-th ballast tank; S_{BTg} is the base area of the g-th ballast tank. Thus, the ballast tank transfer function as the water level control object has the following form

$$W_{BT_g}(s) = \frac{L_{WBT_g}(s)}{Q_{WFRg1}(s) - Q_{WFRg2}(s)} = \frac{1}{T_{BT_gs}}, \tag{8.26}$$

where T_{BTg} is the time constant of the g-th ballast tank transfer function, $T_{BTg} = S_{BTg}$.

Water mass value M_{WBTg} of the g-th ballast tank is calculated by the following equation

$$M_{WBT_g} = \rho_W \int (Q_{WFRg1} - Q_{WFRg2})dt = \frac{\rho_W(Q_{WFRg1} - Q_{WFRg2})}{s}, \tag{8.27}$$

where ρ_W is the ballast water density. Thus, the ballast tank transfer function as the water mass control object has the following form [54]

$$W_{BT_g}(s) = \frac{M_{WBT_g}(s)}{Q_{WFRg1}(s) - Q_{WFRg2}(s)} = \frac{K_{BT}}{s}, \tag{8.28}$$

where k_{BT} is gain coefficient of the ballast tank transfer function, which is the same for all the floating dock's ballast tanks, $k_{BT} = \rho_W$.

In this chapter, it is advisable to use the transfer function given in the equation (8.28) as the floating dock water ballast tank mathematical model.

The floating dock moves in the upright position relative to the surface of the water during submerging and surfacing operations, and accordingly must have required buoyancy at different distances. Buoyancy is due to the effect on the floating dock of the forces of its own weight, together with the weight of the vessel and the pressure of water.

In the static mode, the gravitational force of the P_{FD} acting on the body of the floating dock, which is located in the water, is balanced by the buoyancy force of the D_A [54]

$$P_{FD} \quad D_A \quad \gamma V_{FD}, \tag{8.29}$$

where V_{FD} is the volumetric displacement (volume of immersed parts of floating dock and vessel) and γ is the specific gravity of seawater.

The equation of the uneven vertical motion of a floating dock can be described as follows [54]

$$(M_\Sigma + \lambda)\frac{d^2 H}{dt^2} \quad P_{FD} - R_R - D_A, \tag{8.30}$$

where R_R is the hydrodynamic resistance of water; λ is attached mass of the floating dock; M_Σ is the total mass of the floating dock, which consists of the mass M_{FD} of the body of the floating dock, mass of the vessel M_S, the ballast water of the $M_{\Sigma BT}$ of all tanks; H is the draft of the floating dock.

The hydrodynamic force of the liquid resistance consists of the friction R_{FR} and resistance of the form R_{FO}

$$R_R \quad R_{FR} + R_{FO}. \tag{8.31}$$

Frictional resistance is the sum of the resistance: the resistance, which is due to the curvature and roughness of the surface of the floating dock and the friction resistance of the technically smooth equivalent plate [54]

$$R_{FR} \quad (k_{FR} + k_R)\frac{\rho}{2}\left(\frac{dH}{dt}\right)^2 \Omega, \tag{8.32}$$

where k_{FR} is the dimensionless coefficient of friction resistance of a technically smooth plate; k_R is the dimensionless coefficient of resistance surface roughness of the floating dock; Ω is wetted surface of the floating dock.

Resistance of the form is determined by the formula:

$$R_{FO} = k_{FO} \frac{\rho}{2} \left(\frac{dH}{dt} \right)^2 \Omega, \tag{8.33}$$

k_{FO} is the dimensionless coefficient of resistance of the form.

The coefficient of friction resistance k_{FR} of a technically smooth plate and coefficient of resistance of the form k_{FO} depends on the Reynolds number Re [54]. The dimensionless coefficient of resistance surface roughness of the hull can be $k_{FO} = (0.3\ldots0.6)10^{-3}$. The mathematical model of the floating dock for low-tonnage vessels, which is used in the design process of the FC, consists of the equations (8.24), (8.28) and (8.30–8.33).

The type of the goal function of the draft ACS is selected at *Step 1* of the proposed approach according to the equation (8.2) and has the form of mean integral quadratic deviation $I(t, \mathbf{E}, \mathbf{K}, \mathbf{P}, q)$ of the real transient characteristic of the draft ACS $H_R(t, \mathbf{E}, \mathbf{K}, \mathbf{P}, q)$ from the desired $H_D(t)$

$$I(t, \mathbf{E}, \mathbf{K}, \mathbf{P}, q) = \frac{1}{t_{max}} \int_0^{t_{max}} (H_D(t, \mathbf{E}, \mathbf{K}, \mathbf{P}, q) - H_R(t, \mathbf{E}, \mathbf{K}, \mathbf{P}, q))^2 dt. \tag{8.34}$$

In turn, its optimal value is determined to be equal 1.5 ($I_{opt} = 1.5$) for the given floating dock. The desired transients of the draft ACS of the given floating dock should have a duration of about 3,200 seconds and aperiodic character. The given desired transient characteristic can be obtained on the basis of the reference model of the draft ACS with the transfer function (8.1), parameters of which are defined and have the following values: $T_{RM} = 500$ s, $\nu = 2$.

Initially, the vector $\mathbf{E}$ of the FC input is selected according to the expression (8.4) at *Step 2* of the proposed method. It is presented in the form

$$\mathbf{E} = \{\varepsilon_H, \dot{\varepsilon}_H\}. \tag{8.35}$$

The vector of coefficients of proportionality $\mathbf{K}$ for the input and output variables of the FC of the draft ACS, presented by the expression (8.10), and its initial values $\mathbf{K}_0$ are determined at *Step 3*. In turn, the initial values of the coefficients K_{P0} and K_{D0} for the input variables of the FC ε_H and $\dot{\varepsilon}_H$ are calculated by the formula (8.15) and are equal to 7 and 0.003, respectively ($K_{P0} = 7$; $K_{D0} = 0.003$). The initial value of the coefficient K_{u0} for the output variable u_{FC} is calculated by the formula (8.16) and is equal to 10 ($K_{u0} = 10$). Also, the operating ranges of the input $\mathbf{E}$ and output u_{FC} variables of the draft FC are set from −1 to 1 at *Step 3*.

Initially, the parameter q is set equal to one ($q = 1$) at *Step 4*. Thus, 3 linguistic terms ($m_i = w = 3$) are selected for each input and output variables of the given FC in accordance with the formula (8.17). They are: N – negative; Z – zero; P – positive.

For all of the above terms the triangular type of membership functions is selected at *Step 5*. The vector **P** of the parameters (vertices) of the LTMF of the input and output variables of the FC, that is presented in general form by the expression (8.18), and its initial values P_0 are also determined at this step. In this case, j {1, 2, 3} and k {1, 2, 3}. In turn, the initial values of the vector of the LTMF parameters P_0 are defined in such a way that the linguistic terms of the input **E** and output u_{FC} variables of the draft FC are evenly distributed in their operating ranges. The appearance of the LTMF of the input and output variables of the FC with the specified at this stage parameters is shown in Figure 8.5.

The preliminary synthesis of RB of the draft FC is carried out at *Step 6*, on the basis of the input **E** and output u_{FC} variables, as well as their linguistic terms selected at *Steps 2* and *4*, respectively. The total number of rules s, according to the equation (8.19), in this case is 9. Each rule of the synthesized RB is the linguistic statement, represented by the expression (8.36)

$$\text{IF ``}\varepsilon_H \quad x_1\text{'' AND ``}\dot{\varepsilon}_H \quad x_2\text{'' THEN ``}u_{FC} \quad y\text{''},\qquad(8.36)$$

where x_1, x_2, y, are the corresponding values of the linguistic terms.

The RB of the FC, synthesized at this stage, is presented in Table 8.1.

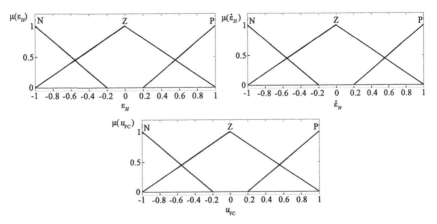

Figure 8.5   FC linguistic terms and their parameters at $m_i = w = 3$.

Table 8.1   Rule base of the FC at $m_i = w = 3$

		Rate of Error Change, $\dot{\varepsilon}_H$		
		N	**Z**	**P**
	N	N	N	N
Error, ε_H	**Z**	N	Z	P
	P	P	P	P

The computational procedures of the fuzzy inference and defuzzification method of this FC are selected at *Step 7*. In this case, the operation "min" is selected as an aggregation operation, the operation "max" is selected for both activation and accumulation operations as well as the gravity center method is chosen as the defuzzification method.

The calculation and estimation of the value of the goal function I of the draft ACS with the current configuration of its FC is carried out at *Step 8*. The simulation of transients of the draft ACS of the floating dock is carried out at all possible operation modes (at submerging and surfacing with and without the vessel) at calculating the value of the goal function at this stage. The goal function value the for the current configuration of the FC $I = 3.2$, that is much larger than the given optimal value I_{opt}. Therefore, the transition to *Step 9* is carried out.

The optimization procedures of the vector of coefficients **K** and vector of the LTMF parameters **P** of the input and output variables of the FC are carried out at *Step 9* on the basis of the methods of gradient descent of numerical optimization in order to increase the quality indicators of control as well as to reduce the value of the goal function I of the draft ACS. The goal function I, presented by the equation (8.34), and its optimal value I_{opt} for the optimization procedures are previously selected at *Step 1*. As the initial values (initial hypothesis) of the vectors **K** and **P** the values **K₀** and **P₀** are selected, which are previously determined at *Steps 3* and *5*, respectively. The boundaries of the operating ranges of the input and output variables of the FC, which are set at *Step 3*, are selected as the constraints for the possible values of the vector **P** of the LTMF parameters. In turn, the given boundaries are selected as the constraints for the possible values of the vector **K**

$$K_P \in [0.05; 10];$$
$$K_D \in [0.001; 0.015];$$
$$K_u \in [0.05; 20];$$

The iterative procedures of optimization of the vectors **K** and **P** are carried out on the basis of the vector equations (8.20) and (8.21) according to the

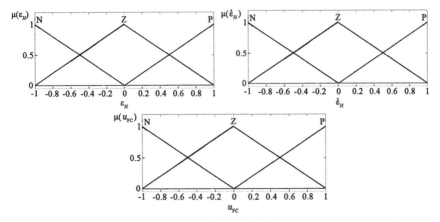

Figure 8.6   FC linguistic terms and their optimized parameters at $m_i = w = 3$.

gradient method of the fastest descent. The optimized values of the vector **K** of the coefficients of proportionality are: $K_P = 6.22$; $K_D = 0.00236$; $K_u = 12.73$.

The appearance of the LTMF of the input and output variables of the draft FC with the parameters optimized at *Step 9* is shown in Figure 8.6.

The computer simulation of transients and comparative analysis of the quality indicators of the ACS of the floating dock draft with the developed FC and optimally tuned conventional PD-controller are carried out to verify the efficiency of the synthesized at this stage draft FC.

The transfer function $W_{PD}(s)$ of the conventional PD-controller is represented by the expression

$$W_{PD}(s) \quad K_{P1} + \frac{K_{D1}s}{T_F s + 1}, \tag{8.37}$$

where K_{P1}, K_{D1} and T_F are the coefficients and time constant of the filter of the conventional PD-controller, which are also found in the process of parametric optimization based on the desired transients using the gradient method of the fastest descent. In turn, $K_{P1} = 2.319$; $K_{D1} = 246.2$; $T_F = 1.125$.

The quality indicators comparative analysis of the ACS of the floating dock draft with the designed FC and optimally tuned conventional PD-controller are presented in the Table 8.2, where t_t is the transient time.

The transients of the draft ACS at the floating dock submerging are presented in Figure 8.7.

Table 8.2   Quality indicators comparative analysis of the ACS of the floating dock draft

Quality Indicators	ACS Quality Indicators Values			
	RM	Optimally Tuned Conventional PD-controller	Non-Optimized FC at $m_i = w = 3$	Optimized FC at $m_i = w = 3$
t_t, s	3100	3480	47800	3270
I	0	2.35	3.2	1.96

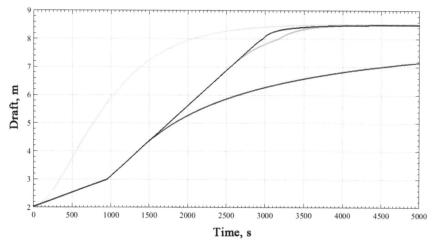

Figure 8.7   ACS transients at the floating dock submerging: yellow line – RM; blue line – with optimally tuned conventional PD-controller; red line – with non-optimized FC at $m_i = w = 3$; green line – with optimized FC at $m_i = w = 3$.

As can be seen from Table 8.2 and Figure 8.7, the ACS with optimized FC at $m_i = w = 3$, developed at this stage, has worse quality indicators than ACS with optimally tuned conventional PD-controller and RM. In turn, the values of the goal function I for three ACSs are much larger than the given optimal value I_{opt}. In turn, the overshoot and static error have zero values. The change in slope of the ACS transients at the floating dock submerging is due to its significant nonlinearity as a control object. Up to a draft of 3 meters ($H < 3$), the dock's submerging is relatively slow, since at this stage a large in volume pontoon is submerging. After that, when the pontoon has already submerged (at $H > 3$), the submerging speed becomes much faster, since the smaller in since towers of the dock are submerging.

At conducting the given above iterative procedures the case (c) of condition of the optimization completion has been fulfilled (when for a certain

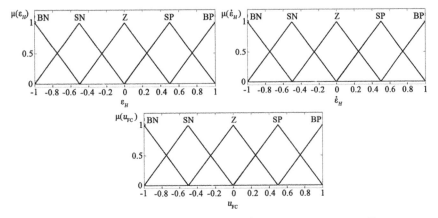

Figure 8.8   FC linguistic terms and their parameters at $m_i = w = 5$.

number of iterations τ the value of the goal function I has not decreased). Thus, the return to *Step 4* is carried out.

At the return to *Step 4*, the parameter q is set equal to two ($q = 2$). Thus, 5 linguistic terms ($m_i = w = 5$) are selected for each input and output variables of the given FC in accordance with the formula (8.17). They are BN – big negative; SN – small negative; Z – zero; SP – small positive; BP – big positive.

For all of the above terms, the triangular type of membership functions is also selected at *Step 5*. The vector **P** of the parameters (vertices) of the LTMF of the input and output variables of the FC, that is presented in general form by the expression (8.18), and its initial values P_0 are also determined at this step. In this case, j $\{1, \ldots, 5\}$ and k $\{1, 2, 3\}$. In turn, the initial values of the vector of the LTMF parameters P_0 are defined in such a way that the linguistic terms of the input **E** and output u_{FC} variables of the draft FC are evenly distributed in their operating ranges. The appearance of the LTMF of the input and output variables of the FC with the specified at this stage parameters is shown in Figure 8.8.

The synthesis of RB of the draft FC is carried out at *Step 6*, on the basis of the input **E** and output u_{FC} variables, as well as their linguistic terms selected at *Steps 2 and 4*, respectively. The total number of rules s, according to the equation (8.19), in this case is 25. Each rule of the synthesized RB is also presented by the linguistic statement, represented by the expression (8.36). The RB of the FC, synthesized at this stage, is presented in Table 8.3.

The computational procedures of the fuzzy inference and defuzzification method of this FC are selected at *Step 7*. In this case, the operation "min"

Table 8.3 Rule base of the FC at $m_i = w = 5$

		Rate of Error Change, $\dot{\varepsilon}_H$				
		BN	SN	Z	SP	BP
	BN	BN	BN	BN	BN	SN
	SN	BN	BN	BN	SN	SN
Error, ε_H	Z	BN	SN	Z	SP	BP
	SP	SP	SP	BP	BP	BP
	BP	SP	BP	BP	BP	BP

is also selected as an aggregation operation, the operation "max" is selected for both activation and accumulation operations as well as the gravity center method is chosen as the defuzzification method.

The calculation and estimation of the value of the goal function I of the draft ACS with the current configuration of its FC is carried out at *Step 8*. The goal function value for the current configuration of the FC $I = 2.95$, which is also larger than the given optimal value I_{opt}. Therefore, the transition to *Step 9* is carried out.

The optimization procedures of the vector of coefficients **K** and vector of the LTMF parameters **P** of the input and output variables of the FC are carried out at *Step 9* fully as in the previous time on the basis of the gradient method of the fastest descent in order to increase the quality indicators of control as well as to reduce the value of the goal function I of the draft ACS.

The optimized values of the vector **K** of the coefficients of proportionality are $K_P = 6.493$; $K_D = 0.00418$; and $K_u = 14.796$.

The appearance of the LTMF of the input and output variables of the draft FC with the parameters optimized at *Step 9* is shown in Figure 8.9.

The computer simulation of transients and comparative analysis of the quality indicators of the ACS of the floating dock draft with the developed FC and optimally tuned conventional PD-controller are carried out to verify the efficiency of the synthesized at this stage draft FC.

In turn, transients and comparative analysis of the quality indicators of the ACS of the floating dock draft with the developed FC and optimally tuned conventional PD-controller are presented in Figure 8.10. and Table 8.4, respectively.

As can be seen from Table 8.4 and Figure 8.10, the value of the goal function is reduced to $I = 1.26$, that is less than the given optimal value I_{opt}, and the draft ACS with optimized FC at $m_i = w = 5$ has high enough quality indicators.

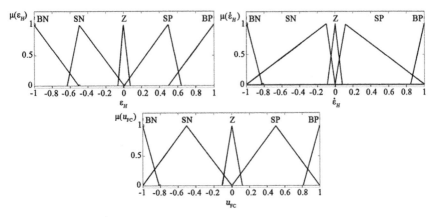

Figure 8.9   FC linguistic terms and their optimized parameters at $m_i = w = 5$.

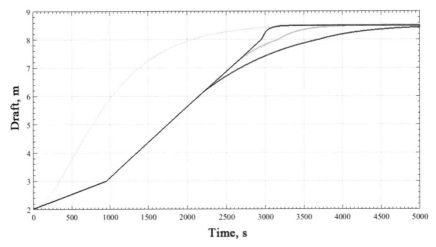

Figure 8.10   ACS transients at the floating dock submerging: yellow line – RM; green line – with optimally tuned conventional PD-controller; red line – with non-optimized FC at $m_i = w = 5$; blue line – with optimized FC at $m_i = w = 5$.

Thus, according to the proposed approach, as the optimal value of the goal function is reached ($I = I_{opt}$), the transition to *Step 10* is carried out for further reducing the RB of the designed draft FC in order to simplify its hardware and software implementation.

The calculating of changing of the truth degree $\mu^R(t)$ and evaluation functional $G[\mu^R(t)]$ of the influence of the RB rules of the FC on the draft control process is carried out at *Step 10* by means of the equations (8.22)

Table 8.4   Quality indicators comparative analysis of the ACS of the floating dock draft

| Quality Indicators | ACS Quality Indicators Values | | | |
	RM	Optimally Tuned Conventional PD-controller	Non-Optimized FC at $m_i = w = 5$	Optimized FC at $m_i = w = 5$
t_t, s	3100	3480	4280	3060
I	0	2.35	2.95	1.26

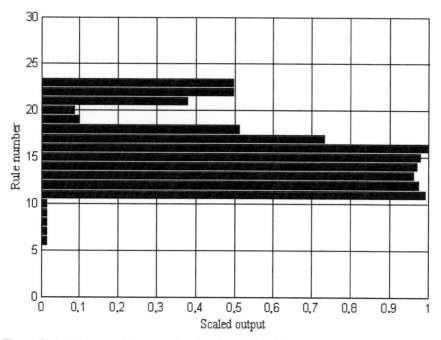

Figure 8.11   Diagram of the evaluation functional of the influence of RB rules of the FC on the control process of the floating dock draft.

and (8.23) at the simulation of the ACS of the floating dock in all possible functioning modes. The diagram of the evaluation functional $G[\mu^R(t)]$ of the influence of all the rules of the FC RB on the control process is presented in Figure 8.11.

The ranking series of rules **R**, which include all the rules of the FC RB located in the order of decreasing of their values of the calculated evaluation functional $G[\mu^R(t)]$ has the form

$$\mathbf{R} \quad \{16, 11, 15, 12, 14, 13, 17, 18, 22, 23, 21, 19, 20, 6, 7, 8, 9, 10,$$
$$1, 2, 3, 4, 5, 24, 25\}.$$

The exclusion of the RB rules, whose influence on the formation of the signal u_{FC} is insignificant, is carried out consistently one by one rule (beginning from the end of the formed ranking series **R**) to a certain optimal value of their number s_{opt}, at which the value of the goal function I does not exceed the optimal value I_{opt}, and the quality indicators of the ACS of the floating dock draft remain in the acceptable limits. Thus, the optimal value of the rules number $s_{opt} = 13$.

The transients and comparative analysis of the quality indicators of the ACS of the floating dock draft with the developed FC with full RB as well as with reduced (optimized) RB are presented in Figure 8.12 and Table 8.5, respectively.

As can be seen from Table 8.5 and Figure 8.12, the draft ACS with optimized FC with reduced at this stage RB has high enough quality indicators

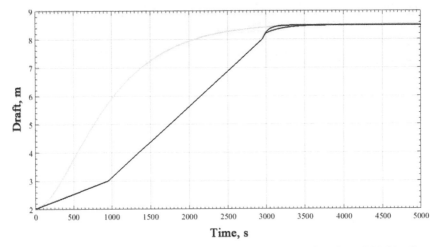

Figure 8.12   ACS transients at the floating dock submerging: yellow line – RM; blue line – with optimized FC at $m_i = w = 5$ with full RB; red line – with optimized FC at $m_i = w = 5$ with reduced RB.

Table 8.5   Quality indicators comparative analysis of the ACS of the floating dock draft with optimized FC with full and reduced RB

| | | ACS Quality Indicators Values | |
Quality Indicators	RM	Optimized FC at $m_i = w$ = 5 with Full RB	Optimized FC at $m_i = w = 5$ with Reduced RB
t_t, s	3100	3060	3130
I	0	1.26	1.43

and the value of its goal function lies within the optimum limits. Also, the software and hardware implementation of the developed and optimized FC at *Step 11* will be simplified by the reduction of the RB, conducted at *Step 10*, and it can be used for further experimental research in the draft ACS of the given floating dock at *Step 12*.

The application of the structural-parametric optimization procedures of the designed FC at *Step 4* and *Step 9* allowed to significantly increase the quality indicators of the draft ACS of the given floating dock and to reduce the goal function I to its optimal value I_{opt}. In turn, further reduction of the RB of the given FC at *Step 10* allowed to reduce the total number of the RB rules up to 13 rules without degrading the quality indicators of the draft ACS and significant increasing of the goal function value. This gives the opportunity to significantly simplify the software and hardware implementation of the designed FC in the draft ACS of this floating dock.

The above studies confirm the high efficiency of the proposed combined approach to design of fuzzy controllers for the floating docks main docking operations. The application of the developed on the basis of this approach FCs will provide high quality indicators of control of the main control parameters of the floating docks operations, high economic and operational indicators, low energy consumption, as well as high quality of the docking operations conducting.

8.5 Conclusion

The combined approach developed by the authors to design of fuzzy controllers for the floating docks main operations ACSs is presented in this chapter.

The proposed approach allows to carry out the synthesis of the structure and parameters of the FCs of Mamdani type for the ACSs of the main control parameters of the floating docks main operations using expert knowledge and assessments in conjunction with certain optimization procedures on the basis of mathematical programming methods. In particular, the structural and parametric optimization of the FCs is performed on the basis of the ACS desired transients and methods of gradient descent of numerical optimization, that allows to provide high quality indicators of control of the docking operations. In turn, the reduction of the FC RB is carried out on the basis of calculating the influence of each rule on the control process and allows to reduce the total number of rules of the RB without deterioration of the quality indicators of the docking operations control. This allows to significantly simplify the FC further software and hardware implementation.

For studying the effectiveness of the proposed approach the design of the Mamdani type FC for the draft ACS of the floating dock for low-tonnage vessels is carried out in this chapter. The designed FC has a relatively simple hardware and software implementation as well as allows to achieve high quality indicators of the draft control, high economic and operational indicators, low energy consumption, as well as high quality of the docking operations conducting, that confirms the high efficiency of the proposed approach.

Thus, the application of the FCs, designed on the basis of the given approach, in the docking operations ACSs of the floating docks of different types and sizes allows to provide high quality indicators of control of the main operating parameters, which implies high economic and operational efficiency, low energy consumption as well as reliability and safety of the docking operations conducting. Also, such FCs have a relatively simple software and hardware implementation.

References

[1] P. Y. Pavlov, A. N. Rogulin, 'Efficiency of operation of docks', Transport, 1987 (in Russian).

[2] A. Apostolidis, J. Kokarakis, A. Merikas: 'Modeling the Dry-Docking Cost: The Case of Tankers', Journal of Ship Production and Design, vol. 28, no. 3, pp. 134–143, 2012.

[3] A. Topalov, O. Kozlov, Y. Kondratenko, 'Control processes of floating docks based on SCADA systems with wireless data transmission', Proc. of the International Conference MEMSTECH, pp. 57–61, Ukraine, 2016.

[4] L. L. Vagushchenko, N. N. Tsymbal, 'Automatic ship movement control systems', Latstar, 2003 (in Russian).

[5] L. L. Vagushchenko, N. N. Tsymbal, 'Ship as an object of automatic control', Latstar, 2002 (in Russian).

[6] I. R. Freidzon, 'Mathematical modeling of automatic control systems on ships', Shipbuilding, 1969 (in Russian).

[7] A. S. Bolshev, E. V. Mikhalenko, S. A. Frolov, 'Mathematical modeling of the behavior of marine floating structures', Retrieved from URL: http://esg.spb.ru/files/content/files/All/AS_2006.pdf, (in Russian).

[8] Y. D. Zhukov, A. T. Ivanchenko, 'The use of simulation in automated controlled systems of floating docks', Scientific Methodical Journal

P. Mohyly Black Sea National University, vol. 61, no. 48, pp. 20–25, 2007 (in Russian).

[9] B. N. Gordeev Yu. D. Zhukov, A. V. Kurakin, S. B. Prikhodko, 'Automated remote control system for floating dock landing parameters', Modern Information and Energy-Saving Technologies of Human Life Support, pp. 258–260, no 5, 1999 (in Russian).

[10] Yu. Zhukov, B. Gordeev, A. Zivenko and A. Nakonechniy, 'Polymetric sensing in intelligent systems', Advances in intelligent robotics and collaborative automation, Y. P. Kondratenko and R. J. Duro (Eds.), vol. 1, pp. 211–234, River Publishers, 2015.

[11] Yu. D. Zhukov, B. N. Gordeev, Y. I. Logvinenko, E. O. Prischepov, 'Computerized maritime polymetric system', Proceedings of the Second International Conference on Marine Industry, pp. 245–252, Varna, Bulgaria, 1998.

[12] Y. P. Kondratenko, N. Y. Kondratenko, 'Reduced library of the soft computing analytic models for arithmetic operations with asymmetrical fuzzy numbers'. Soft Computing: Developments, Methods and Applications, Alan Casey (Ed), Series: Computer Science, technology and applications, NOVA Science Publishers, Hauppauge, pp. 1–38, 2016.

[13] Y. Kondratenko, V. Kondratenko, 'Soft computing algorithm for arithmetic multiplication of fuzzy sets based on universal analytic models', In book: Information and Communication Technologies in Education, Research, and Industrial Application. Communications in Computer and Information Science 469, Ermolayev, V. et al. (Eds): ICTERI'2014, pp. 49–77, Springer International Publishing Switzerland, 2014. Doi: 10.1007/978-3-319-13206-8_3

[14] Y. P. Kondratenko, N. Y. Kondratenko, 'Soft computing analytic models for increasing efficiency of fuzzy information processing in decision support systems' Chapter in book: Decision Making: Processes, Behavioral Influences and Role in Business Management, R. Hudson (Ed.), pp. 41–78, Nova Science Publishers, New York, 2015.

[15] V. Novak, I. Perfilieva et al., 'Mathematical principles of fuzzy logic', Dordrecht: Kluwer, 1999.

[16] I. Perfilieva, V. Novák, A. Dvořák, 'Fuzzy transform in the analysis of data' International Journal of Approximate Reasoning, vol. 48(1), Elsevier, pp. 36–46, 2008.

[17] I. Perfilieva, 'Fuzzy transform: application to reef growth problem', in: R. Demicco, G. J. Klir (Eds.), Fuzzy Logic in Geology, Academic Press, Amsterdam, pp. 275–300, 2003.

[18] J. R. Castro, O. Castillo, P. Melin, A. Rodríguez-Díaz, 'A hybrid learning algorithm for a class of interval type-2 fuzzy neural networks', Information Sciences, vol. 179(13), pp. 2175–2193, 2009.

[19] J. Börcsök, 'Fuzzy control. Theory and industry use', Verlag Technik, 2000 (in German).

[20] R. Hampel, M. Wagenknecht, N. Chaker (Eds.), 'Fuzzy control: theory and Practice', Physika-Verlag, Heidelberg, 2000.

[21] M. Jamshidi, N. Vadiee, T. J. Ross (Eds.), 'Fuzzy logic and control: software and hardware application', vol. 2. Series on Environmental and Intelligent Manufacturing Systems, Prentice Hall, 1993.

[22] A. Piegat, 'Fuzzy modeling and control'. Springer, Heidelberg, 2001.

[23] T. Takagi, M. Sugeno, 'Fuzzy identification of systems and its applications to modeling and control', IEEE Transactions on Systems, Man, and Cybernetics, vol. SMC-15, no. 1, pp. 116–132, 1985.

[24] R. R. Yager, D. P. Filev, 'Essentials of fuzzy modeling and control' Sigart Bulletin, vol. 6, no. 4, pp. 22–23, 1994.

[25] L. A. Zadeh, 'Fuzzy sets, information and control', no. 8, pp. 338–353, 1965.

[26] D. Rutkovskaya, M. Pilinsky, L. Rutkowski, 'Neural networks, Genetic algorithms and fuzzy systems' Hotline – Telecom, 2006, (in Russian).

[27] Y. P. Kondratenko, O. V. Kozlov, O. V. Korobko, 'Two modifications of the automatic rule base synthesis for fuzzy control and decision making systems', Chapter in a book: "Information Processing and Management of Uncertainty in Knowledge-Based Systems. Theory and Foundations", Medina, J., Ojeda-Aciego, M., Verdegay, J. L., Pelta, D. A., Cabrera, I. P., Bouchon-Meunier, B., Yager, R. R. (Eds.), Book Series: Communications in Computer and Information Science, vol. 854, pp. 570–582, Berlin, Heidelberg: Springer International Publishing, 2018. Doi: 10.1007/978-3-319-91476-3.

[28] Y. P. Kondratenko, O. V. Korobko, O. V. Kozlov, 'Synthesis and optimization of fuzzy controller for thermoacoustic plant', Recent Developments and New Direction in Soft-Computing Foundations and Applications, Studies in Fuzziness and Soft Computing, Lotfi A. Zadeh et al. (Eds.), vol. 342, pp. 453–467, Berlin, Heidelberg: Springer-Verlag, 2016.

[29] Y. Kondratenko, V. Korobko, O. Korobko, G. Kondratenko, O. Kozlov, 'Green-IT approach to design and optimization of thermoacoustic waste heat utilization plant based on soft computing', Chapter in a book: "Green IT Engineering: Components, Networks and Systems Implementation", vol. 105, pp. 287–311, Springer, Heidelberg, 2017.

[30] K. Michels, F. Klawonn, R. Kruse, A. Nürnberger, 'Fuzzy control: Fundamentals, stability and design of fuzzy controllers', Studies in Fuzziness and Soft Computing, pp. 1–405, 200, 2006.

[31] V. Novák, I. Perfilieva, J. Močkoř, 'Mathematical Principles of Fuzzy Logic', Kluwer, Boston, 1999.

[32] P. Melin, O. Castillo, A. Pownuk, O. Kosheleva, V. Kreinovich, 'How to gauge the accuracy of fuzzy control recommendations: A simple idea', Advances in Intelligent Systems and Computing, 648, pp. 287–292, 2018. Doi: 10.1007/978-3-319-67137-6_32.

[33] M. L. Lagunes, O. Castillo, J. Soria, 'Optimization of membership function parameters for fuzzy controllers of an autonomous mobile robot using the firefly algorithm', Studies in Computational Intelligence, 749, pp.199–206, 2018. Doi: 10.1007/978-3-319-71008-2_16.

[34] O. R. Carvajal, O. Castillo, J. Soria, 'Optimization of membership function parameters for fuzzy controllers of an autonomous mobile robot using the flower pollination algorithm', Journal of Automation, Mobile Robotics and Intelligent Systems, 12(1), pp. 44–49, 2018. Doi: 10.14313/JAMRIS_1-2018/6.

[35] M. Biglarbegian, W. W. Melek, J. M. Mendel, 'Design of novel interval type-2 fuzzy controllers for modular and reconfigurable robots: Theory and experiments', IEEE Transactions on Industrial Electronics. 58(4), 5464358, pp. 1371–1384, 2011. Doi: 10.1109/TIE.2010.2049718.

[36] A. W. Deshpande, D. V. Raje, 'Fuzzy logic applications to environment management systems: case studies', IEEE International Conference on Industrial Informatics (INDIN) 2003-January, 1300356, pp. 364–368, 2003. Doi: 10.1109/INDIN.2003.1300356.

[37] A. N. Trunov, 'Criteria for the evaluation of model's error for a hybrid architecture DSS in the underwater technology ACS', in Eastern-European Journal of Enterprise Technologies, vol. 6, no. 9(84), pp. 55–62, 2016.

[38] S. D. Shtovba, 'Ensuring accuracy and transparency of mamdani fuzzy model in learning by experimental data', Journal of Automation and Information Sciences, vol. 39(8), pp. 39–52, 2007.

[39] M. Pasieka, N. Grzesik, K. Kuźma, 'Simulation modeling of fuzzy logic controller for aircraft engines', International Journal of Computing, vol. 16, no. 1, pp. 27–33, 2017. Retrieved from http://computingonline.net/computing/article/view/868.

[40] Z. Gomolka, E. Dudek-Dyduch, Y. P. Kondratenko, 'From homogeneous network to neural nets with fractional derivative Mechanism. In: artificial intelligence and soft computing' Lecture Notes in Artificial Intelligence 10245, 16th International Conference ICAISC, pp. 52–63, Springer, 2017.

[41] K. Tanaka, H. O. Wang, 'Fuzzy control systems design and analysis: a linear matrix inequality approach' John Wiley & Sons, 2001.

[42] Y. P. Kondratenko, O. V. Kozlov, L. P. Klymenko, G. V. Kondratenko, 'Synthesis and research of neuro-fuzzy model of ecopyrogenesis multicircuit circulatory system', Advance Trends in Soft Computing, Studies in Fuzziness and Soft Computing, vol. 312, Berlin, Heidelberg: Springer-Verlag, pp. 1–14, 2014.

[43] Y. P. Kondratenko, O. V. Kozlov, O. S. Gerasin, Y. M. Zaporozhets, 'Synthesis and research of neuro-Fuzzy observer of clamping force for mobile robot automatic control system' Proceedings of the 2016 IEEE First International Conference on Data Stream Mining & Processing (DSMP), pp. 90–95, Lviv, Ukraine, 2016.

[44] D. Driankov, H. Hellendoorn, M. Reinfrank, 'An introduction to fuzzy control', Springer Science & Business Media, 2013.

[45] Y. P. Kondratenko, D. Simon, 'Structural and parametric optimization of fuzzy control and decision making systems', In: Zadeh L., Yager R., Shahbazova S., Reformat M., Kreinovich V. (eds), Recent Developments and the New Direction in Soft-Computing Foundations and Applications. Studies in Fuzziness and Soft Computing, vol. 361, Springer, pp 273–289, 2018. Doi: https://doi.org/10.1007/978-3-319-75408-6_22.

[46] B. Jayaram, 'Rule reduction for efficient inferencing in similarity based reasoning', International Journal of Approximate Reasoning, vol. 48, no. 1, pp. 156–173, 2008.

[47] Y. Yam, P. Baranyi, C.-T. Yang, 'Reduction of fuzzy rule base via singular value decomposition', IEEE Transactions on Fuzzy Systems, vol. 7, no. 2, pp. 120–132, 1999.

[48] Y. P. Kondratenko, E. Y. M. Al Zubi, 'The optimization approach for increasing efficiency of digital fuzzy controllers', Annals of DAAAM for 2009 & Proceeding of the 20th Int. DAAAM Symp. "Intelligent

Manufacturing and Automation", Published by DAAAM International, pp. 1589–1591, Austria, 2009.

[49] D. Simon, 'H∞ estimation for fuzzy membership function optimization', International Journal of Approximate Reasoning, pp. 224–242, 2005.

[50] D. Simon, 'Evolutionary optimization algorithms: biologically inspired and population-based approaches to computer intelligence', John Wiley & Sons, 2013.

[51] Y. P. Kondratenko, L. P. Klymenko, E. Y. M. Al Zu'bi, 'Structural optimization of fuzzy systems' Rules Base and Aggregation Models', Kybernetes, vol. 42(5), pp. 831–843, 2013. Doi: http://dx.doi.org/10.1108/K-03-2013-0053.

[52] D. Simon, 'Design and rule base reduction of a fuzzy filter for the estimation of motor currents', International Journal of Approximate Reasoning, no. 25, pp. 145–167, 2000.

[53] H. Ishibuchi, T. Yamamoto, 'Fuzzy rule selection by multi-objective genetic local search algorithms and rule evaluation measures in data mining', Fuzzy Sets and Systems, vol. 141, no. 1, pp. 59–88, 2004.

[54] Y. P. Kondratenko, O. V. Kozlov, O. V. Korobko, A. M. Topalov, 'Synthesis and optimization of fuzzy control system for floating dock's docking operations', Fuzzy Control Systems: Design, Analysis and Performance Evaluation, Wendy Santos (Eds.), Nova Science Publishers, pp. 141–215, 2017.

[55] L. T. Koczy, K. Hirota, 'Size reduction by interpolation in fuzzy rule bases', IEEE Transactions on Systems, Man, and Cybernetics, Part B: Cybernetics, vol. 27, no. 1, pp. 14–25, 1997.

[56] R. Alcalá, J. Alcalá-Fdez, M. J. Gacto, F. Herrera, 'Rule base reduction and genetic tuning of fuzzy systems based on the linguistic 3-tuples representation', Soft Computing, vol. 11, no. 5, pp. 401–419, 2007.

[57] W. Pedrycz, K. Li, M. Reformat, 'Evolutionary reduction of fuzzy rule-based models' Fifty Years of Fuzzy Logic and its Applications, STUDFUZ 326, Cham: Springer, pp. 459–481, 2015.

[58] Y. P. Kondratenko, V. L. Timchenko, 'Increase in navigation safety by developing distributed man-machine control systems' Proc. of the Third Intern. Offshore and Polar Engineering Conference, Singapore, vol. 2, pp. 512–519, 1993.

[59] V. L. Timchenko, Y. P. Kondratenko, 'Robust stabilization of marine mobile objects on the basis of systems with variable structure of feedbacks', Journal of Automation and Information Sciences, vol. 43, no. 6,

New York: Begel House Inc., pp. 16–29, 2011. Doi: 10.1615/JAutomat-InfScien.v43.i6.20

[60] Y. P. Kondratenko, T. A. Altameem and E. Y. M. Al Zubi, 'The optimisation of digital controllers for fuzzy systems design', Advances in Modelling and Analysis, AMSE Periodicals, pp. 19–29, Series A 47, 2010.

[61] T. A. Altameem, E. Y. M. Al Zu'bi, Y. P. Kondratenko, 'Computer Decision Making System for Increasing Efficiency of Ship's Bunkering Process' Annals of DAAAM for 2010 & Proceeding of the 21th Int. DAAAM Symp. "Intelligent Manufacturing and Automation", (20–23 Oct., 2010, Zadar, Croatia), Published by DAAAM International, Vienna, Austria, pp. 0403–0404, 2010.

[62] M. Solesvik, Y. Kondratenko, G. Kondratenko, I. Sidenko, V. Kharchenko, A. Boyarchuk, 'Fuzzy decision support systems in marine practice', In: Fuzzy Systems (FUZZ-IEEE), 2017 IEEE International Conference on, pp. 9–12, July 2017. IEEE. Doi: 10.1109/FUZZ-IEEE.2017.8015471.

[63] J. Kacprzyk, 'Multistage fuzzy control of a stochastic system using a bacterial genetic algorithm', Advances in Intelligent Systems and Computing, 315, pp. 273–281, 2015. Doi: 10.1007/978-3-319-10765-3_32.

[64] D. Filev, T. Larsson, 'Intelligent adaptive control using multiple models', IEEE International Symposium on Intelligent Control - Proceedings, Mexico City; Mexico, pp. 314–319, 2001.

[65] V. Novák, 'Genuine linguistic fuzzy logic control: Powerful and successful control method' Lecture Notes in Computer Science (including subseries Lecture Notes in Artificial Intelligence and Lecture Notes in Bioinformatics), 6178 LNAI, pp. 634–644, 2010. Doi: 10.1007/978-3-642-14049-5_65

[66] E. H. Mamdani, S. Assilian, 'Experiment in linguistic synthesis with a fuzzy logic controller'. International Journal of Man-Machine Studies, vol. 7(1), pp. 1–13, 1975. Doi: 10.1016/S0020-7373(75)80002-2

[67] D. Simon, 'Sum normal optimization of fuzzy membership functions', International Journal of Uncertainty: Fuzziness and Knowledge-Based Systems, no 10, pp. 363–384, 2002.

[68] D. Simon, 'Training fuzzy systems with the extended Kalman filter', Fuzzy Sets and Systems, no 132, pp. 189–199, 2002.

9

Efficiency Control for Multi-assortment Production Processes Taking into Account Uncertainties and Risks

Anatoliy Ladanyuk, Viacheslav Ivashchuk[*] and Jaroslav Smityukh

National University of Food Technologies, Kyiv, Ukraine
*Corresponding Author: ivaschuk@nuft.edu.ua

The article describes the method of efficiency control that authors developed, for monitoring and automation of multi-assortment processes for control of technological complexes what can be presented for a food production. The functional structure and main components of control system for technological complexes that are generalized for the production of multi-assortment products are presented. The approach that authors proposed is based by uses of logical-dynamic models, which adapted for batch process. Procedures for improving the efficiency of multi-assortment production (MAP) of food products using methods of system analysis and modern control theory are considered, which makes it possible to use complex assessments of function of technological complexes (TC) for MAP, as well as necessary mathematical models and implement intelligent systems for optimizing production taking into account energy and resource-efficient technologies, to provide the required quality of products. Main attention was paid to special qualities of the functional structure, software and hardware implementation, as well as the integrated human interface and the machine to maintain the adequacy of the intelligent model of the work process, which presented in the form of a multilayered structure for the control systems of targeted TC.

9.1 Introduction

The development of processing industry is closely related by the demand for final products, which had characteristics that depend on consumer preferences and tend to change by product cycle. Methods for change of product characteristics, which are implemented by the variation of recipes, characteristics of raw materials and parameters of the processing are form the concept of production assortment.

The main structural unit of the MAP is technological complexes, which consist of technological processes and aggregates, where the complex processes of heat and mass transfer, hydrodynamics, physical and chemical transformations of a substance are proceed for finished product or semi-finished product produce. Usually, for this processes the considerable resources of material and energy are depend, the optimal use of which directly affects on quality and cost of products [1]. The decision maker person (DMP) assist for processes of TC, organizational and technological systems, which makes it possible for uses of complex methods for evaluation of processes and the formation of control actions [2], because that the processes of TC is the main source of production efficiency. The article that discusses some aspects of development, analysis of the functioning and creation of a control system for MAP in various industries are present in the technical literature[3–6], but the conditions of full solution for food industry in such work was not performed. The MAP food products had been selected into a separate class of control objects by many features, such as a time of stocks and raw materials processing, short times for storage of finished products, strict demands of product quality, purity and consumer attractiveness, seasonal fluctuations of demand, market situation, etc [7]. As a typical instance of the MAP are dairy [8], baking [9], confectionery, beverage production, etc. Each step for assortment increasing of dairy, vegetable processing, bakery industry have a complex manufacturing and technology development, and the process of its implementation requires the technical re-equipment of production, construction of a separate technological line [10]. This way requires substantive volume of material, the share of which is related to development of technological apartment, increase of production space, which is determined by location of such production, and as a consequence is the limits for production assortment. The reconfiguration of technological units at industrial locality is lead to complications of control systems that do not provide of stability conditions as necessary [11], which is determine to decline in product quality and increase in energy consumption as unforeseen. The evaluation of parameters identity

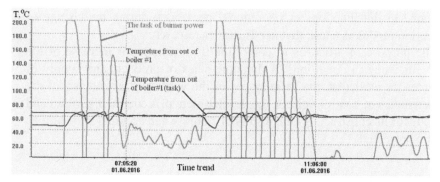

Figure 9.1 The Control Process of Water Boiling by uses of Linear Model.

of processing for changes in modes of equipment is complicated, because the object of control over the structure states and modes of work often can be represented by nonlinear models at circumstances that are pointed [12], with the ambiguous inverse reactions that are involved. Low quality control, for the use of linear type models (Figure 9.1), leads to additional losses in product characteristics.

For multi-assortment technologies, there is a need for the creation of new control methods that would response to nature of variability of processes in TC (technological system), in conjunction with methods that as solve of optimization problems, as an increment of resource and energy efficiency of production in most cost.

The analysis of operation of production control systems [13], which was adapted to assortment products, indicate the absence of any comprehensive methodology that can provide predictable quality control of technological variables in modes for change of operations and characteristics of raw material processing, lack of techniques for the efficient use of resources, as energy for example, in the case of a change in the range of production. It should be noted that classical methods and technical means for increasing the efficiency of function, for systems of control of flexible technological processes, require further development both for the development of the theory and its practical application. The systematic methods are requiring by tasks that had been pointed for processes control of enterprise in the complex. The control systems, in which the flexibility of the tasks is provided by simplification of subsystems that are used, by include of parallel systems [14–16] are widely used now. An additional factor that complicates the implementation of MAP is change in the functions of operational personnel, during the change of production modes as a task that needs further research.

The purpose of article is to determine the methods for increase of the MAP efficiency in conditions of dynamic sales of products and supply of resources. In particular, the main stages of the work are: (a) analysis of the characteristics of control objects; (b) development of a control algorithm and methods for its construct; (c) the results of stability and accuracy of solutions that are presented.

9.2 Decision-Making Under Uncertainty

We define the restriction for alternatives of object states in the class of organization-technological systems for the control of MAP, where its entropy by features that was determined

$$H(y) \quad 0.$$

So as the state of object is evaluated by single criterion y' and set in one of n states by probability $p(y'_1), \ldots, p(y'_n)$, so the dimension of information basis is the adequacy of entropy $H(y)$ as a measure of state uncertainty of object [17]:

$$H(y) \quad -\sum_{i=1}^{n} p(y'_i) log\, p(y'_i), \tag{9.1}$$

that require to reduce entropy by information redundancy:

$$J(y, y') \quad H(y) - H(y/y'), \tag{9.2}$$

where $H(y/y')$ is a conditional entropy for determine of object state after update a new information about y'.

The strategy for control synthesis for the MAP will be determined by the following steps as the entropy approach of obtaining information:

- reduce the number of alternative states $H(y)$ of object to improve the quality of control;
- increase the number of control variables for $H(x)$, which should correspond to the variety of object states by $H(y)$;
- reduce the ambiguity of control variables relative to object states $H(x/y)$ by increase the information about condition and external environment of object, up to constrain the dimension of space for coordinates change of object state, which should be adequate to the dimension of

control variables, which is conditioned by the necessity of satisfy to diversity principle of U. R. Eshbi for to develop a subsystem of decision making support by the implementation of case control [11].

But, the technical implementation of such control as for assessment multi-criteria is technically complicated [18]. So that, the implication of decision maker by limited information basis is an alternative for control of MAP.

The decision to control of MAP is chosen from a group of possible alternatives by development of situations that allow reduce the dimension of alternatives group. The reduction of this group dimension is provided by the method of pairwise comparison as an alternative solution [19]. This method allows resolve the conflict situations in particular by use of criteria, which is characterized as mutually correlative by arguments of criterion for the case increasing of control dimension. For the alternative x from group X_i, so for $x \in X_i$ is the target criterion, which reproduces a partial projection of the goal, when it is defined as a function $q(x)$, where the adjacent alternative X_{i+1} determine the competitive strategy of choice

$$X_{i+1} : (X_i > X_{i+1}) \Rightarrow q(X_i) < q(X_{i+1}). \tag{9.3}$$

Previously, we stipulate that the choice of any alternative leads unambiguously to an effective resolve. Thus, X^* is the best alternative that has the highest value of criterion:

$$X^* \quad arg\ max_{x \in X}\ q(x). \tag{9.4}$$

The numerical method for searching X^* is determined by the type of function $q(x)$: the dimension of arguments, the presence of extremums, points of function discontinuity. The choice of adequate control of MAP is being constructed by a multilayer functional principle [20] of subordination, which requires the formation of a dynamic multi-stage structure of situation assessment (Figure 9.2).

In this way, the evaluation of response for control of MAP as the elements of upper level of hierarchical system are used for increasing the information basis about of MAP objects by to increasing the number of system's behavior aspects.

The further application of the multi-headed principle determines the element of the upper level as a "command" for other elements and the consequences of determining state as the basis of solutions determination for totality approach to state of all subordinate elements to the general purpose

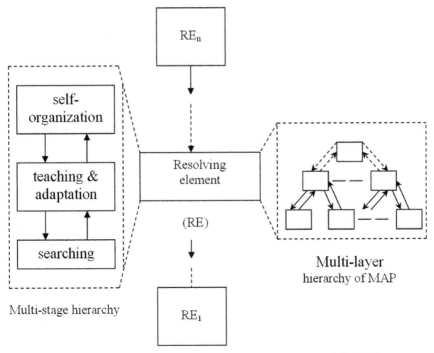

Figure 9.2   The structure of decision-making for control of the MAP.

of control. Moreover, the elements of the upper level leave adequate system behavior over longer time periods.

Therefore, the sample for state determination is estimated by location hierarchy of element and volume of decision risk, where the increase of risk responsible to an increase of training sample size.

The problem of large dimension is solved in article by decreasing the number of alternative variants for X, which is limited by set of H_m that are distinguished at the next m stage of raw materials processing.

9.3 The Estimation of Stability and Accuracy of Presented Solutions

The efficiency of control of technological processes and aggregates is evaluated by a generalized indicator that has interpretation as a cost function, and its value depends of control quality of technological variables and is coinciding with characteristics of energy and resource efficiency.

The reduction of group of partially defined alternative that used for decision-making strategies for control is conditioned by the next state estimation of technological route [20]. Sequence of stages is accompanied by restrictions of predefined risks of efficiency loss and accidents emergence. The group of states of individual processes is limited by technological product card. The vector of process characteristics changes the space of permissible variation for each product. Thus, the space of uncertainty of coordinates decreases and the process state obtained by the number of alternatives $q(x)$, which doesn't allows linearization [21].

The formation of constraints $sub(X)$ to determine the space $X \cap H$ of possible solutions are obtaining the raw material characteristics as coordinates of H, investment flow dI_{pr}/dt for function processing and maximum specific losses dI_{vir}/dF_{sir}. Involving the analysis by a group of arguments allows to provide $max[dI_{vir}/dF_{sir}/(dI_{pr}/dt)]$. This method of construction solutions admits a flexible strategy with relative to characteristics of the TP as an instance of class of dynamic programming.

For example, consider the process of evaporation of fruit flavors by evaporator MZS-320, where the ingredients are sugar syrup - as a preservative, fruit pulp - as the main raw material and the heating steam as the main control effect. Since the multidimensional control involves a loss of static accuracy, we leave the level of exhaustion vacuum in space of evaporator as a mode parameter those changes as a new stage for the task of new product. As an optimization of degree of product processing, is being considered the changeable part from parameters of TP, which varies according to the change of assortment. Thus, in addition to the components of logistics $I_{vup\_res\_j}$, which has been considered as a relatively constant, expenses for basic and optional components of the product according to their cost and defined assortment as a priority, the optimize of volume perfection main TP is:

$$dIdQ^{-1} \quad (sub\,dI - \inf dI)\exp(-Q_{vip}Q^{-1}_{vip\_base}), \tag{9.5}$$

Q_{vip} – degree of evaporation, $Q^{-1}_{vip\_base}$ – the basically evaporation degree, which has been considered for a product as the lowest value of dry components. Thus, for full cost of product obtain:

$$dI \quad dI_{var\_vip}\exp(-Q_{vip}Q^{-1}_{vip\_base})$$
$$-dI_{trans}(m_{base} - Q_{vip\_base}Q^{-1}_{vip}m_{base})$$
$$+dI_{dop\_sir}\exp(-k_{sir}Q_{vip\_base}Q^{-1}_{vip}) \tag{9.6}$$

where k_{sir} as the technological need for control conditions in relation to $m_{sir} m_{dop\_sir}^{-1}$.

For the selected product recipe, which determines the volume of components $\langle x_1, x_2 \rangle$, calculate the optimal degree of its technological perfection:

$$g_{vip} \quad Q_{vip}(Q^*_{vip})^{-1}, \tag{9.7}$$

where for (9.6) obtain:

$$\min_g \; dI(Q_{vip}) \colon dI \quad 0,37 x_1 e^{-g} - 2,3 x_1 x_2 g + 4,1 x_2 e^{k_{sir} g}, \tag{9.8}$$

where the basically components of investments assume a value of dI_{vip} $0,37$; dI_{pered_prod} $2,3$; dI_{dod_sir} $4,1$.

The inverse task of transcendental programming takes the form

$$v_{prod\_i}(dI) \quad (0,37 y_1^{-1})^{y_1} (-2,3 y_2^{-1})^{y_2} (-4,1 y_3^{-1})^{y_3} (y_4^{-1})^{y_4}$$

from which, taking into account the optimization argument from (9.7), obtain:

$$v_{prod\_i}(dI) \quad (0,37 x_1 e^{-g} y_1^{-1})^{y_1} (-2,3 x_1 x_2 y_2^{-1})^{y_2} (-4,1 x_2 e^{k_{sir} g} y_3^{-1})^{y_3}$$

By selecting the variables that determine the content of components by recipe, we obtain:

$$v_{prod\_i}(dI) \quad (0,37 y_1^{-1})^{y_1} (-2,3 y_2^{-1})^{y_2} (-4,1 y_3^{-1})^{y_3}$$
$$\times (x_1 e^{-g})^{y_1} (x_1 x_2)^{y_2} (x_2 e^{k_{sir} g})^{y^3}, \tag{9.9}$$

which should reach the maximum.

Another example is the involvement of algorithm for control a group of water-heating boilers that provide the dynamic load. The destructive variable is the pressure in the water supply network. An additional complication is the necessary regulation of heat power by using group of boilers, introducing variability and dynamic error as a consequence during change of modes and load task by temperature (heating/domestic water). The loss by error (Figure 9.1), which is estimated for the dimensions of coordinates that had been involved for the decision-making task n, justifies the effectiveness of strategy along with the classical methods of coordination of control.

The adding a new product is only available with the support of operator, that determine the signs of an alternative production strategy by stochastic cognitive behavior and a cumulative assessment of persistent situations. In this way it is possible to reduce the entropy of states (Figure 9.3).

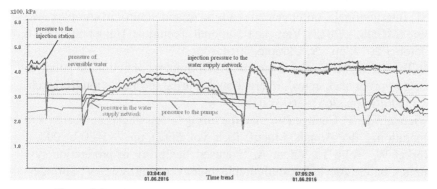

Figure 9.3 The structure of decision-making for control of the MAP.

9.4 Conclusion

The system that had been represented makes it possible to simplify the control of TP of a production complex with minimization of risks provided by violation of technological norms of production. The method that had been involved allows to increase the efficiency of MAP production, by maintaining the adequacy of the functions by which the control decisions are formed. This method ensures minimization of static control errors. The estimation of decisions by the model, which reduces the level of uncertainty, provides minimization of energy costs for processing of the unit of product.The examples that had been used indicate a decrease of the process variability in relation to the application of classical control systems.The next improvement of the MAP relates by the optimization of dimension of control actions to ensure controllability, reduction uses of energy during transients, the enhancement of the number of productivity task and range of assortment.The obtained result can be effectively used in tasks where the function $q(x)$ loses its adequacy during the change of assortment [22, 23].

References

[1] Ivaschuk V., Ladanyuk A. 'Minimize energy losses for implementation of changes of product range', in J. Machinery in agricultural production, industry machine building, automation, vol. 29, pp. 192–196, 2016.

[2] Ivaschuk V., Ladanyuk A. 'Algorithm for Decision Support in the Task of Control System of Industries with Variable Assortment of Products',

XII International Conference measurement and control in complex system (MCCS – 2018) Vinnytsia National Technical University, 14–16 October 2014, pp. 29, Ukraine, 2014.

[3] A. Tomasella, M. Clerici, Sacconi M. 'Optimal reconfiguration policy to react to product changes', International Journal of Production Research, vol. 10(46), pp. 2651–2673, 2008.

[4] Wiendahl H.-P. Changeable manufacturing – classification, design and operation. CIRP Annals: Manuf. Tech, vol. 56(2), pp. 783–809, 2007.

[5] Terkaj W., Tolio T., Valent A. 'Focused flexibility in production systems: Changeable and reconfigurable manufacturing systems', Springer-Verlag, pp. 47–66, 2009.

[6] Lafou M. 'Manufacturing System Flexibility: Product Flexibility Assessment : Research and Innovation in Manufacturing: Key Enabling Technologies for the Factories of the Future'. in Proc. of the 48th CIRP Conference on Manufacturing Systems, vol. 41, pp. 99–104, 2016.

[7] Ivashchuk V., Ladanyuk A. 'Definition of depth for flexibility of technological system' in Ukrainian Food Journal, vol. 3(2), pp. 233–242, 2015.

[8] Ivashchuk V. V., Ladanyuk A. P. 'Automated control of multiassortment manufacture of dry milk products'. in J. Izvestia vuzov. Pishevaya tekhnologia, vol. 1, pp. 85–88, 2016.

[9] Ivashchuk V. V., Ladanyuk A. P., Echkalov D. V. 'Features of controlling the heating of a bread oven for the preparation of assortment production'. in J. Izvestia vuzov. Pishevaya tekhnologia, vol. 5–6, pp. 73–76, 2014.

[10] Ivashchuk V., Ladanyuk A. 'Features of control for multi-assortment technological process' in book of abstracts "Systems, Control and Information Technology- 2016" Industrial Research Institute for Automation and Measurements PIAP, Warsaw, Poland, pp. 40, 18–22 April 2016.

[11] Ivashchuk V., Ladanyuk A. 'Use of the models for sliding control mode under conditions of assortment production'. Scientific Works of National University of Food Technologies, vol. 22(2), pp. 7–14, 2016.

[12] Panneerselvam R. Operations research/PHI Learning Pvt. Ltd.,p. 620, 2006.

[13] Matta A., Semeraro Q. Design of Advanced Manufacturing Systems. Kluwer Academic Pub, p. 267, 2005.

[14] Stecke K. E., Raman N. FMS planning decisions, operating flexibilities, and system performance. IEEE Technology and Engineering Management Society, vol. 42, no. 1. pp. 82–90, 1995.

[15] Parshall J., Lamb L. B. 'Applying S88: batch control from a user's perspective Publisher ISA, p. 155, 2011.

[16] Ivashchuk V., Ladanyuk A., Yechkalov D. 'Features of temperature regulation as a task for assortment of bread baking'. In J. of Food and Packaging Science. Technique and Technologies. vol. 3, pp. 102–105, 2014.

[17] Ivashchuk V. V. 'Developing criteria for optimization of multiproduct processes for task of control of regimes'. In Book of Abstracts 8th Central European Congress on Food, Kyiv, p. 57, 23–26 May 2016.

[18] Ivashchuk V. V. Development of control systems for multi-product technological plant. In material II International Scientific-Technical Conference 'Modern methods, information, software and technical equipment of control system of organizational and technological complexes'. Kyiv, pp. 107–108, 25 November 2015.

[19] Ladanyuk A., Ivashchuk V. Algorithm for decision support as the tool for control system of industries with variable assortment of products In Proc. SPIE 9816, Optical Fibers and Their Applications, Bellingham; USA: Spie-Int Soc Optical Engineering, p. 5, 2015.

[20] Ladanyuk A., Ivashchuk V., Boyko R., Savchuk O. Methods of situational control of multipurpose production. Scientific Works of National University of Food Technologies, vol. 22(3), pp. 25–30, 2016.

[21] Tondel P., Johansen T. A., Bemporad A. Evaluation of piecewise affine control via binary search tree. In J. Automatica, vol. 39, pp. 945–950, 2003.

[22] Vlasenko L. O., Ladanyuk A. P., Ivaschuk V. V. 'Sub-system for support of decision making for control of technological complexes' Patent. 30556 Ukraine: MPC (2006) G05B 13/04, No. 200713857. accept. 10.12.07, public. 25.02.08, vol. 4. p. 2.

[23] El Ghaoui L., Boyd S. Stability Robustness of Linear Systems to Real Parametric Perturbations. Proceedings of the 29th IEEE Conference on decision and control, pp. 1247–1248, 1990.

10

On the Coordinate Determination of Space Images by Orbital Data

Dmitriy V. Lebedev

International Research and Training Center for Information
Technologies and Systems, Ukraine
Corresponding Author: ldv1491@gmail.com

The problem of coordinate determination of objects located on space images only by orbital data is investigated. Taking into account foreground role of in-flight geometric calibration of opto-electronic equipment of remote sensing satellites in the procedure of image binding from space, considerable attention is given to discribing of in-flight geometric calibration algorithms designed by the author. Characteristics of accuracy of coordinate binding of satellite images, which is realized with usage of proposed algorithms of in-flight geometric calibration, are investigated. Computer modeling confirms the effectiveness of the proposed algorithmic solutions.

10.1 Introduction

Remote sensing of the Earth is one of the important and rapidly developing areas of practical cosmonautics.

The term "remote sensing" generally refers to the use of satellite-based (or aircraft-based) sensor technologies to detect and classify objects on Earth, including on the surface and in the atmosphere and oceans. Remote sensing is used in numerous fields, including geography, land surveying and most Earth Science disciplines (for example, hydrology, ecology, oceanography, glaciology, geology); it also has military, intelligence, commercial, economic, planning, and humanitarian applications [1].

The value and practical usefulness of space images is the higher, the higher their resolution and the accuracy of the coordinate fixing of the survey objects, i.e., accuracy of determining the location of the objects of interest in a coordinate system associated with the Earth or with a reference ellipsoid that approximates the shape of the Earth's surface.

Functioning of the SC of ERS according to his function occurs when Earth's surface images of regions of interest, done by them, can be bound by orbital data (data of on-board sources of information of SC) with accuracy enabling performing further image processing. The mention processing can include also the procedures of carrying the accuracy of CB of surveyed objects to the level, declared in demands to final product.

Basic accuracy of CB of images for such spacecrafts as Cartosat-1 and Cartosat-2 amounts to 200 m [2], and for SC Sich-2 it is 250–2000 m [3]. Accuracy of determination of plane coordinates of points only by orbital data of satellite GeoEye-1 is about 3 m [4].

It is evident that for accuracy, which is provided by data from satellites Cartosat-1, Cartosat-2 or Sich-2 further image processing without refinement of orbital data becomes problematic.

Rapid development and improvement of element base of opto-electronic equipment of SC of ERS promote the development technologies of Earth surface images CB which would provide the accuracy appropriate to the accuracy of special resolution of survey equipment.

Of all disturbing factors the one that has the greatest influence on accuracy of space images CB is the residual uncertainty of mutual orientation of CS of shooting camera and CS of SS. As mentioned in [5] the deviations of the relative positions of the above CS from the initial position determined due to pre-flight geometric calibration are caused by action of some disturbing factors like thermal and vibration loading occurred under transportation, SC launching and under its functioning on the obit and also weightlessness, aging of materials in SC structure and others.

High accuracy of CB of images is feasible due to application of space platforms with high stability (the possibility of long-term maintenance of admissible values of mutual orientation parameters of survey equipment and measurement means of control systems of SC orientation and stabilization) and improved accuracy of determining orientation of ERS satellite. Without such possibility there arises necessity of periodic FGC.

At least there are known two ways of solving in-flight calibration problem different in applied sources of external information: calibration by stars and calibration in observation mode of SP with known reference objects (for example, point landmarks). Calibration by ground point landmarks can be conducted with the use of both toporeferenced landmarks and a priori unknown ones (see, for example, [6–13]).

Taking into account foreground role of FGC in the procedure of image binding from space, as well as not the least of the factors of operability and labor content of this calibration (including expenses of onboard energy sources on SC orientation control in the process of shooting) main attention in the present chapter is paid to substantiation of methods and algorithms of calibration for minimal number of topographically bound point references and shooting frames of SP.

On computer modeling it is investigated characteristics of accuracy of CB of satellite images, which is realized with usage of proposed algorithms of FGC.

10.2 Systems of Coordinates. Problem Statement of In-flight Calibration

Let SC of ERS with installed high resolution camera, SS and GPS-receiver revolves around the Earth in the orbit which is nearly circular. Without violating a rigor it is assumed that the mentioned equipment is installed in the mass center O of the spacecraft.

Introduce the required further right orthogonal CS and bases composed by orts of axes of these CS:

- $X_I Y_I Z_I$ (**I**) is inertial CS with the vertex in the Earth center, the axis X directed to the vernal equinox, and the axis Z oriented along the axis of the world in the direction of Polar star;
- $X_G Y_G Z_G$ (**G**) is geographical CS, WGS-84. In WGS-84 the ellipsoid with the equatorial radius $R_1 = 6378137$ m and the polar radius $R_2 = 6356752,3142$ m is taken as the basis. The angular velocity of the Earth rotation is 7,292115 10^{-5} rad/s; [14]
- $X_O Y_O Z_O$ (**O**) – orbital SC with the vertex at the center of mass of the SC (the axis X_O lies in the plane of the orbit and is oriented toward the motion of the space vehicle, the axis Z_O is directed along the geocentric vertical to the zenith);
- $X_T Y_T Z_T$ (**T**) is topocentric CS whose vertex axis at the point O_T of the Earth surface with the coordinates φ_* and λ_* (latitude and longitude correspondingly). The axis Y_T coincides with the tangent to the meridian and is directed to the North and the axis Z_T – vertically upwards.

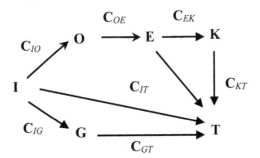

Figure 10.1   Connection between the bases **I, G, O, T, K** and **E**.

- $X_K Y_K Z_K$ (**K**) is CS, fixed with shooting camera such that its vertex is in the center of camera projection, Z_K is directed along optical axis of camera oppositely to shooting object and is orthogonal to plane of light sensitive element of image plant.
- $X_E Y_E Z_E$ (**E**) is CS, fixed with SS is a way, such that the axe Z_E collinear to optical axis of sensor.

Connection between the bases **I, G, O, T, K** and **E** is illustrated by scheme in Figure 10.1, where C_{ij} is rotation matrix transforming the vectors from the basis i in the basis j.

We give the expressions for the matrices of direction cosines C_{IO}, C_{IT}, C_{OE}, characterizing mutual orientation of the bases **I** with bases **O** and **T** and the bases **E** with bases **O** and **K** necessary for the presentation of the material of the chapter:

$$C_{IO} = \begin{bmatrix} -c_u c_i s_\Omega - s_u c_\Omega & c_u c_i c_\Omega - s_u s_\Omega & c_u s_i \\ s_i s_\Omega & -s_i c_\Omega & c_i \\ -s_u c_i s_\Omega + c_u c_\Omega & s_u c_i c_\Omega + c_u s_\Omega & s_u s_i \end{bmatrix}, \qquad (10.1)$$

where Ω – the longitude of the ascending node of the orbit, u – the latitude argument, i – the inclination of the orbit, $c_\alpha = \cos\alpha, s_\alpha = \sin\alpha (\alpha = u, i, \Omega)$;

$$C_{IT} = \begin{bmatrix} \cos(\lambda_* + \lambda_G)\cos\varphi_* & \sin(\lambda_* + \lambda_G)\cos\varphi_* & \sin\phi_* \\ -\sin(\lambda_* + \lambda_G) & \cos(\lambda_* + \lambda_G) & 0 \\ -\cos(\lambda_* + \lambda_G)\sin\varphi_* & -\sin(\lambda_* + \lambda_G)\sin\varphi_* & \cos\varphi_* \end{bmatrix},$$
$$(10.2)$$

where λ_G is the current value of the angle between the direction to the point of the vernal equinox and the Greenwich meridian.

For a sequence of turns $\vartheta \rightarrow \varphi \rightarrow \psi$ (angles of yaw $\rightarrow$roll$\rightarrow$pitch), the matrix C_{OE} is given by relation

$$C_{OE} = \begin{bmatrix} c_\vartheta c_\psi + s_\vartheta s_\varphi s_\psi & c_\varphi c_\psi & -s_\vartheta c_\varphi + c_\vartheta s_\varphi s_\psi \\ -c_\vartheta s_\psi + s_\vartheta s_\varphi c_\psi & c_\varphi c_\psi & s_\vartheta s_\psi + c_\vartheta s_\varphi c_\psi \\ s_\vartheta c_\varphi & -s_\varphi & c_\vartheta c_\varphi \end{bmatrix}. \tag{10.3}$$

where $c_\beta = \cos \beta$, $s_\beta = \sin \beta (\beta = \vartheta, \varphi, \psi)$.

If the matrix C_{EK} is parameterized by the coordinates of the orientation vector $\boldsymbol{\theta}_E = \{\theta_1, \theta_2, \theta_3\} = \mathbf{e}\theta$, or the Euler vector ($\mathbf{e}$ – the unit vector of the Euler rotation axis, $\theta = \|\boldsymbol{\theta}_E\|$ – the rotation angle with respect to this axis), then its connection with coordinates of the vector $\boldsymbol{\theta}_E$ is given by the relations [15]

$$C_{EK} = E_3 + \frac{\sin\theta}{\theta}\Phi(\boldsymbol{\theta}_E) + \frac{1-\cos\theta}{\theta^2}\Phi^2(\boldsymbol{\theta}_E),$$

$$\Phi(\boldsymbol{\theta}_E) = \begin{bmatrix} 0 & -\theta_3 & \theta_2 \\ \theta_3 & 0 & -\theta_1 \\ -\theta_2 & \theta_1 & 0 \end{bmatrix},$$

where E_3 – the third-order unit matrix; $\Phi(\boldsymbol{\theta}_E)$ is (3×3)– matrix of the vector multiplication operator by the vector $\boldsymbol{\theta}_E$. For small values of the angle θ, when $\theta^2 \ll 1$, the matrix C_{EK} can be approximated by the expression

$$C_{EK} = E_3 + \Phi(\boldsymbol{\theta}_E). \tag{10.4}$$

Formulate the in-flight geometric calibration problem. Let be known the matrix C_{IT}, location and orientation of SC in the inertial CS – radius vector $\mathbf{r}$ and the matrix C_{IE}, and radii vectors $\mathbf{r}_i (i = \overline{1, n})$ topographically referenced point landmarks $M_i (i = \overline{1, n})$ being within the field of the camera view. By the available information and images of the ground polygon received from the orbit of SC flight it is necessary to clarify the value of the matrix C_{EK} defining mutual orientation of bases $\mathbf{E}$ and $\mathbf{K}$ of SS and the camera.

10.3 Coordinate Binding of Space Images

We will bring two procedures of coordinate determination of objects of space survey.

10.3.1 Estimation of the Point Landmark Coordinates by Stereoshooting Results

Consider two positions of SC in the orbit – point 1 and 2 in Figure 10.2, – on which images of the same fragment of the Earth surface are formed.

Let, then, the reference point with the number i be chosen and a one-to-one correspondence be established between the images of this landmark.

Using stereo pair data, we estimate the coordinates of the reference point in question. In doing so, we will use the data obtained from on-board information sources.

In the general case, the directions to the landmark, characterized by the vectors ρ_1 and ρ_1 in Figure 10.2, do not intersect. Vectors $\rho_1 = \rho_1 e_1, \rho_2 = \rho_2 e_2$ and $d = de_d$ (ρ_1, ρ_2 and d are the norms of these vectors) correspond to the situation when the distance d between directions is minimal. Let us consider what point in this case it is convenient to take as the point of "intersection" of the vectors mentioned.

Suppose that there are three vectors **a, b** and **c**. Their mixed product is written in the form $(\mathbf{a} \times \mathbf{b})^T \mathbf{c} = \mathbf{abc}$. Consider the equation

$$\mathbf{a}x + \mathbf{b}y + \mathbf{c}z = \mathbf{d}, \tag{10.5}$$

here **a, b, c** and **d** are known vectors; x, y and z are unknown scalar quantities.

The solution of equation (10.5) has the form [16]

$$x = \frac{\mathbf{dbc}}{\mathbf{abc}}, y = \frac{\mathbf{adc}}{\mathbf{abc}}, z = \frac{\mathbf{abd}}{\mathbf{abc}}. \tag{10.6}$$

We use equation (10.5) and its solution (10.6) to calculate the constants ρ_1, ρ_2 and d. From Figure 10.2 implies the equality

$$\rho_1 \mathbf{e}_1 - \rho_2 \mathbf{e}_2 - d\mathbf{e}_d = \Delta_{\mathbf{r}}, \tag{10.7}$$

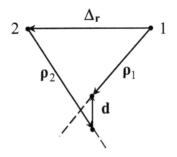

Figure 10.2   Forming of the stereopair.

where $\mathbf{e}_d = (\mathbf{e}_1 \times \mathbf{e}_2)/\|\mathbf{e}_1 \times \mathbf{e}_2\|$

Considering (10.7) as an equation with respect to ρ_1, ρ_2 and d, we obtain the following result of the solution of this equation:

$$\rho_1 = \frac{\mathbf{e}_d^T(\Delta_\mathbf{r} \times \mathbf{e}_2)}{\|\mathbf{e}_1 \times \mathbf{e}_2\|}, \rho_2 = \frac{\mathbf{e}_d^T(\Delta_\mathbf{r} \times \mathbf{e}_1)}{\|\mathbf{e}_1 \times \mathbf{e}_2\|}, d = \frac{\mathbf{e}_d^T \Delta_\mathbf{r}}{\|\mathbf{e}_1 \times \mathbf{e}_2\|}.$$

Naturally, in the absence of perturbations in measurements (calculations), the point of intersection of the vectors ρ_1 and ρ_2 is uniquely determined. In the case considered above, for the purposes of subsequent calculations, it is necessary to agree on what is meant by the "intersection" point of these vectors. Usually, the middle of vector $\mathbf{d}$ is chosen as this point [16]. Then the vector

$$\mathbf{R}_i = \mathbf{r}_1 + \rho_1^{(i)} - \mathbf{d}^{(i)}/2 = \mathbf{r}_2 + \rho_2^{(i)} + \mathbf{d}^{(i)}/2 = (\mathbf{r}_1 + \rho_1^{(i)} + \mathbf{r}_2 + \rho_2^{(i)})/2 \quad (10.8)$$

can serve as an estimate of the location of the i-th landmark in the chosen coordinate system.

10.3.2 Geo-referencing of Frame Elements

Under the geo-referenced images of the Earth from space, we will understand the procedures for processing a picture of the earth's surface, as a result of which each element of the image of interest is associated with its geographical coordinates (latitude and longitude).

Let the point P with the coordinates x, y and $-F$ be chosen in the image of the earth's surface in the coordinate system $\mathbf{K}$ associated with the camera. It is necessary to calculate the coordinates of the preimage P' of the same point in the coordinate system, $X_G Y_G Z_G$.

From the coordinates of the point P' we form the unit vector $\mathbf{e}_K$. After the procedure of in-flight calibration has been performed, the unit vector $\mathbf{e}_{P'}$ of the direction to this point from the survey point O, represented in the CS $X_G Y_G Z_G$, is determined by the equality (Figure 10.1)

$$\mathbf{e}_{P'} = \{l, m, n\} = C_{KG}\mathbf{e}_K = C_{EG}(E_3 + \Phi(\boldsymbol{\theta}_E))\mathbf{e}_K. \quad (10.9)$$

As coordinates of the point P' in the WGS-84, we take the coordinates of the point of intersection of the straight line passing through the survey point O with the coordinates X_G^*, Y_G^*, Z_G^* parallel to the directing vector $\mathbf{e}_{P'}$, with the ellipsoid of rotation adopted in WGS-84 [14]. Thus, the required coordinates are a solution of the system of equations

$$\frac{X_G - X_G^*}{l} = \frac{Y_G - Y_G^*}{m} = \frac{Z_G - Z_G^*}{n},$$

$$\frac{X_G^2}{R_1^2} = \frac{Y_G^2}{R_1^2} = \frac{Z_G^2}{R_2^2} = 1, \tag{10.10}$$

in which R_1 and R_2 are respectively the equatorial and polar radii of the Earth's ellipsoid of rotation of the Earth's shape adopted in the WGS-84.

This system of equations has two solutions:

$$X_{G1} = \alpha Z_{G1} + \beta, Y_{G1} = \gamma Z_{G1} + \delta, Z_{G1} = -p + \sqrt{p^2 - q},$$
$$X_{G2} = \alpha Z_{G2} + \beta, Y_{G2} = \gamma Z_{G2} + \delta, Z_{G2} = -p - \sqrt{p^2 - q}. \tag{10.11}$$

In the expressions (10.11) the following notations are introduced:

$$\alpha = l/n, \beta = X_G^* - \alpha Z_G^*, \gamma = m/n, \delta = Y_G^* - \gamma Z_G^*,$$
$$p = (\alpha\beta + \gamma\delta)D, q = (\beta^2 + \delta^2 - R_1^2)D, D = R_2^2/(R^2 + R_2^2(\alpha^2 + \gamma^2)).$$

From two solutions of the system (10.3), one is chosen that corresponds to the point closest to the point $O(X_G^*, Y_G^*, Z_G^*)$, a point P'. Its radius vector is denoted by $\mathbf{P'} = \{X_G, Y_G, Z_G\}$.

In order to calculate the geocentric coordinates of a point P', it is sufficient to determine the coordinates of the unit vector $\mathbf{e}_G = \{e_X, e_Y, e_Z\} = \mathbf{P'}/\|\mathbf{P'}\|$ and use the equalities

$$e_X = \cos\varphi_P \cos\lambda_P, e_Y = \cos\varphi_P \sin\lambda_P, e_Z = \sin\varphi_P.$$

To the sources of methodical errors of the variant of estimation of geographical coordinates of points of the object of the survey given here, one should include the influence of the height of the real object on the ellipsoid of the Earth's figure taken in the WGS-84 system with the deviation of the line of sight from the local vertical. Naturally, when shooting in a nadir this error is absent.

10.4 Factors Affecting the Accuracy of the Images Coordinate Reference. Assessment of the Accuracy of the Binding

The error in the coordinate reference is primarily due to the deviation of the line of sight of the camera from the true direction to the survey object due to the presence of disturbing factors, such as the errors in determining the position of the spacecraft and the orientation of the Earth observation equipment, the instability of the angular position of the spacecraft (during the survey use in the shooting equipment of CCD-matrices), distortion of the optical system of surveying equipment, atmospheric refraction, aging of materials in space flight conditions and other perturbations.

If we characterize this deviation by an angle ρ, then it is related to the coordinates of the vector $\boldsymbol{\theta}_E$ by the relation [13]

$$\sin \rho = \frac{\|\boldsymbol{\Phi}^2(\mathbf{e}_K)\boldsymbol{\theta}_E\|}{\sqrt{1 - \boldsymbol{\theta}_E^T \boldsymbol{\Phi}^2(\mathbf{e}_K)\boldsymbol{\theta}_E}} \approx \|\boldsymbol{\Phi}^2(\mathbf{e}_K)\boldsymbol{\theta}_E)\|, \tag{10.12}$$

where $\boldsymbol{\Phi}(.)$ is 3×3 – matrix of the vector multiplication operator, $\mathbf{e}_K$ is the unit vector of the line of sight.

Let's estimate the value ρ at a small angle of view of the camera, typical for high and ultra high resolution of ERS spacecrafts. In this case, the coordinates of the unit vector $\mathbf{e}_K$ satisfy the system of inequalities $|e_{3K}| \gg |e_{1K}|, |e_{3K}| \gg |e_{2K}|$. In addition, an analysis of the accuracy of the in-flight geometric calibration by algorithms formed on the basis of various schemes of using the photogrammetric conditions of collinearity and coplanarity (see, for example, [9–11]) indicates a poor estimate of the coordinate θ_3 of the vector $\boldsymbol{\theta}_E$ As a result of this, the errors $\delta\theta_3$ in estimating the coordinate θ_3 greatly exceed the errors $\delta\theta_1$ and $\delta\theta_2$ of estimating the coordinates θ_1 and θ_2 of the small rotation vector $\boldsymbol{\theta}_E : |\delta\theta_3| \gg |\delta\theta_1|, |\delta\theta_3| \gg |\delta\theta_2|$.

Taking into account the structure of the matrix $\boldsymbol{\Phi}^2(\mathbf{e}_K)$ in (10.12) and the above inequalities for the coordinates of the vectors $\mathbf{e}_K$ and $\delta\boldsymbol{\theta}_E : [\delta\theta_1 \delta\theta_2 \delta\theta_3]^T$, the angle ρ is calculated from the formula

$$\rho \approx \Sigma_{i=1}^3 \sqrt{1 - e_{iK}^2}|\delta\theta_i| \cong |\delta\theta_1| + |\delta\theta_2| + \sqrt{e_{1K}^2 + e_{2K}^2}|\delta\theta_3|. \tag{10.13}$$

If the angle ρ is known, then it is not difficult to estimate the error δR_M in determining the coordinates of the point M on the planes $X_G Y_G$ or $Z_T = 0$. Indeed, let the SC photographs the earth's surface in the mode of route survey at the angle of roll $\gamma = \gamma_* = const$ and zero values of the angles of pitch ϑ and ψ. Then, as shown in [13], the following estimate of the error of the coordinate determination of the point M on the plane $X_T Y_T$ of the topocentric coordinate system takes place:

$$\delta R_M \cong \sqrt{(\delta X_M)^2 + (\delta Y_M)^2} \leq \frac{Z_O}{\cos^2 \gamma_*} \rho, \tag{10.14}$$

where Z_O is the coordinate of the projection center of the optical system at the time of shooting.

Using the sensitivity function

$$\frac{\partial \delta R_M}{\partial |\delta \theta_3|} = \frac{Z_O}{\cos^2 \gamma_*} \frac{\partial \rho}{\partial |\delta \theta_3|} = \frac{Z_O}{\cos^2 \gamma_*} \sqrt{e_{1K}^2 + e_{2K}^2},$$

it is possible to determine the contribution to the magnitude of the error of the topographic binding of the point M of uncertainty of the value of the angle θ_3 that occurs after the implementation of the in-flight calibration procedure.

As an example, consider the situation when a polygon size $1000 \times 2000\ m$ survey is made from a circular orbit 680 km high in a mode in which the sub-satellite point coincides with the top of the topocentric coordinate system. Let all landmarks of the polygon fall into the shot frame, and the spacecraft is stabilized in the orbital coordinate system. Then relation (10.13) takes the form

$$\rho \leq |\delta \theta_1| + |\delta \theta_2| + 1,64.10^{-3}|\delta \theta_3|.$$

It follows from this inequality and formula (10.14) that under the conditions of the example considered, the error of the coordinate reference of the image is approximately 610 times less sensitive to variations in the coordinate θ_3 of the vector $\boldsymbol{\theta}_E = [\theta_1 \theta_2 \theta_3]^T$ in comparison with the variations of the remaining coordinates of this vector.

The above-mentioned specificity of the Earth observation satellite's surveying equipment leads to the fact that the error $\delta \theta_3$ in estimating the coordinate θ_3 of the vector $\boldsymbol{\theta}_E$ does not play a decisive role in the formation of the error of the snapshot coordinate determination. This increases the requirement for the ability of in-flight calibration algorithms to minimize the residual uncertainty of the values of the coordinates θ_1 and θ_2 of the mutual orientation vector of the bases $\mathbf{E}$ and $\mathbf{K}$.

Analysis of expression (10.13) allows, on the one hand, to evaluate the contribution of each component of the vector $\delta \boldsymbol{\theta}_E$ to the error of binding the survey objects, and on the other hand, to justify the possibility of correctly constructing an algorithm of flight geometric calibration for coordinates θ_1 and θ_2 of the vector $\boldsymbol{\theta}_E$ that are well estimated.

10.5 Algorithms of In-flight Geometric Calibration

The variety of possible algorithms for solving the problem of the flight geometric calibration of the ERS spacecraft imagery equipment is associated with various schemes for the presentation of photogrammetric conditions for collinearity and coplanarity. Some calibration algorithms based on these conditions are given in this section.

10.5.1 Calibration Algorithms Based on Photogrammetric Equations

Let us characterize the position of the spacecraft in the inertial space (in the coordinate system $X_I Y_I Z_I$) by the radius vector $\mathbf{R} = [X, Y, Z]^T$ of its center of mass O and the matrix $C_{IE} = \{c_{ij}\}(i, j = 1, 2, 3)$ of the direction cosines. On the camera image plane, we define the coordinate system $O_1 x_\pi y_\pi$ with the origin at the main point O_1 of the camera (the point of intersection of the optical axis with the sensitive area) and the axis $O_1 x_\pi, O_1 y_\pi$ parallel to the corresponding axes of the coordinate system $X_K Y_K Z_K$.

Let there is a system of N point landmarks with known coordinates. Then the relationship between the coordinates x_i and y_i the mapping of the i-th landmark on the image plane $O_1 x_\pi y_\pi$ with the coordinates X_i, Y_i, Z_i of this reference point in the inertial space is given by the direct equations of photogrammetric [17]

$$X_i = X + (Z_i - Z)\frac{c_{11}x_i + c_{21}y_i - c_{31}F}{c_{13}x_i + c_{23}y_i - c_{33}F},$$
$$Y_i = Y + (Z_i - Z)\frac{c_{12}x_i + c_{22}y_i - c_{32}F}{c_{13}x_i + c_{23}y_i - c_{33}F}, \tag{10.15}$$

and the inverse equations

$$x_i = -F\frac{c_{11}(X_i - X) + c_{12}(Y_i - Y) + c_{13}(Z_i - Z)}{c_{31}(X_i - X) + c_{32}(Y_i - Y) + c_{33}(Z_i - Z)},$$
$$y_i = -F\frac{c_{21}(X_i - X) + c_{22}(Y_i - Y) + c_{23}(Z_i - Z)}{c_{31}(X_i - X) + c_{32}(Y_i - Y) + c_{33}(Z_i - Z)}, \tag{10.16}$$

where F is the focal length of the camera.

10.5.1.1 Algorithm 1

To form the calibration algorithm, we use equations (10.15). In these equations, the coordinates X_i, Y_i, Z_i of the i-th reference point are known a priori, the values X, Y, Z of the coordinates of the projection center (the center of mass of the SC) come from the GPS-receiver, elements of the matrix C_{IE} form the star sensor. Under x_i, y_i, z_i we understand the coordinates of the landmark mapping, given in the CS with the basis **E**.

Since the measurements of the coordinates x_i', y_i' of the mapping of the i-th landmark are made in the coordinate system $X_K Y_K Z_K$ connected with the camera, then in order to use the equations (10.15), the measurement results must be recalculated in the coordinate system $X_E Y_E Z_E$.

Characterizing the estimated discrepancy between the bases **E** and **K** by the vector $\boldsymbol{\theta}_E = [\theta_1, \theta_2, \theta_3]^T$, we define the required transformation by the relation

$$\begin{bmatrix} x_i \\ y_i \\ z_i \end{bmatrix} = C_{EK}^T \begin{bmatrix} x_i' \\ y_i' \\ -F \end{bmatrix} \approx (E_3 - \Phi(\boldsymbol{\theta}_E)) \begin{bmatrix} x_i' \\ y_i' \\ -F \end{bmatrix}, \tag{10.17}$$

Taking into account equality (10.17) and notation

$$C_j = c_{3j} y_i' + c_{2j} F, \, S_j = c_{3j} x_i' + c_{ij} F,$$
$$V_j = c_{2j} x_i' - c_{1j} y_i', \, W_j = c_{1j} x_i' - c_{2j} y_i' - c_{3j} F \quad (j = 1, 2, 3)$$
$$\rho_{Xi} = \frac{X_i - X}{Z_i - Z}, \rho_{Yi} = \frac{Y_i - Y}{Z_i - Z},$$

equations (10.15) are reduced to the form

$$\mathbf{q}_i = G_i \boldsymbol{\theta}_E \tag{10.18}$$

The two-dimensional vector $\mathbf{q}_i$ and the (2×3) – matrix G_i in the equation of measure (10.18) are calculated from formulas

$$\mathbf{q}_i = [W_1 - \rho_{Xi} W_3 \quad D_2 - \rho_{Yi} W_3]^T,$$
$$G_i = \begin{bmatrix} C_1 - \rho_{Xi} C_3 & -(S_1 - \rho_{Xi} S_3) & V_1 - \rho_{Xi} V_3 \\ C_2 - \rho_{Yi} C_3 & -(S_2 - \rho_{Yi} S_3) & V_2 - \rho_{Yi} V_3 \end{bmatrix}.$$

The evaluation of the vector $\boldsymbol{\theta}_E$ is obtained as a solution of all the generated measurement equations (10.18). It is sufficient to compose and solve with respect to $\boldsymbol{\theta}_E$ the system of normal equations of the method of least squares corresponding to i equations (10.18):

$$B_i \boldsymbol{\theta}_E = \mathbf{Y}_i,$$

$$B_i = \sum_{k=1}^{i} G_k^T G_k = B_{i-1} + G_i^T G_i, \, y_i = \sum_{k=1}^{i} G_k^T y_i = y_{i-1} + G_i^T \mathbf{q}_i,$$

$$B_0 = 0, \mathbf{y}_0 = 0, \tag{10.19}$$

where the summation is over all indices k, indicating the number of visual contact with the landmark [9].

10.5.1.2 Algorithm 2

The basis of this algorithm is the relation (10.17) and equation (10.16). The results of measurements of the coordinates x_i' and y_i' of mapping the i-th landmark to the plane of the camera image are compared with the results of the estimation x_i and y_i the same coordinates obtained from (10.16) on the basis of data from the stellar sensor, the GPS-receiver and a priori information about the coordinates of the landmark in an inertial coordinate system.

When forming estimates x_i and y_i of coordinates x_i' and x_i', in formulas (10.16) elements of the matrix C_{IK} are used instead of matrix elements $C_{IK} = \{c_{ij}\}$. The measurement equations obtained in this way have the structure of expression (10.18), in which the following relations are understood as a vector $\mathbf{q}_i$ and a matrix G_i:

$$q_i = \begin{bmatrix} x_i - x_i' \\ y_i - y_i' \end{bmatrix}, G_i = \begin{bmatrix} 0 & F & -x_i' \\ -F & 0 & y_i' \end{bmatrix}.$$

The desired vector θ is estimated using the above computational procedure.

10.5.2 Calibration for a Priori Unknown Point Landmarks

At the present time the main method of in-flight calibration of the spacecraft optical-electronic complex is the calibration based on observations of topographically bound landmarks. In the practical implementation of this approach, it becomes necessary to attract the opportunities of specially organized terrestrial infrastructures - sub-satellite polygons. Successful implementation of this approach will reduce the dependence of SC from sub-satellite polygons, simplify and reduce the cost of their content and make the calibration process more reliable and operational.

Consider two positions of SC in the orbit – point 1 and 2 in Figure 10.3, – on which images of the same fragment of the Earth surface are formed.

In the Figure 10.3, we denote by $\mathbf{r}_1$ end $\mathbf{r}_2$ the geocentric radius vectors of the center of mass of the SC at points 1 and 2, respectively; $\mathbf{e}_1$ and $\mathbf{e}_2$ are orts of directions to the i-th landmark.

In the absence of perturbations in measurements (calculations) of vectors $\Delta \mathbf{r} = \mathbf{r}_2 - \mathbf{r}_1, \mathbf{e}_1$ and $\mathbf{e}_2$ the condition of coplanarity of these vectors is fulfilled

$$\Delta \mathbf{r}^T (\mathbf{e}_1 \times \mathbf{e}_2) = 0. \tag{10.20}$$

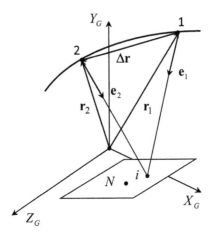

Figure 10.3   Shooting formation.

Write the equality (10.20) in the coordinate system $X_G Y_G Z_G$ (basis **G**). At the points 1 and 2 of the flight trajectory of the spacecraft, the radius vectors $\mathbf{r}_{I1}$ and $\mathbf{r}_{I2}$ and the orientation of the shooting camera in the inertial space (the matrixes C_{IE1} and C_{IE2}) are known from the data of the GPS- receiver and the stellar sensor.

According to indications of the camera $x_k^{(i)}, y_k^{(i)}$ (k – the picture number, i – the landmark number), the unit vectors $\mathbf{e}_{Kk}^{(i)}$ of directions to the i-th landmark for both images are computed in the bases **K** by the formulas

$$\mathbf{e}_{Kk}^{(i)} = \mathbf{p}_k^{(i)} / \|\mathbf{p}_k^{(i)}\|, \ \mathbf{p}_k^{(i)} = \{x_k^{(i)}, y_k^{(i)}, -F\}. \tag{10.21}$$

The representations of the vectors $\mathbf{e}_{Kk}^{(i)}$ and the difference of vectors $\mathbf{r}_1$ and $\mathbf{r}_2$ in the basis **G** are defined by the equalities (see Figure 10.3)

$$\mathbf{e}_{KGk}^{(i)} = C_{KGk}\mathbf{e}_{Kk}^{(i)}(k = 1, 2), \Delta \mathbf{r}_G = \mathbf{r}_{G2} - \mathbf{r}_{G1} = C_{IG2}\mathbf{r}_{I2} - C_{IG1}\mathbf{r}_{I1}. \tag{10.22}$$

Approximating the matrix C_{EK} by $C_{EK} = E_3 + \Phi(\boldsymbol{\theta}_E)$ and taking into account the formulas (10.21) and (10.22), the condition of coplanarity (10.20) for i - th landmark can be written as [10]

$$A_i\boldsymbol{\theta}_E = b_i$$

$$A_i = \Delta \mathbf{r}^T[\Phi(C_{EG1}\mathbf{e}_{K1}^{(i)})C_{EG2}\Phi(\mathbf{e}_{K2}^{(i)}) - \Phi(C_{EG2}\mathbf{e}_{K2}^{(i)})C_{EG1}\Phi(\mathbf{e}_{K1}^{(i)})] \tag{10.23}$$

$$b_i = \Delta \mathbf{r}^T(C_{EG1}\mathbf{e}_{K1}^{(i)} \times C_{EG2}\mathbf{e}_{K2}^{(i)})$$

Analyzing a couple of snapshots of a fragment of the earth's surface, we select N point landmarks on them, for which a one-to-one correspondence is established. We now represent the set of equations (10.23) in the form of the matrix equation

$$A\theta_E = b,\qquad(10.24)$$

where A is $(N \times 3)$ – matrix, $\mathbf{b} = \{b_i\}$ – N – dimensional vector. Solving it by the method of least squares, we arrive at the relation $\theta_E = A^+b$, in which $A^+ = (A^T A)^{-1} A^T$ is the matrix that is pseudoinverse for the matrix A.

The estimate of the unknown magnitude of the discrepancy vector θ_E between the basis K and E obtained in this way depends significantly on the random errors of the stellar sensor realized in measuring the orientation of the basis E with respect to the inertial space at the points 1 and 2 of the flight trajectory of the SC. The influence of these errors can be mitigated in many ways, one of them is the averaging of the value of a vector θ_E by processing data from M pairs of images of some sections of the earth's surface:

$$\hat{\theta}_E = \frac{1}{M}\sum_{j=1}^{M}\theta_j.$$

The vector $\hat{\theta}_E$ is used later to refine the coordinate determination.

10.5.3 Calibration Using a Virtual Reference System

Let on SC flight route (or near it) there be a sub-satellite polygon with coordinately fixed landmarks. Let, further, flying over the polygon, the survey equipment of the satellite forms $k = \overline{1, S}$ the frames of shooting of this range.

From n known points (landmarks) in a topocentric coordinate system $O_T X_T Y_T Z_T$ we choose a point M_i. We then choose an arbitrary point $V_j (j = \overline{1, m})$ whose coordinates in the same coordinate system can be given arbitrarily, a virtual point (Figure 10.4).

For known vectors $\mathbf{r}, \mathbf{r}_{M_i}$ and $\mathbf{r}_{V_j}$ (Figure 10.4), we calculate the unit vectors $\mathbf{e}_{M_j}, \mathbf{e}_{V_j}$ of the directions OM_i and OV_j, and also the vector $\Delta\mathbf{r}_{ij} = \mathbf{r}_{V_j} - \mathbf{r}_{M_i}$. In the basis of the algorithm for the refinement of the value of the matrix C_{EK}, we set the condition for the coplanarity of the vectors $\Delta\mathbf{r}_{ij}, \mathbf{e}_{M_i}$ и $\mathbf{e}_{V_j}$:

$$\Delta\mathbf{r}_{ij}^T (\mathbf{e}_{V_j} \times \mathbf{e}_{M_i}) = 0.\qquad(10.25)$$

Write condition (10.25) in the coordinate system $O_T X_T Y_T Z_T$. From the camera readings and the known focal length F of the optical system, we form

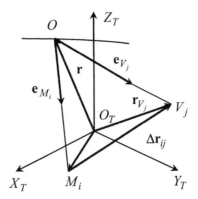

Figure 10.4   Position of a virtual point.

the unit vector $\mathbf{e}_{M_iK}$ of the direction to the i-th landmark. As follows from the scheme in Figure 10.4, in the topocentric coordinate system it is defined by the relation

$$\mathbf{e}_{M_iT} = C_{KT}\mathbf{e}_{M_iK} \tag{10.26}$$

From the same scheme we find the expressions necessary for calculating the matrix C_{KT}:

$$C_{KT} = C_{ET}C_{EK}^T, C_{ET} = C_{IT}C_{IE}^T.$$

Then equation (10.26), taking into account the formula $C_{EK} = E_3 + \Phi(\boldsymbol{\theta}_E)$, can be written in the form

$$\mathbf{e}_{M_iT} = C_{ET}(E_3 - \Phi(\boldsymbol{\theta}_E))\mathbf{e}_{M_iK} \tag{10.27}$$

For known vectors $\mathbf{r}$ and $\mathbf{r}_{V_j}$ ort $\mathbf{e}_{V_j}$ of direction from the point O to the chosen virtual point V_j is determined by the formula

$$\mathbf{e}_{V_j} = (\mathbf{r}_{V_j} - \mathbf{r})/\|\mathbf{r}_{V_j} - \mathbf{r}\|. \tag{10.28}$$

Taking into account equalities (10.27) and (10.28), it possible to write the coplanarity condition (10.25) as

$$B\boldsymbol{\theta}_E = q, \tag{10.29}$$

where the following notations are introduced:

$$B = \Psi\Phi(\mathbf{e}_{M_iK}), q = \Psi\mathbf{e}_{M_iK}, \Psi = \Delta\mathbf{r}_{ij}^T\Phi(\mathbf{e}_{V_j})C_{ET}.$$

On processing information about all a priori known and virtual reference points the number of equations (10.29) of type is equal to mn.

As the result, we have an overdetermined system of linear algebraic equations

$$\mathbf{P}\boldsymbol{\theta}_E = \mathbf{Q} \tag{10.30}$$

with a matrix $\mathbf{P}$ of size $mn \times 3$ and mn-dimensional vector $\mathbf{Q}$.

The estimate of the desired values of coordinates of the vector $\boldsymbol{\theta}_E$ is extracted from the solution of the system of equations (10.30).

Thus, the procedure for refining the mutual orientation of the bases $\mathbf{E}$ and $\mathbf{K}$ the stellar sensor and camera is reduced to solving the linear least-squares problem.

On the presence of several frames of sub-satellite polygon it is expedient to use for every frame its own system of virtual reference points, formed randomly from the given range of the variations of their coordinates; then according to the above scheme it is necessary to calculate the values of vectors $\hat{\boldsymbol{\theta}}_{E,k}(k = \overline{1,S})$. Then as estimate $\hat{\boldsymbol{\theta}}_E$ of the current misalignment of the bases $\mathbf{E}$ and $\mathbf{K}$, the mean value of estimates $\hat{\boldsymbol{\theta}}_{E,k}(k = \overline{1,S})$ is taken:

$$\hat{\boldsymbol{\theta}}_E = \{\hat{\theta}_1, \hat{\theta}_2, \hat{\theta}_3\} = \frac{1}{S} \sum_{k=1}^{S} \hat{\boldsymbol{\theta}}_{E,k}.$$

10.5.4 Reduction of the Flight Calibration Algorithm

Let us consider some computational aspects of the solution of the system of algebraic equations (10.30).

The disadvantage of the convenient method of calculating the normal equations for solving the linear least-squares problem is the bed condition numbers of the resulting system of equations [19]. Indeed, conditionality factor of the matrix $\mathbf{P}^T\mathbf{P}$ of normal equations and the matrix $\mathbf{P}$ of the system of equations (10.30) ($cond(\mathbf{P}^T\mathbf{P})$ and $cond(\mathbf{P})$ respectively) are related by equality $cond(\mathbf{P}^T\mathbf{P}) = cond^2(\mathbf{P})$. Nevertheless, the usage of normal equations for small dimension of the vector of desired variables gives a solution that is satisfactory in accuracy [19]. This method of solving of the system of equations (10.30) is used below in the modeling of the calibration algorithm, considered as a basic algorithm for flight geometric calibration.

It was noted in [20] that the singular numbers of the matrix of a linear over determined system of algebraic equations can serve as quantitative measures of the estimation of the corresponding variables. If the singular number is small, then it makes sense to take zero as an estimate of the corresponding variable, thereby performing a reduction – a decrease in the dimension of the vector of the parameters being evaluated. It also indicates that for those state variables that can be estimated with satisfactory accuracy, it is advisable to build separately algorithms for their evaluation.

If we refuse to estimate the coordinate θ_3 of the vector $\boldsymbol{\theta}_E$ due to the insignificant influence of the error in its estimation on the accuracy of the coordinate reference of the images (see Section 10.4), then to evaluate the coordinates θ_1 and θ_2 of the mentioned vector it is necessary to find the solution of the following system of algebraic equations:

$$\hat{\mathbf{P}}\hat{\boldsymbol{\theta}}_E = \mathbf{Q}, \tag{10.31}$$

where $(mn \times 2)$ – matrix $\hat{\mathbf{P}}$ is obtained from the matrix $\mathbf{P}$ in the algorithm (10.30) by deleting the third column, and the vector $\hat{\boldsymbol{\theta}}_E$ contains two coordinates: $\hat{\boldsymbol{\theta}}_E = [\theta_1 \theta_2]^T$.

Let us agree to continue to call the procedures of flight geometric calibration by algorithms (10.30) and (10.31) as B- and R-calibration respectively.

10.6 Modeling of Algorithms of Calibration and Topographic Binding

The simulation was performed for the purpose of analyzing the accuracy of the coordinate binding of satellite images realized with the help of the above algorithms from data obtained from on-board information sources. On computer modeling the influence of the number N_K of shooting frames of a sub-satellite polygon, the number N_3 of topographically biding landmarks and the accuracy characteristics of the GPS-receiver, the stellar sensor, and topographical reference of landmarks on the accuracy of the flight calibration and the binding of space images were analyzed.

The calibrations of the optical-electronic complex of the SC and the coordinate determination of the images were carried out in the presence of a system of parametric perturbations formed in the form of normally distributed centered random variables with the corresponding root-mean-square deviations.

Simulation was performed when the satellite was moving in a circular orbit with the following parameters: altitude of the orbit – 680 km, inclination of the orbit –98°, longitude of the ascending node – 142°.

The perturbation system included the following parameters:

- additive errors with root-mean-square deviations $\sigma_X = \sigma_Y = \sigma_Z = \sigma_{GPS}$ arising when the coordinates of the center of mass of the SC are formed from information from the global positioning system;
- stellar sensor errors $\delta_i (i = 1, 2, 3)$ (standard deviation σ_i), which correspond to the rotation of the basis $\mathbf{E}$ around the axes X_E, Y_E, Z_E in its calculated position;
- the error of the focal length δF of the camera;
- errors of topographical reference of landmarks with standard deviation σ_P;
- reading errors of coordinates x_k, y_k of the k-th reference point on the plane of the camera image due to of their pixel-wise representation. They are subject to uniform distribution law with standard deviations σ_k.

Coordinates of the vector $\boldsymbol{\theta}_*$ characterizing the initial uncertainty of mutual orientation of the bases $\mathbf{E}$ and $\mathbf{K}$ are centralized normally distributed random variables with the same root-mean-square deviations σ_{θ_*}.

Given the statistical nature of the errors inherent in the above sources of information, it is natural to use the method of statistical tests to evaluate the accuracy of calibration algorithms. It is important to note that a large number of tests are required to confidently evaluate the accuracy of the algorithms. So that the relative error of estimating the error variance of the calibration algorithm with probability 0, 9973 does not exceed 1%, it is necessary to conduct at least 45,000 tests [9]. In this connection, the modeling of the calibration process should be organized in such a way that the required estimates can be obtained in a reasonable time.

10.6.1 Calibration Algorithms 1 and 2

In modeling, it was assumed that landmarks are located on 28×28 km area. The location N of landmarks on the polygon was specified using a system of pseudo-random numbers with a uniform distribution law. Statistical tests were carried out for the case when the camera was directed to the nadir when surveying landmarks.

The mathematical expectation and the standard deviation of the error estimation of the vector $\boldsymbol{\theta}$ were determined by processing the statistical information obtained during the computational experiment.

The contribution of each of the perturbing factors to the magnitude of the root-mean-square deviation of the error in estimating the coordinates of

Table 10.1   Results of Algorithm 2 simulation

Variants	Coordinates of Vector θ_E, ang., min			Error of Estimation, ang., sec		
1	−22,33	−26,87	−9,33	8,82	2,04	−3,09
2	10,96	8,66	1,02	−11,52	2,95	−0,44
3	−3,78	−0,67	2,27	−2,08	−7,57	4,11

the vector will be determined using the matrix (vector) of the sensitivity coefficients.

The presence of information on the sensitivity coefficients makes it possible not only to estimate the contribution of each of the perturbing factors to the resulting error of the alignment process, but also to formulate the requirements for the level of parametric perturbations in order to align the "camera and stellar sensor" system with the required accuracy.

During the simulation it was established that algorithms 1 and 2 give almost the same accuracy. In connection with this, algorithm 2, which has less computational complexity, was subsequently investigated.

We will illustrate the results of statistical tests performed for number of landmarks $N = 7$, focal distance $F = 2.2$ m, field of view of the camera − $3.5° \times 3.5°$. For the case when the root-mean-square deviations in the system of disturbing factors were equal, respectively

$$\sigma_{\delta 1} = \sigma_{\delta 2} = \sigma_{\delta 3} = 4'', \sigma_P = 0.1\ m, \sigma_{GPS} = 15\ m,$$
$$\sigma_K = 3 \cdot 10^{-6}\ m, \sigma_{\theta_0} = 10' \tag{10.32}$$

the mathematical expectations of errors in determining the coordinates θ , θ_y, θ_z of the vector θ were hundredths of the angular seconds, and the root-mean-square deviations of the errors were estimated by the values (angular sec):

$$\sigma_{\theta_x} = 6.09, \quad \sigma_{\theta_y} = 6.09, \quad \sigma_{\theta_z} = 8.36, \quad \sigma_{\theta}^{\Sigma} = 11.98.$$

In none of the 45,000 implementations, the values of the errors in the determination of angles did not exceed the absolute values of the three standard deviations of the errors of the corresponding estimated variables.

To judge the level of residual uncertainty in the relative orientation of the bases E and K, as an example in Table 10.1 shows the calculation data, performed according to Algorithm 2 for three variants of generating a system of random perturbations with parameters (10.32).

10.6.2 Calibration Algorithms B and R

Of greatest interest were the characteristics of the accuracy of image binding, performed taking into account the results of B- and R-calibrations with a minimum number N_3 of control points on the sub-satellite polygon and the only frame of his shooting. (In the implementation of the B-calibration, the minimum value N_3 is two, while for the R-calibration, it is one.). Note that for both calibration the number N_V of virtual reference points must be at least two (simulation was performed at $N_V = 5$).

The ground polygon is the size 1000×2000 m along the KA flight path. It contains topographically binding landmarks whose height is in the range ± 50 m. Geographic coordinates of the center of the polygon – vertices O_T of the topocentric coordinate system – $\varphi_* = 50°$, $\lambda_* = 30.5°$. Virtual points with coordinates X_T and Y_T within ± 5000 m are selected on the plane $O_T X_T Y_T$. The coordinates of n known and m virtual landmarks are formed using a system of pseudo-random uniformly distributed numbers.

Simulation is performed when there are one or three images of a ground polygon. A single snapshot is formed when the sub-satellite point coincides with a point O_T – the top of the topocentric coordinate system. With three frames of the polygon shooting, the above picture is supplemented by two pictures separated in time from the moment of passing the sub-satellite point through the point O_T to the values $\tau_1 = -10c$ and $\tau_2 = 10c$, respectively. The SC orientation control system functions in the stabilization mode of the line of sight in the direction of the point O_T, providing stabilization with a standard deviation equal to 0.05 degrees for each of the degrees of freedom of the satellite. The simulation scheme is described in detail in [13].

Simulation was performed for three variants of specifying the parameters in the system of disturbing factors:

option I $- \sigma_{GPS} = 15\ m$, $\boldsymbol{\sigma}_\delta = \{2'', 2'', 20''\}$, $\sigma_T = 2\ m$;
option II $- \sigma_{GPS} = 5\ m$, $\boldsymbol{\sigma}_\delta = \{1, 1, 10''\}$, $\sigma_T = 1\ m$;
option III $- \sigma_{GPS} = 1\ m$, $\boldsymbol{\sigma}_\delta = \{1, 1, 10''\}$, $\sigma_T = 1\ m$;

Data of the flight calibration and the coordinate reference of the images are placed in Table 10.2. In it $\sigma_j(i = 1, 2, 3)$ are standard deviations of errors in the estimates of the vector $\boldsymbol{\theta}_E$ coordinates that take place after the corresponding calibration is performed.

Analyzing the results of the simulation, we note that for the case when the geometric calibration in the flight task of the satellite is not provided,

Table 10.2 Results of B- and R-algorithms simulation

Variant	Algorithm	N_3	N_K	σ_1	σ_2	σ_3	σ_X	σ_Y	σ_Z
				ang., sec			m		
I	B	2	1	11,53	10,02	2963,3	54,41	57,25	39,59
		2	3	5,86	6,39	1853,9	32,29	37,48	25,56
		5	1	5.44	5.42	414.4	21,01	23,88	15,99
	R	1	1	5,07	5,07	599,3	21,09	23,96	16,10
		1	3	2,92	2,91	602,3	17,74	20,17	13,54
		5	1	5.02	5.01	600.2	20,97	23,89	15,98
II	B	2	1	6,36	5,21	1786,6	29,78	32,63	22,21
	R	1	1	2,00	1,96	599,3	10,76	12,28	8,20
III	B	2	1	3,78	3,77	1067,1	18,04	20,42	13,90
	R	1	1	1,30	1,24	599,3	9,05	10,34	6,88

the standard deviations of the images coordinates for the variant I of the perturbation parameters are characterized by the following values:

$$\sigma_X = 1644, 1 \ m, \sigma_Y \approx 1881, 4 \ m, \sigma_Z \ 1256, 3 \ m.$$

With a change from 1 to 10 in the number N_3 of ground reference points, the errors of in-flight calibration and the coordinate determination of space images take the following values:

variant I : $\sigma_1 \approx \sigma_2 = 5, 0''$; $\sigma_X \approx 21, 0 \ m$, $\sigma_Y \approx 23, 9 \ m$, $\sigma_Z \approx 16, 0 \ m$;

variant II : $\sigma_1 \approx \sigma_2 = 1, 9''$; $\sigma_X \approx 10, 7 \ m$, $\sigma_Y \approx 12, 2 \ m$, $\sigma_Z \approx 8, 2 \ m$;

variant III : $\sigma_1 \approx \sigma_2 = 1, 2''$; $\sigma_X \approx 9, 0 \ m$, $\sigma_Y \approx 10, 3 \ m$, $\sigma_Z \approx 6, 9 \ m$.

It is significant that high accurate winding of images by orbital data is possible on the basis of R-calibration performed from information on the position of a single landmark in one picture of the sub-satellite polygon.

10.7 Conclusion

In this chapter, as applied to shooting facilities with narrow field of view, peculiar for observation SC of the Earth with high and extremely high resolution the problem of exact coordinate determination of space images only by orbital data is investigated.

Since the greatest influence on the accuracy of the coordinate reference of space images is caused by the residual uncertainty of the mutual orientations of the stellar sensor and shooting camera, and an effective geometric calibration is used to calculate it, a set of calibration algorithms based on photogrammetric conditions of collinearity and coplanarity (algorithms that use direct and inverse photogrammetric equations, virtual landmarks, algorithms for a priori unknown point landmarks), has developed by the author.

The analysis of expression (10.13), which establishes the connection between the error in the orientation of the survey frame and the parameters of the relative orientation of the bases $\mathbf{E}$ and $\mathbf{K}$ the stellar sensor and the survey equipment (coordinates of the vector θ_E, on the one hand, allows us to estimate the contribution of each component of the vector θ_E to the binding error of the survey objects, and on the other hand – to substantiate the possibility of correctly constructing an algorithm of flight geometric calibration for those coordinates of the vector θ_E that are well estimated.

Two variants of the in-flight geometric calibration procedure are proposed: the so-called B- and R-calibrations. The first option (basic) involves the construction of an algorithm for estimating all the coordinates of the vector θ_E, the second – a reduced algorithm – specifies the values of only the coordinates θ_1 and θ_2 of the specified vector. Both calibration options are based on the photogrammetric condition of coplanarity, a system of known landmarks and landmarks chosen arbitrarily – virtual reference points.

Accurate coordinate binding of images on the orbital data is possible on the basis of R-calibration, performed from information on the position of a single landmark in one picture of the sub-satellite polygon. The accuracy of the coordinate fixing of the survey objects for various variants of specifying the numerical values of the parameters in the system of disturbing factors is much greater than the accuracy of the binding obtained under the same conditions from the results of the B-calibration and two terrestrial landmarks.

References

[1] https://en.wikipedia.org/wiki/Remote_sensing
[2] T. P. Srinivasan, B. Islam, S. K. Singh, B. Copala Crishna, P. K. Srivastava. In-flight geometric calibration – an experience with Cartsat-1 and Cartsat-2 //Archives of the Photogrammetry and Remote Sensing and Spatial Information Sciences. XXXVII. Part B1, pp. 83–88, Beijing, 2008.
[3] O. N. Kravchenko, M. S. Lavrenyuk, N. N. Kussul. Algorithms of geographical binding of images of satellite "Sich-2". – Naukovi pratsi DonNU, Seriya: "Informatyka, кібернетыка та obchislyuvalna tekhnika", no. 2(18), pp. 71–79, 2013.

[4] http://www.gisa.ru/54663.html.

[5] S. Yu. Samoilov. Ways of reduction of errors of geographical binding from spacecrafts the Earth remote sensing // Vestnik FGUP NPO imeni S.A. Lavochkina., no. 2, pp. 19–22, 2012.

[6] D. V. Lebedev, A. I. Tkachenko. Calibration of information measuring complex of spacecraft for the Earth surface survey // Problemy upravleniya i informatiki≫, no. 1, pp. 101–120, 2004.

[7] E. I. Somov, S. A. Butyrin, V. K. Technology of processing the accompanying measurement information for highly accurate coordinate referencing of space images, Izvestiya Samarskogo nauchnogo tsentra RAN, 11, no. 56, pp. 156–163, 2009.

[8] I. A. Pyatak. Problems of coordinate referencing of images made by SC, Visnyk Dnipropetrovskogo universytetu, Ser., "Raketno-kosmichna tekhnika", no. 14, 116–122, 2011.

[9] D. V. Lebedev, A. I. Tkachenko. Parametric adjustment of the complex ≪camera and star sensor≫ installed on low-orbiting spacecraft // Izvestiya RAN. Teoriya i sistemy upravleniy, no. 2, pp. 153–165, 2012.

[10] D. V. Lebedev. In-flight geometric calibration of optoelectronic equipment of remote sensing satellite by unknown landmarks //Mezhdunarodnyi nauchno-tekhnicheskiy zhurnal "Problemy upravleniya i informatiki", no. 5, pp. 114–125, 2013.

[11] A. I. Tkachenko. On in-flight adjustment of optoelectronic complex of spacecraft // Izvestiya RAN. Teoriya i sistemy upravleniya, no. 6, pp. 122–130, 2013.

[12] A. I. Tkachenko. On coordinate biding of ground objects by space images // Космічна наука і технологія, vol. 2(21), pp. 65–72, 2015.

[13] D. V. Lebedev. On the coordinate determination of space images by orbital data // Mezhdunarodnyi nauchno-tekhnicheskiy zhurnal "Problemy upravleniya i informatiki", no. 6, pp. 116–126, 2016.

[14] WGS84, http://ru.wikipedia.org/wiki/WGS-84.

[15] I. N. Bronstein, K. A. Semendyaev. Handbook of Mathematics, [in Russian], edition, GITTL. Moscow, 1957.

[16] S. K. Ghosh. Fundamentals of computational photogrammetry, New Delhi, 2005.

[17] A. N. Lobanov. Photogrammetry [in Russian], Nedra, 1984.

[18] J. E. Bortz A new mathematical formulation for strapdown inertial navigation, IEEE Trans. on Aerospace and Electron. Syst., AES-7, no. 1, pp. 61–66, 1971.

[19] S. A. Belov, N. Yu. Zolotykh. Numerical methods of linear algebra. Laboratory practice [in Russian], Nizhegorodskiy gosunivesitet imeni N. N. Lobochevskogo, Nizhnii Novgorod, 2005.

[20] A. A. Golovan, N. A. Parusnikov. Mathematical fundamentals of navigation systems. Part II: Usage of methods of optimal estimation to navigation problems [in Russian], 2nd edition, MAKS Press, Moscow, 2012.

11

Algorithms of Robotic Electrotechnical Complex Control in Agricultural Production

**Vitaliy Lysenko, Igor Bolbot, Yuriy Romasevych*,
Viatcheslav Loveykin and Valeriy Voytiuk**

National University of Life and Environmental Sciences of Ukraine, Kyiv,
Ukraine
*Corresponding Author: romasevichyuriy@ukr.net

Mobile robotic electrotechnical complex, able to move greenhouse area, by the means of technological guideways, provides monitoring of the main parameters of the atmosphere of greenhouses, as well as phytomonitoring, including the quality of products, while identifying its zones. The strategies of controlling electrotechnical complexes, which provide the growth technology, are formed on the basis of information coming from the robotic electrotechnical complexes, maximizing the profit of production at the current moment. Such a complex should be characterized by individual features of intelligence and operate according to determined management algorithms which allow: to evaluate the quality of products, to monitor the phytosanitary and atmospheric conditions, to move around the area, minimizing the traversed path, avoiding obstacles, etc.

11.1 Introduction

Information on the condition of atmosphere and plants (phyto-condition) in closed-ground structures is extremely important for ensuring guaranteed yields [1]. Growing of plant products in closed-ground buildings – modern greenhouse facilities is accompanied by significant energy expenditure (in the structure of the cost of production their share totals 70%). Moreover, the products should be provided in the maximum quantity of the appropriate quality. Here are the main factors that influence the profit of a modern enterprise, which specializes on the cultivation of plant products in closed-ground facilities. The reduction of energy costs for the production of technologically justified quantities of products and provision of their appropriate quality may be achieved through the development of strategies for managing the production of plant products based on the use of the results of monitoring phytosanitary and atmospheric conditions via intelligent robotic electrotechnical complex, which will ultimately maximize the profit of production.

The practical realization of a robotic electrotechnical complex for the indicated purposes requires the use of microcontrollers, high-performance drive units and sensors. The most complete use of the available resources of such complexes is related to the regulation of their parameters of motion, which allows obtaining qualitative information on phytosanitary condition, providing energy resources saving and preventing the occurrence of abnormal (sometimes even emergency) situations in the production.

The task of automatic control of the speed of the robotic electrotechnical complex is solved at the lowest – executive level of the control system. Access to the fixed speed of robot movement, which moves on the object of agrarian production (for example, in a greenhouse), is accompanied by fluctuations. A similar situation may arise when the stochastic effects influence the robot movement (for example, the unevenness of the track it moves on, unpredictable obstacles on its way, etc.). The above-mentioned fluctuations reduce the energy efficiency of the robot platform drive, which is especially important with a limited battery charge, and reduces the accuracy of the positioning of the platform, and hence the quality of the information collected. Unfortunately, the above-mentioned factors cannot be eliminated in advance. Therefore, for the qualitative realization of the given (programmed) movement of the robotic complex, it is necessary to complete the synthesis of the system of automatic regulation of its speed.

11.2 An Overview of Recent Findings

Information on the phytosanitary and atmospheric conditions (phyto-condition) in closed-ground structures is extremely important for ensuring guaranteed yields. In Ukraine, the production of such equipment is extremely insufficient. The use of phyto-monitoring as the main method for ensuring the technological parameters of plant development in closed-ground structures is discussed in findings by Camilla Barattoa, Guido Fagliaa, Matteo Pardoa, Marco Vezzolia, Luca Boarinob, Massimo Maffeic, Simone Bossic, Giorgio Sberveglieria. The sensors for monitoring the development of plants are in the focus of the research. The disadvantage of such technical means is bind to one plant. Solving this problem requires the use of mobile platforms that can monitor several plants along the trajectory of their movement. In this case, the quality control of the platform movement, which is usually solved with the use of PI regulators, is a critical issue. The quality of regulation at the same time depends on the method of adjusting the coefficients of proportional and integral components. To date, dozens of such methods are known, each of which has its disadvantages and advantages [1]. However, the choice of the method of setting the PI regulator is not the only way to improve the quality of regulation. One of them is the synthesis of nonlinear PI regulators [2–4], which, in addition, can provide an acceptable quality of regulation of substantially nonlinear objects. In this key one can consider the modification of the structure of the controller, which is used in paper [5] to regulate the first order object. However, the feasibility of using this modification for robotic complexes in agroindustrial production, which is usually presented by systems of the second and higher orders, needs to be proved.

11.3 The Purpose of the Research

The purpose of the research is to develop algorithms for controlling robotic electrotechnical complexes in agricultural production, which provides monitoring of the atmospheric and phytosanitary conditions in closed-ground structures for the formation of management strategies that maximize the profit of production. The achievement of this purpose is related to the development of algorithms for controlling the work of robotic electrotechnical complexes at the upper (strategic) and lowers (executive) levels. For the upper level of control, the development of algorithms provides a general strategy for controlling the production of plant products based on the use of the results of monitoring the phytosanitary and atmospheric conditions by means of robotic electrotechnical complex. For the lower level, it is necessary to perform the synthesis of the nonlinear structure of the PI regulator to control the movement of the platform of the robotic electrotechnical complex and to prove the expediency of its use for the modes of adjustment of the set point and the neutralization of external stochastic disturbances.

11.4 Statement of Basic Materials

11.4.1 Development of Upper-Level Control Algorithms

Localization and development of a terrain map is a key issue that needs to be considered when designing a mobile robot. The main way to solve this problem is to use SLAM (Simultaneous Localization and Mapping) algorithms. This type of algorithms is considered within the framework of this research, since it has the high accuracy of localization and sufficient performance for real-time operation.

At the moment, there are many different SLAM algorithms that differ in both the type of input information, representation of the surrounding space in the form of a map, and the methods of processing this information. Let's introduce the classification of localization algorithms in terms of the size of the mapped space: two-dimensional localization on the plane (2D-SLAM); three-dimensional localization in space (3D-SLAM); color localization for R, G, B components of the image (Color-SLAM); color space three-dimensional localization (6D-SLAM).

These specifications depend directly on the sensor used. With the use of the simplest laser range finders, the input information for the algorithm is a two-dimensional horizontal section of the terrain of the surrounding objects, respectively, for processing we use 2D-SLAM. It should be noted that the localization algorithms used by us combine 2D-SLAM and Color-SLAM i.e., it is expanded to color three-dimensional localization in space (6D-SLAM).

The mobile robot sensory system collects information about the coordinates of its location. It includes two subsystems of motion sensors and a system of technical vision. The movement algorithm of the electrotechnical complex of phytomonitoring in a greenhouse with the use of 6D-SLAM is presented in Figure 11.1.

Another feature of SLAM algorithms of this type is the ability to close the cycle, that is, recognition of the already traversed map area followed by the relaxation of the entire map along the trajectory of the robot. Sequential localization defines a change in the position of the robot between two consecutive scans.

Algorithms of this type tend to give a more accurate result of determining the position of a mobile robot, compared to global SLAM algorithms, but must be performed in real time. However, if this algorithm didn't converge at least once, its further use is impossible without additional amendments. To achieve the shortest result in terms of accuracy and reliability of localization, it is necessary to use both algorithms simultaneously.

In addition to the phytoclimatic parameters, which are obtained from the sensors aboard the robot, the image is filmed and transferred from the video camera for determining the degree of maturity of tomatoes and their amount.

The vision of a mobile robot can be defined as the process of isolating, identifying and transforming information obtained from three-dimensional

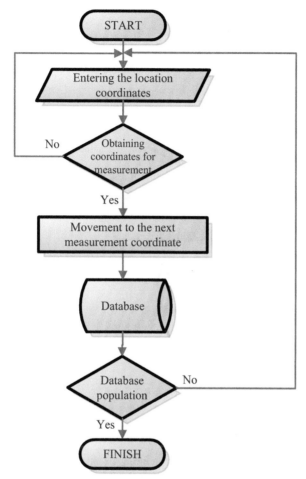

Figure 11.1 The movement algorithm of the electrotechnical complex of phytomonitoring in a greenhouse with the use of 6D-SLAM.

images. Visual information is transformed into signals using a video sensor, performed by the camcorder. After spatial sampling and quantization in amplitude, these signals give a digital image. The image is recognized by the software installed on the server, which is responsible for the strategic level. The algorithm for measuring the parameters of phytosanitary and atmospheric conditions is shown in Figure 11.2.

A special role is assigned to the computing system of the robot, which relies on the processing of signals coming from the sensor system, recognition

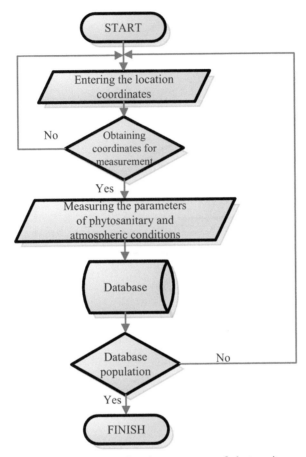

Figure 11.2   The algorithm for measuring the parameters of phytosanitary and atmospheric conditions.

of the environment conditions, determination of the desired trajectory of the robot, calculation of the deviations of the current configuration and the speed of the robot from the desired values and the revaluation of deviations into managerial influence. The use of traditional methods and algorithms of management is a challenge for solving such problems, there arises a necessity to use special trajectory traffic management strategies applying the principles of adaptation and self-study.

A situational approach may serve as an alternative method for solving a problem. This method is based on the identification of situations with a predetermined number and the adoption of managerial decisions, associated with situations. Discrete-event models of various types, including probabilistic automata, are used to describe the situation of transitions. Probabilistic machines are now increasingly used in various programming areas. Their main advantages are simplicity and visibility. The most elaborate question of the use of probabilistic automata is a parsing analysis in various kinds of translators of algorithmic languages, they are also used in the field of logical control and object-oriented programming, used in the programming of protocols, games and schemes of programmable logic. When using this approach, the mobile robot is considered as an independent system. Such systems respond to the flow of events by changing states and performing actions while transiting from one state to another or during the transition of actions in states. The general control algorithm of the mobile robot with the aim of monitoring phytosanitary and atmospheric conditions in closed-ground structures is presented in Figure 11.3 [6, 7].

The implementation of the multi-level hierarchical structure of controlling the electrotechnical complex of phytomonitoring in the greenhouse, which includes: strategic, tactical and executive levels, requires the use of non-similar software.

The implementation of the mobile robot control system is based on the integrated Arduino environment with the corresponding hardware complex. The Arduino Integrated Environment is a multi-platform Java application that includes code editor, compiler, and firmware modules for the board. The development environment requires the use of the Processing programming language, which is similar to the Wiring language. Applications are processed using a preprocessor, and are then compiled using the AVR-GCC.

The program for managing the intelligent robot for monitoring the phytosanitary and atmospheric conditions in the closed ground facilities is written in the programming language C or C++. The Arduino development environment includes a Wiring program library that allows you to easily perform many standard I/O operations.

Programming is done through its own software shell (IDE), available free of charge on the Arduino site and does not require a license purchase. This shell contains a text editor, a project manager, a preprocessor, a compiler, and tools for downloading the program to a microcontroller. The shell is written in Java based on the Processing project, operates on the basis of Windows, Mac OS X and Linux.

The Arduino IDE project manager has a non-standard mechanism for adding libraries, which, in the form of source code in the standard C++ language, are added to a special folder in the IDE working directory. The name

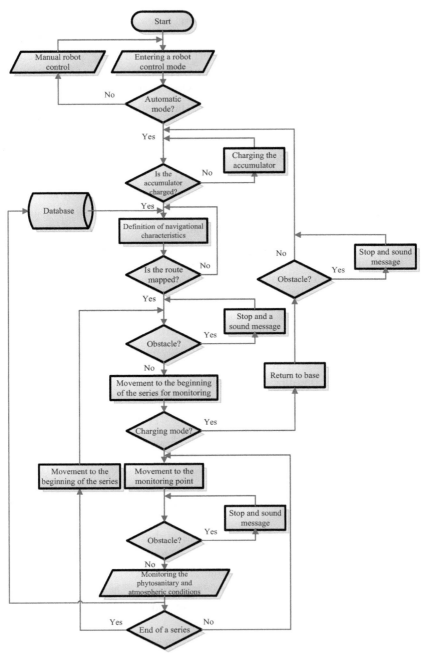

Figure 11.3 Flow diagram of the control algorithm of the mobile robot with the aim of monitoring phytosanitary and atmospheric conditions in closed-ground structures.

of the library will be added to the list of libraries in the IDE menu. Only needed libraries are selected and they are added to the compilation list.

Downloading the program into the Arduino microcontroller is through a pre-programmed special bootloader (all Arduino microcontrollers go basically with this bootloader). The downloader is based on the Atmel AVR ApplicationNote AN109 and is capable of functioning using RS-232, USB, or Ethernet interfaces depending on the peripherals of a specific processor board. In some cases, a separate adapter is required for programming.

Since the Arduino platform cannot process information and is limited in its technical specifications, we faced the problem of collecting data, processing them and displaying them for further processing. Using a laptop on board a robot would be the best solution. But we faced the technical problems of the increase in the mass of the robot and its energy consumption. Therefore, the Raspberry pi 3 – miniature portable computer was used. Its small size and energy consumption allow it to be placed on board a robot. It was coupled up with functionality, namely the establishment of a fully functional operating system, the availability of GPIO and the ability to remotely implement settings in the program code without making significant structural changes made it possible to more broadly approach this issue.

The operating system Raspberry – an analogue of Linux, designed specifically for this mobile robot, was installed. This operating system has software designed for software and computer engineering.

The Node-RED software environment, which has an open license, was used for the work. This program allows you to create complex software products in the form of flow diagrams. This speeds up software development time, not only for communication with devices like the Arduino, but also provides the ability to develop a WEB interface for the correct operation with them.

The complexity of software development was to use several programming languages. Thus, for the next step you need to use JavaScript. This programming language is essential for working with Node-Red.

We have installed NPM – package management system – for complete operation. This software package is installed over Node-Red, to extend its functionality. Flow diagram of content control is shown in Figure 11.4 [8].

A flow diagram for output of data to the screen and storing information in a file is also developed. See Figure 11.5.

It should be noted that these blocks are a ready-made code in the JavaScript language, which introduces adjustments for the operability of the entire program.

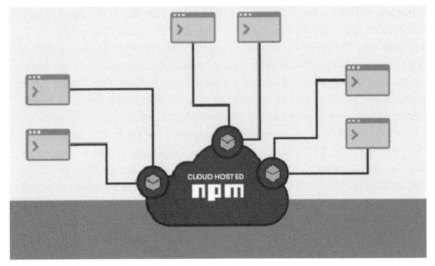

Figure 11.4   Flow diagram of content management.

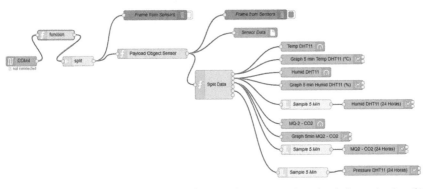

Figure 11.5   A flow diagram for output of data to the screen and storing information in a file.

The first unit is responsible for extracting information from the COM port. Arduino Mega is connected to Raspberry via USB, and this unit is responsible for the technical aspect of the information transfer. It is followed by the functional unit with a code that decodes the information received. We made the appropriate adjustments to the program code for the correct operation of this scheme at the beginning of Arduino programming. This flow diagram is important because it performs the role of a function of this program and it streams the information from the data stream into separate components and displays it further.

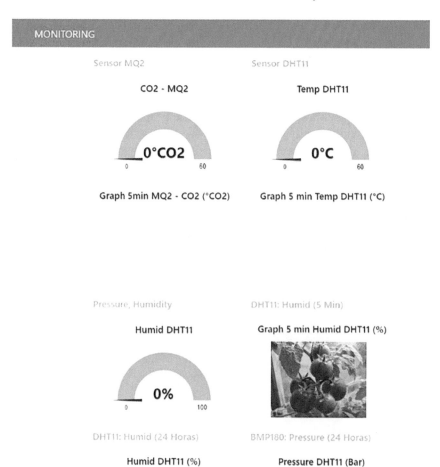

Figure 11.6 Interface of the control system of the electrotechnical complex of the strategic level.

Only at the final stage the streamed data are displayed on a specially designed WEB-interface and stored in a file for use in the calculations. WEB-interface of the control system of the electrotechnical system for phytomonitoring in the greenhouse of the strategic level is presented in Figure 11.6.

The hardware implementation of the control system by an intelligent mobile robot provides:

- the speed corresponding to the requirements for moving an intelligent mobile robot in real time;

- typical requirements for onboard systems (it is compact, reliable and consumes little energy).

The implementation of the task as for measuring the phytometric and phytoclimatic parameters of greenhouse crops by mobile robot for phytomonitoring in a greenhouse containing the frame on which the chassis and the mechanism of fixing the trolley are mounted, according to the proposed solution, is that the mobile robot additionally has a rotary tower, a camcorder, retractable mechanisms, measurement items of phytometric and phytoclimatic parameters, control unit and a pair of double wheels, bearing in mind that – front inside wheel has a diameter of 19 cm, the near-side wheel – 8 cm, rear inside wheel – 8 cm, rear carrying wheel – 5 cm.

The mobile phytomonitoring robot (Figure 11.7) is used jointly with the intelligent microclimate control system, artificial lighting and mineral

Figure 11.7   The appearance of a mobile robot.

fertilization of plants. The new solution to the problem is its mobility, self-sustainability, the presence of the unit, which ensures its connection with the existing control system [9].

In order to ensure a good positioning of the mobile robot platform at given points, increasing the energy efficiency of using the battery of its power supply, increasing the performance of monitoring and eliminating undesirable phenomena during its movement (fluctuation of structural elements, possible abnormal situations, etc.), it is necessary to implement a synthesis of algorithms of motion control at the executive level.

11.4.2 Development of Control Algorithms at the Lower Level

To ensure the above-mentioned requirements, it is necessary to modify the structure of the PI regulator. Its essence can be illustrated by the following equation:

$$u = k_p e + k_I f(e, e_{\partial on}, \int_0^t edt), \qquad (11.1)$$

where u – system moving control function; k_P and k_I – coefficients of adjustment of the proportional and integral components respectively; e and $e_{\partial on}$ – current and permissible error of the angular speed regulation of the robot drive motor $(e = w_y - w, e_{\partial on} = 0.05 w_y)$ respectively; w_y and w are the given and actual values of the angular speed of robot drive motor respectively; f – nonlinear discontinuous function of its arguments.

To illustrate the suggested modification of the PI regulator structure, let's do calculations for the robotic complex, the parameters of which are given in Table 11.1. The drive voltage control is performed by pulse-width modulation of the signal coming from the PI regulator.

Taking into account the data shown in Table 11.1, we calculated the process transfer function on the channel "power supply voltage – angular speed of the drive". It may be presented as follows:

$$G(s) = \frac{1}{2,30 \cdot 10^{-2} + 6,17 \cdot 10^{-3}s + 8,20 \cdot 10^{-5}s^2}. \qquad (11.2)$$

In order to determine the effectiveness of the proposed modification (11.1), let's perform an estimate of the quality of automatic control of the robot speed for two modes: set point mode (acceleration of the robot to the steady speed) and neutralizing the external disturbances (providing the fixed speed under stochastic disturbances on the robot).

Table 11.1   Parameters of the robotic system

Parameter	Parameter Value
Mechanical parameters	
Robot weight, kg	$1.51 \cdot 10^1$
Robot wheel radius, m	$1.50 \cdot 10^{-1}$
Gear ratio of the drive	$4.98 \cdot 10^1$
Drive efficiency	$7.10 \cdot 10^{-1}$
The coefficient of static resistance of the robot motion	$5.00 \cdot 10^{-2}$
Electrical parameters	
Battery voltage, V	$1.20 \cdot 10^1$
Carrier frequency PWM, Hz	$5.00 \cdot 10^1$
Armature windings resistance, Ohm	$7.10 \cdot 10^{-1}$
Armature inductance, Gn	$9.43 \cdot 10^{-3}$
Coefficient of the engine torque, Nm/A	$2.30 \cdot 10^{-2}$

The selected duration of the simulation of the first mode is equal to one second ($T_{SPM} = 1$ s). For evaluating the quality of control during the set point mode, we use the following indicators:

1. integral error

$$\bar{e}_{SPM} = T_{SPM}^{-1} \int_0^{t_p} |e| dt, \qquad (11.3)$$

where t_p – response time;

2. integral control quality index

$$\bar{u}_{SPM} = T_{SPM}^{-1} \int_0^{t_p} |u| dt; \qquad (11.4)$$

3. overshoot

$$e_{max \cdot SPM} = \left(1 - \frac{\max(e)}{\omega_y}\right) \cdot 100\%; \qquad (11.5)$$

4. response time (time of the output angular speed of the drive in the area where the error does not exceed the permissible value)

$$t_p = \arg(\omega(t) \leq \omega_y \pm e_{\partial on}). \qquad (11.6)$$

The lower index of indicators (11.3–11.5) denotes the set point mode.

The indicators (11.3–11.6) were calculated for the set point mode of the angular speed of the drive for the methods of setting the PI regulator: Ziegler-Nichols [10], Kappa-Tau [11], AMIGO (Approximate M-constrained integral gain optimization) [12], Chien-Hrones-Reswick [13], Cohen-Coon [14], Lambda [15], Skogestad [16] (Table 11.2).

Table 11.2 Automatic control quality indicators for set point processing

Method of Setting the PI Regulator	The Structure of the PI Regulator							
	Classical				Modified			
	$\bar{e}_{SPM}$, rad/s	$\bar{u}_{SPM}$, V	$e_{max.SPM}$, %	t_p, s	$\bar{e}_{SPM}$, rad/s	$\bar{u}_{SPM}$, V	$e_{max.SPM}$, %	t_p, s
Ziegler-Nichols	24.75	3.16	65.83	0.47	9.66	1.32	4.92	0.11
Kappa-Tau	19.42	2.78	53.31	0.39	10.54	2.00	8.36	0.25
AMIGO	19.22	2.93	47.98	0.41	9.97	1.46	7.68	0.15
Chien-Hrones-Reswick	22.34	2.93	62.67	0.42	9.66	1.32	4.92	0.11
Cohen - Coon	26.32	3.19	67.89	0.46	9.66	1.32	4.92	0.11
Lambda	29.52	3.62	70.44	0.58	9.66	1.32	4.92	0.11
Skogestad (SIMC)	22.10	2.95	62.14	0.43	9.66	1.32	4.92	0.11

Analysis of the results presented in Table 11.2 shows that the modified regulator has significant advantages over the classical one. This applies to all values. The $\bar{e}_{SPM}$ indicator for the modified structure of the regulator decreased by 1.84–3.05 times, which allows to provide a more qualitative exit of the system to established mode of movement. The $\bar{u}_{SPM}$ indicator decreased by 1.39–2.39 times. This signifies that the use of the modified structure of the PI regulator provides a less stressful mode of operation of the battery. The advantage of the PI regulator with a modified structure also lies in the significant reduction of overshoot $e_{max.SPM}$ and time of regulation t_p by 1.56–5.27 times (Figure 11.8).

The graphs shown in Figure 11.8, reflect the function of the relative power supply voltage $\tilde{U}$ of the motor (black line) and the relative angular speed $\tilde{\omega}$ of the electric drive robot (gray line). They are built for the method of setting the Ziegler-Nichols PI regulator, which is one of the most widespread in practice. The graphs shown in Figure 11.8 confirm that the modification of the structure of the PI regulator allows for much better control quality, a less stressful mode of battery operation and a significantly shorter duration of the transient operation of the electric drive. The latter factor makes it possible to increase the energy efficiency of its operation. In addition to the set point mode, calculations were made for the mode of neutralization of disturbances. In this operating mode, the PI regulator is used to maintain the robot steady speed under stochastic disturbances on the robot motion. In the framework of this study, the disturbance function is physically represented as a force acting on the platform of the robot, diverting its speed from the given one.

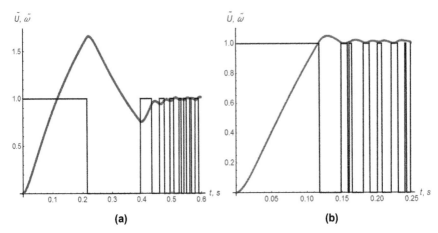

Figure 11.8 Graphs of the function obtained by simulating the mode of working out the set point of the PI regulator with: (a) classical; (b) modified structure.

Table 11.3 Integrated parameters of the random disturbance function*

Parameter	Value
Maximum value	9.80
Minimum value	−7.59
Median RMS value	4.87
Universal mean	0.00

*the numerical values of the combined parameters, which are presented in Table 11.3, are obtained as the ratio of the actual value of the parameter to the fixed value of the moment of resistance of the robot platform movement.

The disturbance function is identical for all calculations performed, the values of its aggregated parameters are given in Table 11.3. The duration of simulation of the disturbances neutralization mode is selected at one minute ($T_{DNM} = 60$ s).

The following indicators were used to assess the quality of the regulation process during the disturbances neutralization mode:

1. integral error

$$\bar{e}_{DNM} = T_{DNM}^{-1} \int_0^T |e| dt; \tag{11.7}$$

2. integral control quality index

$$\bar{u}_{DNM} = T_{DNM}^{-1} \int_0^T |u| dt; \tag{11.8}$$

Table 11.4 Quality indicators of automatic regulation for the disturbances neutralization mode

Method of Tuning the PI Regulator	The Structure of the PI Regulator							
	Classical				Modified			
	$\bar{e}_{DNM}$, rad/s	$\bar{u}_{DNM}$, V	$e_{max.DNM}$, %	$e_{min.DNM}$, %	$\bar{e}_{DNM}$, rad/s	$\bar{u}_{DNM}$, V	$e_{max.DNM}$, %	$e_{min.DNM}$, %
Ziegler-Nichols	7.22	8.76	51.42	−92.49	5.39	8.67	51.98	−24.77
Kappa-Tau	5.99	8.76	51.82	−84.11	7.29	8.61	52.22	−22.66
AMIGO	6.21	8.76	51.79	−81.62	6.75	8.62	51.11	−23.10
Chien-Hrones-Reswick	7.06	8.76	52.08	−89.34	5.81	8.65	51.45	−22.95
Cohen-Coon	7.50	8.76	51.49	−95.07	5.47	8.67	51.84	−23.07
Lambda	8.26	8.76	54.62	−95.08	4.56	8.69	50.97	−23.14
Skogestad (SIMC)	6.99	8.76	50.88	−90.70	5.37	8.67	51.97	−24.29

3. maximum relative error

$$e_{max \cdot DNM} = \left(1 - \frac{max(e)}{w_y}\right) \cdot 100\%; \qquad (11.9)$$

4. minimum relative error

$$e_{min \cdot DNM} = \left(1 - \frac{min(e)}{w_y}\right) \cdot 100\%. \qquad (11.10)$$

The lower index of the indicators (11.7–11.10) denotes the disturbances neutralization mode. The indicators (11.7–11.10) were calculated for the disturbances neutralization mode by the system for a number of tuning methods of PI regulator (Table 11.4).

Comparative analysis of the data presented in Table 11.4 shows that the use of the modified structure of the PI regulator allowed reducing some undesirable indicators of regulation. For example, the minimum relative errors for a PI regulator due to a modified structure decreased by 3.53–4.10 times compared with those obtained for the classical structure of the regulator. The integral error estimate for some of the methods of setting the PI regulator has decreased, and for some it has increased. However, the use of a modified structure does not significantly reduce unwanted indicators of $\bar{u}_{DNM}$ and $e_{max.DNM}$.

Thus, in the framework of this study is expedient to use the modified structure of the PI regulator with purpose of neutralizing disturbances only for such methods of setting the PI regulator: Ziegler-Nichols, Chien-Hrones-Reswick, Cohen-Coon, Lambda and Skogestad (SIMC).

11.5 Conclusion

1. The present research offers the control algorithms of robotic electrotechnical complex in closed-ground structures, which provides monitoring of the basic parameters of the atmosphere, phytomonitoring, including the quality of products, while identifying its zone. The prototype of the intelligent robotic electrotechnical complex for monitoring the phytosanitary and atmospheric conditions in the closed ground structures was developed and the algorithm of its control on the basis of color three-dimensional localization in the space of 6D-SLAM was proposed.
2. The chapter suggests a modification of the structure of the PI regulator, which consists of introducing a nonlinear functional dependence between the integral and the proportional component of the PI regulator. In this case, the modified structure of the PI regulator allows the use of standard methods for its configuration. The modified structure of the PI regulator improves the speed control parameters of the robot platform for set point processing modes (speed, overshoot, integral error estimation and control) and neutralization of disturbances (the minimum relative error of regulation). This indicates the expediency of its use for objects of regulation of this type.

References

[1] A. O'Dwyer 'Handbook of PI and PID controller tuning rules (3rd edition)'. Ireland: Imperial College Press, p. 623, 2009.
[2] A. Andreev, O. Peregudova 'Non-linear PI regulators in control problems for holonomic mechanical systems', Systems Science & Control Engineering, vol. 6(1), pp. 12–19, 2018.
[3] A. Balestrino, V. Biagini, P. Bolognesi, E. Crisostomi 'Advanced variable structure PI controllers' IEEE Conference on Emerging Technologies and Factory Automation (ETFA), Mallorca, Spain, Sept. 2009.
[4] A. Balestrino, A. Brambilla, A. Landi, C. Scali 'Non linear standard regulators' IFAC World Congress Proc., vol. 8, pp. 20–25, 1990.

[5] Ю. О. Ромасевич, В. С. Ловейкін 'Модифікація структури ПІ-регулятора' Збірник тез доповідей міжнародної конференції АВТОМАТИКА – 2017", pp. 177–178 (in Ukrainian).

[6] В. П. Лисенко, І. М. Болбот, Т. І. Лендєл, 'Фітотемпературний критерій оцінки розвитку рослини' Енергетика іавтоматика, 3, pp. 122–128, 2013 (in Ukrainian).

[7] V. Lysenko, I. Bolbot, T. Lendiel 'Vejvlet-analiz v fitometrii rastenij. Sbornik nauchnyx trudov ≪aktualnye voprosy sovremennoj nauki', pp. 163–173, 2014.

[8] В. П. Лисенко, І. М. Болбот, Т. І. Лендєл, І. І. Чернов 'Програмно-апаратне забезпечення системи фітомоніторингу в теплиці' Вісник Харківського національного технічного університету сільського господарства імені Петра Василенка, vol. 154, pp. 42–45, 2014.

[9] В. П. Лисенко, І. М. Болбот 'Роботи та робототехнічні системив агропромисловому комплексі' Науковий вісник НУБіП України, 153, pp. 105–110, 2010.

[10] J. G. Ziegler, N. B. Nichols 'Optimum Settings for Automatic Controllers' Transaction of the ASME, vol. 64. pp. 759–768, 1942.

[11] K. J. Åström, T. Hägglund 'PID Controllers: Theory, Design and Tuning'. Instrument Society of America' NC.: Research Triangle Park, 2 edition., p. 344, 1995.

[12] K. J. Åström, T. Hägglund 'Revisiting the Ziegler-Nichols step response method for PID control' Journal of Process Control, 14, pp. 635–650, 2004.

[13] K. L. Chien, J. A. Hrones, J. B. Reswick 'On the automatic control of generalized passive systems' Transaction of the ASME, vol. 74, no.2, pp. 175–185, 1952.

[14] G. H. Cohen, G. A. Coon 'Theoretical Consideration of Retarded Control' Transaction of the ASME, vol. 75, pp. 827–834, 1953.

[15] L. Eriksson 'Control Design and Implementation of Networked Control Systems. Licentiate thesis' Department of Automation and Systems Technology, Helsinki University of Technology. p. 118, 2008.

[16] S. Skogestad 'Simple analytic rules for model reduction and PID controller tuning' J. Process Control, vol. 13(4), pp. 291–309, 2003.

12

Information Support of Some Automated Systems of Remote Monitoring of Planted Areas State

Vitaliy Lysenko, Serhii Shvorov, Oleksii Opryshko, Natalia Pasichnyk and Dmytro Komarchuk*

National University of Life and Environmental Sciences of Ukraine, Kyiv, Ukraine
*Corresponding Author: dmitruyk@gmail.com

The reviewed problem deals with providing biogas reactors with plants raw materials as a material for anaerobic fermentation. It is proposed to use biomass of planted areas, the further cultivation of which is economically inexpedient for cold spring climate regions.

The low quality of such crops is primarily predicted because of the lack of nitrogen to be advisable evaluated with the use of specialized sensors on the UAV platform.

The reviewed problem deals with the metrological support of sensory equipment using cameras with standard and infra-red lenses. The spectral channels laboratory evaluation was carried on the example of the Go PRO HERO 4 camera.

The prospective vegetative indices for some crops were obtained exampling maize and wheat sowing. It is proved the expediency of using integral indexes while creating stress indices for UAVs with taking into account either spectral or dimensional parameters of grain crops planted areas. Besides, it was reviewed the problem to choose the best routes for harvesting field plants on the base of electronic maps to be prepared with the use of UAVs.

12.1 Introduction

The problem to use local resources for energy supply of consumers in rural areas is extremely relevant because of the high gas price as well as the risks to be caused by political or even military factors.

Traditional raw materials for biogas reactors exampling livestock and poultry waste may not be enough to get the required amount of energy due to the reduction of the stock population.

As additional sources for biogas production plant raw materials are used in the EU countries, namely energy crops and crop residues to be shown in works by A. Meyer et al. (2017) [1] and P. Schröder et al. (2018) [2].

Energy crops sowing areas are mostly planned taking into account the unsuitable cropping land exampling peatland (K. Laasasenaho et al., 2017) [3] as well as logistics for existing biogas reactors.

The technologies to be elaborated to get biogas from the processing plots accumulated waste are adapted to specific raw materials (R. Ciccoli et al., 2018) [4], which limits its application for random raw materials, and therefore for the rural areas with their gas consumers.

In the paper by K. Sahoo et al. (2018) [5], it is shown that plant margins have a certain economic potential for biogas production, but it is still the problem of monitoring the volumes of this raw material and optimizing logistics for its transportation to reactors.

The problem of biomass transportation optimizing within the region was reviewed by J. Höhn et al. (2014) [6] for Finland as well as in the paper by V. Burg et al. (2018) [7] – for Switzerland. However, the places of the promising stationary biogas reactors building were mostly estimated.

The papers showed that every year the location of biomass sources is different. It complicates the decision about whole year biomass state operational monitoring to optimize logistics.

In the abovenamed papers from the EU countries, the main attention was paid to such crops as wheat, barley and maize because according to the statistics by L. C. Ferreira et al. (2014) [8], straw is a classic raw material for anaerobic digestion- it has a low operating cost on a large production scale.

While choosing raw materials to get biogas, the seasonal factors concerning the availability of raw material should also be taken into account: there are difficulties with the temperature regime to be in need for sufficient anaerobic digestion in cold climate regions.

According to C. Mao et al. (2015) [9], the mesophilic temperature regime to be actively used in t biogas production requires a temperature of 37°C.

Therefore, for economic reasons, it is expedient to use the entire warm period when there is no need in additional heating to start and to operate a biogas reactor.

According to the technology by C. Rodriguez et al. (2017) [10] to provide the biogas reactor with raw materials in spring grass can be used conditioning straw lack as well as the green mass of damaged or underdeveloped cultivated plants, which further cultivation is inappropriate.

The use of plant crops to be in the initial stages of vegetation as a raw material for biogas reactors may be caused by economic reasons exampling the forecast of low yield due to a significant lag in growth and development to be caused by acute shortage of nutrients, damage from harmful organisms or other stress factors.

Remote monitoring of agricultural planted areas state may serve as an efficient tool in need to provide biogas reactors with plant raw material.

Such examination should indicate the amount of plant biomass and its exact location, plant state exampling the level of mineral nutrition, which became the purpose of this chapter.

12.2 Formulation of the Problem. Scheme of the Method

12.2.1 Investigated Objects

Experimental studies in the monitoring of grain crops mineral nutrition state were conducted on the fields of the VN NULES of Ukraine, "Agronomic Research Station", in the long term field station of the Department of Agrochemistry and Quality of Crop Production to be in the status of National Heritage and on the productive fields of this farm.

Viewing the plant stress state to be caused by moisture provision, for production testing there were selected the areas with no puddle formation after snow melting Field experiment is located in the village Pshenychne, Vasylkivskyi district, Kyiv region., GPS coordinates of the field 50° 4′28″ N, 30° 13′20″ E.

The soil of the experimental site is meadow-chernozem carbonate rough-pew-lime-gravel on loessed loam.

To study different fertilizers norms influence the following variants of experiment with winter wheat were selected: 1) without fertilizers (control); 2) P; 3) RK; 4) NAR (recommended norm); 5) NAR (1,5 recommended norms). Fertilizers were introduced in the form of ammonium nitrate, ammophos and potassium chloride. Determination of the nitrogen content in the dry matter was carried out in laboratory conditions by photometric method.

Viewing the resolution of UAV photo shots, unlike satellite imagery it is possibility to fix individual parts of plants, for single-crop crops (wheat and barley) nitrogen content was determined separately in the three upper leaves and in the whole terrestrial part of the plants.

The decision about the appropriateness of further grain crops growing to get grain or to be the raw material for biogas reactors should be taken during the adequate periods of vegetation while fertilizing. For wheat this is the stage of "getting into the tube," for maize – "4–6 leaves" because it is the time for plants to form phytomass and to lay the reproductive organs. Taking into account the cost of technological operations and calculated products, adequate decision about the appropriateness of feeding should be made.

12.2.2 Hardware for Spectral Monitoring

The specificity of crop monitoring to detect the state of nitrogen supply is a limited time for decision-making.

This causes the choice of technical means to conduct such tests.

Standard solutions to use satellites have certain limitations – because of the complexity of clouds and atmospheric correction to be described in the papers by T. Mannschatz et al. (2014) [11] and A. Ibrahim et al. (2018) [12], taking into account the variations in lighting to be presented in the papers by L. S. Bernstein et al. (2005) [13]. UAVs for monitoring tasks on the status of nitrogen nutrition has significant advantages because it does not require atmospheric correction and may be used even in dense clouds.

Due to these advantages UAV is considered to be promising for use in agricultural production. It led to the creation of a specialized range of sensory equipment such as MAPIR Kernel Camera Array, Parrot Sequoia, Tetracam ADC-micro, and SlantRang.

In addition to specialized cameras to use optical and infrared ranges, standard RGB cameras are also used for monitoring purposes.

Involving lighting changes for the UAV spectral equipment may be carried out either with the help of techniques to be developed for satellites or hardware with specialized anti-aircraft sensors to be described by J. Zhou et al. (2016) [14], as well as software with data from the built-in camera exposure meter.

Correction of illumination with data from the exiff file was proposed by M. M. Saberioona et al. (2014) [15]. This technique has been improved by V. Lysenko et al. (2017) in paper [16].

To create a mineral nutrition monitoring system which should function even at variable cloud when the hardware and software correction are ineffective, there were proposed approaches to assess the lighting state with the use of neural networks in the paper by V. Lysenko et al. (2017) [17].

Vegetation indices are used to monitor the stress state of plants, the review of which is presented in papers by R. Main et al. (2011) [18] and G. Maire et al. (2004) [19].

Vegetation indices were mostly created for satellites, which determined the use of infrared spectrum for measurements, which ensured the identification of soil and vegetation from a long distance.

For the UAV, it is expedient to create specialized indices because of the low cost of the high-resolution images: efficient soil filtration is possible in an optical range too.

The households' choice of touch-sensitive equipment for the RGB range is caused by the need in versatile equipment to be used for various purposes all year round.

Becides, the possibility to use the optical range itself in order to determine the plants' mineral nutrition state was described in the paper by T. M. Shadchina (1999) [20].

Vegetation indices except satellites are also used on ground equipment exampling GreenSeeker, CropSpec, Yara N-Sensor, etc. for precision farming. There is an experience to be described in the paper by Q. Cao et al. (2014) [21] concerning the creation of special vegetation indices to be adapted specifically for such instrument where either optical or infrared (IR) channels were used.

In need to receive IR channel, in some models of RGB range cameras, it is possible to install special infrared lens instead of a stationary one.

Basing on farm needs, the RGB camera PHANTOM VISION FC200, the DGI DGI Phantom 3+ standard camera and the more versatile GoPro HERO4 Standard Edition camera to be installed on the UAV to allow the use of IR lenses, were used for research purposes.

In order to check the correspondence with the obtained measurements results from different cameras, comparative studies were conducted in field conditions. Soil road and experimental wheat fields with different nitrogen feed state were used as the objects for research. Data processing was carried

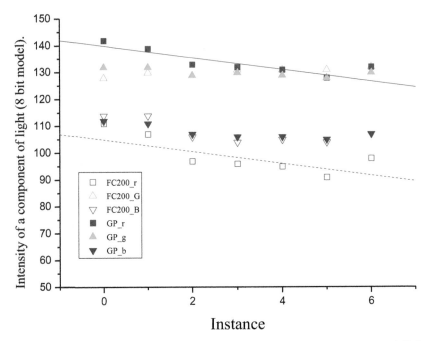

Figure 12.1 Results of objects spectral indicators measurements with FC200 and GoPro HERO4 cameras use.

out according to the method to be described by V. Lysenko et al. (2017) [22], and the results of measurements are presented in Figure 12.1.

According to our research the existing data discrepancy for green and blue components may be explained by measurement errors but discrepancy for red component is caused by differences in the corresponding spectral equipment channel.

Because of the official spectral channels numerical values data absence while using MP lens on GoPro HERO4 IR camera special experimental studies were conducted.

The studies were carried out according to the method to be described in the European Machine Vision Association (EMVA) -1288 standard on the laboratory equipment to be presented in the paper by Yu. A. Hizhnyi et al. (2004) [23]. To take an object image in light of a certain wavelength, a modified dual monochromator DFS-12 was used, which, under certain experimental conditions, allowed to separate the light with a λi wavelength on an accuracy level of $\Delta\lambda$ <0.8 nm. The obtained results are presented in Figure 12.2.

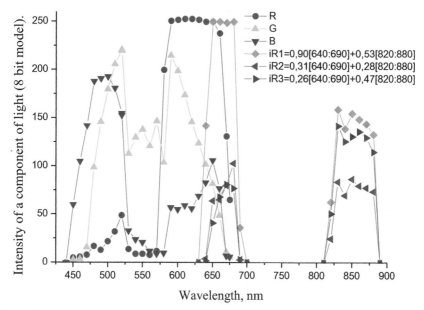

Figure 12.2 Dependence of intensity of components of color from wavelength for GoPro HERO4 camera with standard lens (channels R, G, B) and IR lens MR (channels iR1, iR2, iR3).

It was experimentally found that while using IR lens GoPro HERO4 camera captures three different channels - iR1, iR2, iR3, which using standard lens would correspond to red, green, and blue colours. While using an IR lens, the channels had two components - red and infrared.

At the same time, their "width" was almost identical, but the intensity was different.

We used this feature while creating vegetation indices to be presented in the paper by V. Lysenko and others (2017) [24]. Determination of camera spectral channels while using a standard lens was carried out to allow an assessment of the suitability of other camera models basing on their spectral channels.

12.3 Obtained Results and Discussion

12.3.1 Continuous Sowing Crops – Wheat

The study of the influence of the nitrogen supply state on the spectral indices of the Limarivna winter wheat was carried out in the phase "out of the tube".

Figure 12.3   The photos of wheat crops on the experimental stationary field to be received from the standard lenses and IR lenses (2017.05.19).

The photographs of the experimental areas with usage of standard and infrared lenses are shown in Figure 12.3.

The parameters of digital camera settings for research: Exposure Time when shooting in the optical range - 0.00069 s (FC200); Exposure Time when shooting with an IR lens - 0.00252 (GoPro); Aperture Value - 2.8; Light Source - Fine Weather.

In order to avoid the chance for frame to fix other objects except plants leaves, they carried out "withdrawal" of results that differed from the average value by more than 10%.

According to the calculation results, it was found that due to the extraction of parts with shown soil and other random objects, the component color intensity did not change by more than on 3 units. The results are graphically shown in Figure 12.4.

By approximating the amount of nitrogen in plants up to the technologically recommended value (up to 4.5%), the intensity of the color component in the optical range is being reduced, but using IR lens it is contrary increasing.

The most common indices exampling NDVI are based on the principle that the maximum absorption or reflection of radiation occurs for plants in certain ranges of electromagnetic waves lengths.

Perspective vegetative indices for precision farming technologies are developed taking into account their application exampling differential fertilizer application. For the automation of fertilizer processes, the most suitable is the linear nature of the relationship between the established vegetation index and the experimental parameters of the plant.

For the "exit to the tube" phase, nutrition is carried out, primarily, with nitrogen fertilizers.

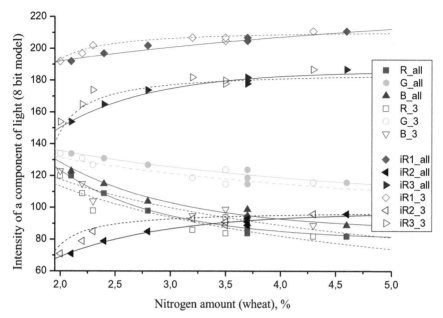

Figure 12.4   Dependence of the components color intensity on nitrogen amount of dry matter for the plants above-ground part (x_all) and three upper leaves (x_3).

Table 12.1   Perspective spectral indices for winter wheat

№	Range	Name	Equation	R^2 Above-Ground Part	3 Leafs
1	RGB	R	–	0.83	0.72
2	–	G	–	0.81	0.75
3	–	B	–	0.85	0.76
4	–	GR	$GR = (G - R)/(G + R)$	0.83	0.68
5	–	GB	$GB = (G - B)/(G + B)$	0.85	0.73
6	NIR	iR1	–	0.92	0.81
7	–	iR2	–	0.88	0.75
8	–	iR3	–	0.86	0.72
9	–	iR	$iRb = (iR1 - iR2)/(iR1 + iR2)$	0.88	0.76

Therefore, the obtained results were approximated in the form of linear dependence, for which the determination coefficient (R2) was calculated. The obtained results are presented in Table 12.1.

The obtained results show that the determination coefficient in the form of linear approximation has approximately the same values while using either optical or infrared lenses.

Thus, for the wheat nitrogen supply state, monitoring the use of IR channel is not obligatory. Regardless of the spectral range, the determination coefficient for approximation in the form of linear dependence is higher on 10–30%, while nitrogen is determined for the whole terrestrial part of plants (in comparison with the three upper leaves), which should be taken into account while creating spectral indices.

12.3.2 Growing Crops

Maize: The research was conducted on maize Pioneer P 8816 hybrids, in "4-6 leaves" and "8–10 leaves" phases. The shooting was done using optical and infra-red lenses (Figure 12.5).

The peculiarity of spectral monitoring of propagated crops, in comparison with crops of continuous sowing, is the presence on the photo of too large area parts with soil, organic residues, etc.

Thus, the approach to remove additional objects based on filtration of area parts which color intensity rates are different from the average on certain percentage is unacceptable. For crops, it is better to apply the approach based on filters cascade usage, while the removal of soil image was carried out based on its spectral parameters.

On an industrial-scale application of the system, it is expedient to determine the adjusting filters' parameters viewing the ground roads to be available on the fields.

In this case, it is primary need in focusing, on the roadside, because there the soil is not squeezed by the wheels and has the closest spectral indicators similar to field's ones. In addition to the plants spectral indices, it was also determined how the percentage of pixels of the experimental area corresponds precisely to the plants. The obtained results are presented in Table 12.2.

Figure 12.5   The photos of maize on the experimental stationary field to be received from the standard lenses and IR lenses (2017.05.24).

Table 12.2 Spectral indices and dimensions of maize plants

N, %	R	G	B	Plants, %	iR1	iR2	iR3	Plants, %
2,7	129	131	128	5,8	206	109	172	16,2
2,9	133	138	128	18,9	207	113	175	23,6
3,1	133	145	128	20,6	201	94	165	37,2
3,7	135	148	126	24,5	200	92	164	38,7
3,5	137	150	128	16,3	201	95	166	41,3
4,2	138	150	129	23,9	201	92	164	43,8

In the optical range, the number of pixels to be recognized as plants is significantly lower than in the red-infrared range, due to the adjustment in a more refined filter.

The IR lens is better due to the greater difference between soil and plants than in the individual optical range.

Taking into account the noted facts, a decision about the primary expediency of using IR lens was adopted.

For the amount of nitrogen in the range 2.7–4.2%, the maximum difference in the maize component color intensity was observed for green channel - the intensity varied in the range of 131–150 units (low sensitivity).

It is an interesting fact that with increasing content in nitrogen plants, their vegetative mass was increasing too, as well as percentage of the area under plants as shown in the photo. We have recorded such changes - from 6 to 24%. It is advisable to use this method while assessing the mineral nutritional status of maize either individually or in combination with the assessment of the components color intensity.

Figure 12.6 shows the dependence of the index, which takes into account either component color intensity or the percentage of the plants on the experimental plot.

The use of integral indices namely those that take into account either the intensity of radiation reflected by plants or the area under them, may be promising for crops viewing sensitivity and selectivity (insensitivity to the color of the soil).

It has been found that iR (RED + NIR) spectral channels have a sufficient sensitivity in the range up to 3% nitrogen and the geometric component – in the range of 3 to 4%. The combination of these two indices into the integral index, namely the component color intensity (spectral component) and the plant dimensions (geometric parameter), allowed to achieve high sensitivity in the critical range of maize nitrogen supply – up to 3.5–4%. We consider

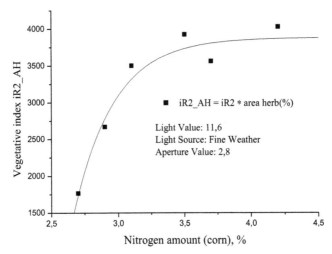

Figure 12.6   Dependence of the IR2_PR integral index on the amount of nitrogen in the maize ground part.

such stress index to be more acceptable for the automation of differentiated fertilizing processes.

12.3.3 Determination of Optimal Movement Routes for Harvesters to Harvest Energy Crops and for Necessary Transport Equipment to Transport Them

One of the important stages of the technological cycle of biogas production is a harvesting campaign for planting or specially grown energy crops (EC) for biogas plants. Non-compliance of the technology to harvest biomass on the complicated area landscape (Figure 12.7) may lead to significant costs exceeding the profit from fertilizer application, the introduction of new varieties of energy crops and the use of intensive technologies for their cultivation.

The main problem concerning these tasks deals with the multicriteria, with the lack of one feature to select decisions or to choose the best one. The main ways to overcome multicriteria is to engage the decision maker.

In this case, the quality of the decision-making process is directly dependent on the completeness of accounting for all factors that are significant for the consequences of the taken decisions. For efficient problem solving, it is in need to apply decision-support systems (SPDS) to be widely used today to formulate proposals for harvesting plant and energy crop residues basing on monitoring the current state of the fields from unmanned aerial vehicles data.

Figure 12.7 Photo of the complicated area landscape with UAV.

With the help of SPPR, the harvesting projects of the EC are being worked out while the decisions about the distribution of machinery (harvesters and vehicles) on the fields are being grounded.

Depending on the availability of the machinery and the predicted conditions (up) of the harvesting campaign, a plurality of options {V} for the execution of harvesting is generated. Among the existing set of such options rational v^P is determined to ensure the maximum profit (P) from the implementation of biomethane:

$$P(v^P) \quad W - (B + Z) \to \max, when \; u^P \epsilon \; \cup, \quad (12.1)$$

where W is the forecasted revenue from the sale of EC or biomethane to be derived from them, B – fuel consumption by unmanned harvesting equipment (BZT) while implementing the second variant of the EC, Z – the costs of monitoring the state and EC transportation by vehicles to get biomethane, UAH, U – conditions of the harvest campaign [25].

Maximization of profit (P) is achieved involving an optimal planning of harvesting, reducing, the cost of EC state monitoring (due to UAVs data processing) and reducing unmanned harvesting equipment (BZT) fuel consumption due to application of the method to synthesize compromise-optimal traffic routes for unmanned harvesters.

The statement of the problem of the compromise-optimal route for unmanned harvesters movement is the next one to be represented further.

The known information is the coordinates of the area where the EC biomass is located, the initial location of each BZT, and its route destination, the coordinates of the passive obstacles and the coordinates of the areas without biomass to be obtained by the UAV. The task is to find the minimum length routes of the BZT with detour without biomass and various obstacles (high-voltage pillars, roads, pits, barriers, etc.).

Definition of compromise-optimal routes of unmanned combine movement is reduced to the initial problem of discrete type and includes the following operations:

1. The length of the path of the BZT movement from level $j - 1$ to level j is determined by the following formula:

$$D^{j,m}_{j-1,i} \quad \sqrt{(x_m - x_i)^2 + (y_i - y_{j-1})^2}, \qquad (12.2)$$

where (x_i, y_{j-1}) – the rectangular coordinates of BZT at the level of the network; x_m, y_j – rectangular coordinates of the permissible point at the same level of the network.

2. To assess the danger of unmanned combine harvesters approaching to passive obstacles (H), the following formula applies:

$$H(\rho) \quad K \cdot e^{-\alpha \rho}, \qquad (12.3)$$

where; $\rho \quad \sqrt{(x - x^*)^2 + (y - y^*)^2}$; (x^*, y^*) – rectangular coordinates of passive obstacles; (x, y) – coordinates of BZT; α and K– coefficients characterizing the degree of danger of unmanned harvesters approaching to passive obstacles [26].

3. The criterion of having unmanned harvesters in the area with no energy crops (θ) is determined by the formula:

$$\theta \quad e^{\eta S_{min}}, \qquad (12.4)$$

where η– positive coefficient and S_{min}– the distance between the BZT and the area with no EC.

4. To determine the compromise-optimal BPT motion routes, a dynamic programming method is used to calculate the total losses $G(j, m)$by the generalized optimality criterion basing on the use of the Bellman equation for each m-th permissible point on each step of j level:

$$G(j, m) \quad \min_{i \in I_{j-1}} [\Delta G^{j,m}_{j-1,i}], j \in [1, J], \qquad (12.5)$$

where m is the number of the permissible point in each step of j level; I_{j-1}– the number of permissible points on the j-1st level of the network; and j – the number of levels of the transition on the network [27, 28].

5. The value $\triangle G_{j-1,i}^{j,m}$ for the generalized criterion (12.5) is calculated in accordance with the methodology of the nonlinear compromise scheme:

$$\triangle G_{j-1,i}^{j,m} \quad \frac{f_1 H_{\max}}{H_{\max} - H_{j,m}} + \frac{f_2 D_{\max}}{D_{\max} - D_{j-1,i}^{j,m}} + \frac{f_3 \theta_{\max}}{\theta_{\max} - \theta_{j-1,i}^{j,m}} \quad (12.6)$$

where H – assessment of the danger of unmanned machinery approaching to passive obstacles; $f1$, $f2$, $f3$ – weight factors, D – length of the unmanned harvesters route; θ –estimation of the probability of unmanned harvesting equipment approach to the area with no energy crops.

6. While distributing vehicles between the fields, the main aim is to minimize downtime for BZT and transport equipment, as well as to obtain minimal expenses for the transportation of the harvested energy crop.

The process of biomass transporting from the harvester to its storage place will be presented as a mass maintenance system. The physical model of biomass harvesting is taken when the intensity of the bunkers filling of λ_{kd} (bunkers/h) is defined as the application of requirements and the body of the vehicle - as a service channel of the system. Based on the fact that unloading biomass from unmanned harvester bunker into one vehicle is practically instantaneous, the mass maintenance system is in one-channel.

Stopping operations due to malfunctions of unmanned harvesting equipment, the lack of vehicle or organizational reasons may be formally considered as a failure of the serving channel and the time intervals without harvesting – as the intervals of the time to recover channel.

It is assumed that the input flow of requirements is the easiest. If the duration of random intervals of repairing time intends to zero, unmanned harvesters will move around the field without long stops and the full bunkers discrete flow at the equal intervals of time will be approximated by the distribution of Poisson.

Depending on the classes of unmanned combine harvesters, they detect the number of harvesters bunkers 0, 1, 2, ... and the corresponding class $\Omega_1 \subset \Omega$ of the vehicle to transport EC from bunkers during one route.

In this case, the average time of vehicle movement from the moment of leaving field up to its return is determined by the formula:

$$\tau \quad \tau^1 + \tau^2 + \tau^3 + \tau^4, \quad (12.7)$$

where τ^1 – average time of movement of the vehicle from the field to the point of unloading, h; τ^2 – average time of movement of the vehicle from the point of unloading to the field, h; τ^3 – average time of transport being on an unloading point, h; τ^4 - average time of transport being on the field, h.

7. The average intensity of vehicle arrival to the first field is based on the formula:

$$\varpi_l \quad \frac{1}{\tau_l}, \tag{12.8}$$

8. The average intensity of vehicles arrival to the field of in the quantity of $c_{\beta k}$ equals:

$$\overline{\varpi}_l \quad c_{\beta l} \cdot \varpi_l, \tag{12.9}$$

where $c_{\beta k}$ – the number of vehicles to be in need to maintain β class harvesters or:

$$\overline{\varpi}_l \quad \lambda_{\beta l}/\mathrm{m}, \tag{12.10}$$

where $\lambda_{\beta k}$– the average intensity of unmanned harvesters of a certain class (bunkers/h).

The total number of vehicles to be in need for the continuous operation of the β-class BZT on the first field is determined in the following ways:

if m 1, the number of 1 class vehicles equals:

$$c_{\beta l}^1 \quad [\lambda_{\beta l \cdot \tau_1^u}] + 1, \tag{12.11}$$

while $m > 1$ the number of m class vehicles equals:

$$c_{\beta l}^1 \quad [\lambda_{\beta l} \cdot \tau^u/\mathrm{m}], \tag{12.12}$$

While the small part of the expression corresponds to the following condition:

$$\{\lambda_{\beta l} \cdot \tau^u/\mathrm{m}\} > \varepsilon_0, \tag{12.13}$$

additional 1st class vehicles are engaged in the $c_{\beta k}^1$ quantities:

$$c_{\beta l}^1 \quad [\{\lambda_{\beta l} \cdot \tau^u/\mathrm{m}\} \cdot \mathrm{m}] + 1, \tag{12.14}$$

Similarly, this technique is used to determine the routes of unmanned harvest vehicles and adequate vehicles, taking into account the number of additional vehicles routes $c_{\beta l}^1$.

As a rule, own vehicles of the enterprise are in use, and in case of their lack of leased machinery.

Figure 12.8 Results of simulation of the routes of harvesting and transport vehicles for the first experimental field at the enterprise TDV ≪Terezone≫.

The evaluation of the efficiency of the elaborated methodology was carried out during the harvesting campaign on 18 experimental fields of company ≪Terezyne≫ of Bila Tserkva district in Kyiv region. As an example for the first experimental field in Figure 12.8 and in Table 12.3 shows the results of modeling the movement of harvesting and transport equipment.

Actual fuel consumption for a harvest campaign for each field with and without the application of the proposed methodology to be implemented in the research support and decision-making system are shown in Figure 12.9 and in percentile measure in Figure 12.10.

Thus, the considered method to determine the routes of harvesting and transport equipment movement is an important component of decision support system to be in need for the efficient solution of the tasks to organize harvesting work.

The practical application of the developed methodology allowed to reduce the excessive length of the BZT traffic routes and the total cost of harvesting campaign on 12–15% due to the operative determination of the harvest, volumes, the planning, and implementation of compromise-optimal routes of harvesting and transport equipment movement to transport biomass toward its storage points.

Proceeding from this, the company's profit increased by more than 12%. Besides, as to be shown by results of the practical application of the proposed method to be implemented in the SPPR, the time expenditures for making

Table 12.3 Data of modeling the routes of harvesting and transport equipment for the first experimental field

№	The Movement Lenth, m	Fuel Consumption	The Harvested Plant Raw Material	Coordinates Start	End
1	1235	9,38	5,85	49.822805, 30.191600	49.817307, 30.202221
2	1233	9,17	5,35	49.817250, 30.202197	49.822773, 30.191557
3	1231	9,35	5,64	49.822740, 30.191514	49.817193, 30.202173
4	1233	9,09	5,40	49.817137, 30.202149	49.822708, 30.191471
...	...	...	...	...	...
89	1584	12,51	7,92	49.819957, 30.187812	49.810304, 30.202272
90	1568	12,38	7,84	49.810296, 30.202158	49.819924, 30.187769
91	1581	10,94	6,46	49.819892, 30.187726	49.810287, 30.202044
...	...	...	...	...	...
173	824	6,14	3,57	49.817238, 30.184197	49.809586, 30.192713
174	828	6,12	3,65	49.809577, 30.192600	49.817206, 30.184154
175	824	6,08	3,64	49.817173, 30.184111	49.809569, 30.192486

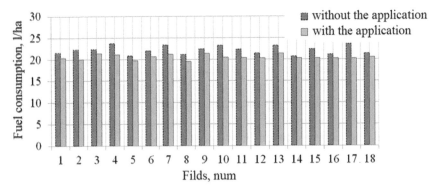

Figure 12.9 Actual fuel consumption of fuel for a harvest campaign for each field using and with and without the application of the proposed method to be implemented in the SPPR.

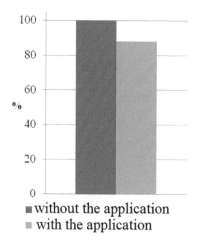

■ without the application
■ with the application

Figure 12.10 Total fuel consumption of fuel for a harvest campaign for each field using and with and without the application of the proposed method to be implemented in the SPPR.

informed decisions are significantly reduced due to the processing of large volumes of information to be obtained from the UAV.

12.4 Conclusion

While using UAVs for spectral monitoring of nitrogen nutrition, it is advisable to develop and use specialized stress indices based upon the usage of images with a high level of planting detailization.

For growing crops stress indices with the ability to take into account either the spectral component or the geometric dimensions of planting areas are considered to be the most promising.

In future, such integral indices may be useful for continuous cropping crops as well.

The obtained information based upon the results of UAV's usage makes an efficient possiblity to determine crop area, the area of crop yields and obstacles in the way of unmanned harvesting machinery, which provides an optimal planning of harvesting campaign taking into account perspective crop rotation.

The consequence of the proposed methodology to determine the routes of unmanned harvest machinery and the quantity of transport equipment is the higher efficiency and accuracy to drive harvesting and transport equipment, as well as the reduction of fuel consumption.

References

[1] A. K. P. Meyer, E. A. Ehimen, J. B. Holm-Nielsen. Future European biogas: Animal manure, straw and grass potentials for a sustainable European biogas production. Biomass and Bioenergy. 2017 https://doi.org/10.1016/j.biombioe. 2017.05.013.

[2] P. Schröder, B. Beckers, S. Daniels, F. Gnödinger, E. Maestri, N. Marmiroli, M. Mench, R. Millan, M. M. Obermeier, N. Oustriere, T. Persson, C. Poschenrieder, F. Rineau, B. Rutkowska, T. Schmid, W. Szulc, N. Witters, A. Sæbø. Intensify production, transform biomass to energy and novel goods and protect soils in Europe—A vision how to mobilize marginal lands Science of the Total Environment, vol. 616–617, pp. 1101–1123, 2018.

[3] K. Laasasenaho, A. Lensub, J. Rintalac Planning land use for biogas energy crop production: The potential of cutaway peat production lands. Biomass and Bioenergy, vol. 85, pp. 355–362, 2016.

[4] R. Ciccolia, M. Sperandei, F. Petrazzuolo, M. Broglia, L. Chiarini, A. Correnti, A. Farneti, V. Pignatelli, S. Tabacchioni Anaerobic digestion of the above ground biomass of Jerusalem Artichoke in a pilot plant: Impact of the preservation method on the biogas yield and microbial community. Biomass and Bioenergy, vol. 108, pp. 190–197, 2018.

[5] K. Sahoo, S. Mani, L. Das, P. Bettinger GIS-based assessment of sustainable crop residues for optimal siting of biogas plants Biomass and Bioenergy, vol. 110, pp. 63–74, 2018.

[6] J. Höhn, E. Lehtonen, S. Rasi, J. Rintala A Geographical Information System (GIS) based methodology for determination of potential biomasses and sites for biogas plants in southern Finland. Applied Energy, vol. 113, pp. 1–10, 2014.

[7] V. Burg, G. Bowman, M. Erni, R. Lemm, O. Thees. Analyzing the potential of domestic biomass resources for the energy transition in Switzerland. Biomass and Bioenergy, vol. 111, pp. 60–69, 2018.

[8] L. C. Ferreira, A. Donoso-Bravo, P. J. Nilsen, F. Fdz-Polanco, Pérez-Elvira. Influence of thermal pretreatment on the biochemical methane potential of wheat straw. Bioresource Technology, vol. 143, pp. 251–257, 2013.

[9] C. Mao, Y. Feng, X. Wang, G. Ren. Review on research achievements of biogas from anaerobic digestion. Renewable and Sustainable Energy Reviews, vol. 45, pp. 540–555, 2015.

[10] C. Rodriguez, A. Alaswad, K. Y. Benyounis, A. G. Olabi. Pretreatment techniques used in biogas production from grass. Renewable and Sustainable Energy Reviews, vol. 68 (Part 2), pp. 1193–1204, 2017.

[11] T. Mannschatz, B. Pflug, E. Borg, K.-H. Feger, P. Dietrich. Uncertainties of LAI estimation from satellite imaging due to atmospheric correction. Remote Sensing of Environment, vol. 153, pp. 24–39, 2017.

[12] A. Ibrahim, B. Franz, Z. Ahmad, R. Healy, K. Knobelspiesse, B. Gao, C. Proctor, P. Zhai Atmospheric correction for hyperspectral ocean color retrieval with application to the Hyperspectral Imager for the Coastal Ocean (HICO). Remote Sensing of Environment, vol. 204, pp. 60–75, 2018.

[13] L. S. Bernstein, S. M. Adler-Golden, R. L. Sundberg, R. Y. Levine, T. C. Perkins, A. Berk. Validation of the QUick Atmospheric Correction (QUAC) Algorithm for VNIR-SWIR Multi- and Hyperspectral Imagery // Proceedings of SPIE, vol. 5806, pp. 668–678, 2005.

[14] J. Zhou, L. R. Khot, H. Y. Bahlol, R. Boydston, P. N. Miklas. Evaluation of ground, proximal and aerial remote sensing technologies for crop stress monitoring // IFAC-PapersOnLine, vol. 49, no. 16, pp. 22–26, 2016.

[15] M. M. Saberioona, M. S. M. Amina, A. R. Anuarb, A. Gholizadehc, A. Wayayokd, S. Khairunniza-Bejoda. Assessment of rice leaf chlorophyll content using visible bands at different growth stages at both the leaf and canopy scale // International Journal of Applied Earth Observation and Geoinformation, vol. 32, pp. 35–45, 2014.

[16] V. Lysenko, O. Opryshko, D. Komarchuk, N. Pasichnyk, N. Zaets, A. Dudnyk Usage of Flying Robots for Monitoring Nitrogen in Wheat Crops. The 9th IEEE International Conference on Intelligent Data Acquisition and Advanced Computing Systems: Technology and Applications, 21–23 September 2017, Bucharest, Romania, vol. 1, pp. 30–34.

[17] V. Lysenko, D. Komarchuk, O. Opryshko, N. Pasichnyk, N. Zaets Determination of the not uniformity of illumination in process monitoring of wheat crops by UAVs / Scientific-Practical Conference Problems of Infocommunications. Science and Technology (PIC S&T), 2017 4th International. Kharkov, Ukraine. Doi: 10.1109/INFO-COMMST.2017.8246394 http://ieeexplore.ieee.org/document/8246394/.

[18] R. Main, M. A. Cho, R. Mathieu, M. M. O'Kennedy, A. Ramoelo, S. Koch An investigation into robust spectral indices for leaf chlorophyll estimation. ISPRS Journal of Photogrammetry and Remote Sensing, vol. 66, pp. 751–761, 2011.

[19] G. le Maire, C. Franc, E. Dufrêne Towards universal broad leaf chlorophyll indices using PROSPECT simulated database and hyperspectral reflectance measurements. Remote Sensing of Environment, vol. 89, pp. 1–28, 2004.

[20] T. M. Shadchina Elaboration of theoretical bases and methods of the remote sensing of winter wheat crops using the high resolution spectrometry.-Manuscript. Thesis for Dr. Sci (Biol.) by speciality 03.00.12-Plant Physiology.-Institute of Plant Physiology and Genetics, National Academy of Sciences of Ukraine, Kyiv, 1999.

[21] Q. Cao, Y. Miao, G. Feng, X. Gao, F. Li, B. Liu, S. Yue, S. Cheng, S. L. Ustin, R. Khosla. Active canopy sensing of winter wheat nitrogen status: An evaluation of two sensor systems. Comput. Electron. Agric. (2014), http://dx.doi.org/10.1016/j.compag.2014.08.12.

[22] V. Lysenko, O. Opryshko, D. Komarchyk, N. Pasichnyk. Drones camera calibration for the leaf research. Scientific Bulletin of the National University of Life and Environmental Sciences of Ukraine, no. 252, pp. 61–65, 2016.

[23] Yu A. Hizhnyi, S. G. Nedilko, V. P. Chornii, M. S. Slobodyanik, I. V. Zatovsky, K. V. Terebilenko Electronic structures and origin of intrinsic luminescence in Bi-containing oxide crystals BiPO4, K3Bi5(PO4)6, K2Bi(PO4)(MoO4), K2Bi(PO4)(WO4) and K5Bi(MoO4)4. Journal of Alloys and Compounds, vol. 614, pp. 420–435, 2014.

[24] V. Lysenko Remote sensing of grain crops for the programming of the crop [Text]: monograph / V. Lysenko, D. Komarchuk, N. Pasichnyk, O. Opryshko; National University of Bioresources and Natural Resources of Ukraine. - K.: CP "Komprint", p. 364, 2017, ISBN 978-966-929-609-2.

[25] S. Shvorov Methodical bases for constructing the system of intellectual decision support for the organization of collection and processing of energy crops in biogas complexes / [Shvorov S. A., Komarchuk D. S., Okhrimenko P. G., Chirchenko D. V., Vasyamovich O. V.] // Power engineering and automatics, no. 2, pp. 144–155, 2016.

[26] S. Shvorov Method of planning of BZT traffic routes using UAV. / Shvorov S. A., Komarchuk D. S., Lukin V. E., Chirchenko D. V. // Scientific Bulletin of NUBiP, no. 261, series "Engineering and Power Engineering of AIC", p. 104–110, 2017.

[27] Synthesis of compromise-optimal trajectories of mobile objects in conflict environment// Voronin, A. N.,Yasinsky, A. G.,Shvorov, S. A. // 2002 // Journal of Automation and Information Sciences. https://www.scopus.com/authid/detail.uri?authorId=7801642066.

[28] N. Kiktev, H. Rozorinov, M. Masoud. Information Model of Traction Ability Analysis of Underground Conveyors Drives // 2017 XIII-th International Conference MEMSTECH, Lviv, 2017, pp. 143–145. http://ieeexplore.ieee.org/document/7937552.

13

Synthesis of an Optimal Combined Multivariable Stabilization System for Adsorption Process Control

Sergey Osadchy, Valentina Zubenko* and Marja Yakoreva

Central Ukrainian National Technical University, Kropivnitsky, Ukraine
*Corresponding Author: zub_valya@ukr.net

The article is devoted to the increasing efficiency of short-cycle adsorption of impurities from hydrogen, which is obtained by the conversion method. The principle of increasing efficiency is based on the approximation of the complete thermodynamic work to the minimum required one due to the introduction of a combined control system. The task of designing a combined control system for the adsorption of carbon oxides from food hydrogen is reduced to the synthesis of the optimal combined multivariable stochastic stabilization system. This stabilization system consists of two control circuits. Signals about changes in carbon oxide concentration are the input signals of the first circuit and the second circuit's input signal is equal to the change of a hydrogen pressure in a receiver.

13.1 Introduction

Recently, the conversion method has become widely used for hydrogen production as the least explosive one. The main final stage of such production is reduction of carcinogenic impurities concentration in hydrogen to admissible standards. Such reduction process is carried out with the help of a short-cycle adsorption (SCA) [1]. Adsorption takes place continuously under the influence of impurities concentration fluctuations at the inlet of the absorber and changes in the finished hydrogen consumption during the hydrogenation of vegetable oil. When separating the gas mixture with a help of two valves, there is a recurrent connection of absorbers to the pressure reservoir and the receiver. While one absorber works at a high pressure, the other works at low one.

The effectiveness of SCA is defined as a ratio of a minimum thermodynamic work required for the release of hydrogen to full thermodynamic work performed by the SCA. The minimum thermodynamic work is determined by a composition of the mixture, its temperature and flow at the entrance to the column SCA. The full thermodynamic work is determined by the design of the column, the type of sorbent, the method of control and the intensity of the finished product's consumption. In conditions of concrete production and fixed suppliers of the mixture, it is difficult to influence the minimum thermodynamic work. Full thermodynamic work at the given design of SCA installation, at the fixed type of sorbent and the receiver structure, depends on the quality of control system.

The purpose of the research is to improve the efficiency of carbon oxide adsorption from food hydrogen by optimizing the accuracy of stabilization its composition. So, the main objectives of this article can be listed as follows: design the hardware structure of automatic control system; formulation a synthesis problem of optimum stabilization system; substantiation of the synthesis algorithm; synthesis of optimal automatic control system; and analysis of its quality.

13.2 Optimal Stabilization System Synthesis Problem Formulation

Examining the existing equipment design (Figure 13.1) shows that the control of carbon oxide adsorption is performed by the operator's O commands using the actuator A. The operator O manages adsorption as a result of commands exchange with the controller C. To generate commands, data from a set of sensors is used: the carbon dioxide concentration sensor GA_1, the carbon monoxide sensor GA_2, and the hydrogen pressure change sensor PS. These data are transmitted to the controller C via the interface I1. Interface I2 serves for the transmission of operator commands to the executive motor M. A change in a speed of the engine M rotation ω affects the actuator A, which is formed by valves. The actuator A changes an adsorption period T_{ads}. Modification of the period T_{ads} causes changes in concentrations of carbon oxide γ_{co} and carbon dioxide γ_{co_2} at the output of the receiver.

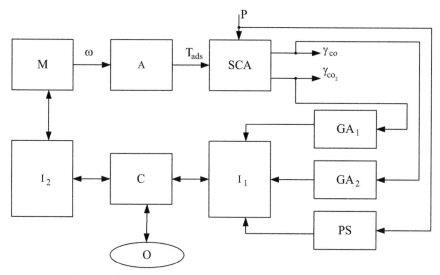

Figure 13.1 The functional structure of the system for adsorption process control.

To achieve the research objectives, it is proposed to exclude the operator O from the control loop and to synthesize the optimal automatic control system. The basis for the optimal control system design is a structural scheme of the automatic stabilization system (Figure 13.2).

It consists of two circuits. The first circuit creates a feedback on the deviation. The vector y_1 acts on the input of this circuits controller with the transfer functions matrix W_1. This vector is formed by signals at the outputs

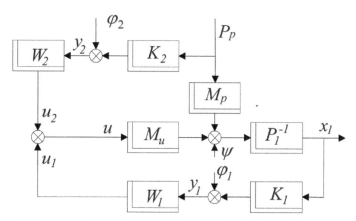

Figure 13.2 The structural scheme of the combined stabilization system.

of the sensors GA_1, GA_2. We will assume that their dynamics is characterized by the matrix of transfer functions K_1, and the measurement noise φ_1 is a two-dimensional centered stationary random process with the known matrix of spectral densities $S_{\varphi_1\varphi_1}$. The second circuit forms feedback on the controlled disturbance P_p. At the input of this circuits' controller with the transfer function W_2, the signal y_2 from the output of the sensor PS operates. In future, we will assume that the dynamics of this sensor is described by the transfer function K_2, and the measurement errors refer to a centered stationary random process with the known spectral density $S_{\varphi_2\varphi_2}$. Taking into account the results of the SCA process structural identification [2], we can assume that the linearized model of this control object dynamics is characterized by the system of ordinary differential equations, transformed according to Fourie, of the form

$$P_1 x_1 = M\,u + M_p P_p + \psi, \tag{13.1}$$

where P_1, M and M_p are polynomial matrixes of a complex argument $s = j\omega$ of 2×2, 2×1, 2×1 dimension correspondingly; x_1 is 2-dimensional vector of changes in carbons' oxide and dioxide concentrations

$$x_1 = \begin{bmatrix} \gamma_{co} \\ \gamma_{co_2} \end{bmatrix};$$

u is a control signal, which is equal to T_{ads}; ψ is a two-dimensional vector of disturbances, which represents two dimensional centered random stationary process with known spectral densities matrix $S\psi\psi$. Let the vector of control signals u be defined by the equation

$$u = W(x + \varphi); \tag{13.2}$$

where W is the desired matrix of transfer functions of a controller such as

$$W = [W_1\ W_2]; \tag{13.3}$$

φ is the vector of measurement noises, which represents centered two-dimensional random stationary process with the known matrixes spectral and mutual spectral densities $S_{\varphi\varphi}$, $S_{\varphi\psi}$.

Then the problem of regulator synthesis is to find the transfer functions matrix of the regulator W, which ensures the stability of the stabilization system (Figure 13.2) and minimizes a following quality criterion on the known matrices P_1, M, M_p, K_1, K_2, $S_{\psi\psi}$, $S_{\varphi\varphi}$, $S_{\psi\varphi}$:

$$J = \frac{1}{j} \int_{-j\infty}^{j\infty} tr\left\{ S'_{x_1x_1}(s)R_0 + S'\ (s)C \right\} ds, \tag{13.4}$$

where S'_{x1x1} is a transposed spectral densities matrix of the vector x_1; R_0, C are positively defined weight matrixes; S' is a transposed spectral densities matrix of the vector u; j is an imaginary one; tr is a symbol of the matrix trace operation.

13.3 Justification of Choice and Description of the Algorithm for an Optimal System's Synthesis

Studying perturbation dynamics models acting at CCA showed that these random time functions cannot be replaced by white noises, since the spectral densities of the pressure variation in the receiver and the vector of uncontrolled disturbances differ substantially from the constant values. Thus, the algorithm for regulator synthesis must be sought among the methods given in the sources [3–6].

The comparison of these methods made it possible to choose the synthesis algorithm described in [6]. For its application, it is necessary to take into account the transfer functions of the sensors GA_1, GA_2, PS (Figure 13.1) and determine the matrices K_1, K_2 (Figure 13.2) in the form

$$K_1 = \begin{bmatrix} 1 & 0 \\ 0 & 1 \end{bmatrix}, K_2 = 1, \tag{13.5}$$

after this, introduce the extended perturbation vectors ζ such that

$$\zeta = \begin{bmatrix} E_2 & M_\psi \\ 0_{1\times2} & 1 \end{bmatrix} \begin{bmatrix} \psi \\ P_p \end{bmatrix}, \tag{13.6}$$

and define the measurement noise's vector φ

$$\varphi = \begin{bmatrix} \varphi_1 \\ \varphi_2 \end{bmatrix}, \tag{13.7}$$

where E_2 is a unit matrix of dimension 2×2, $0_{1\times2}$ is a zero matrix of dimension 1×2. As a result, the structural diagram (Figure 13.2) can be rebuilt to the form shown in Figure 13.3.

In the scheme shown in Figure 13.3, it was introduced as the notation x for the extended vector of the system's output signals, equal to

$$x = \begin{bmatrix} x_1 \\ P_p \end{bmatrix}, \tag{13.8}$$

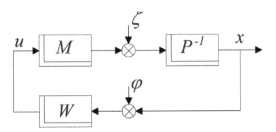

Figure 13.3   The structural diagram of the stabilization system.

P is a polynomial matrix which characterizes dynamics of the extended control object

$$P = \begin{bmatrix} P_1 & 0_{2\times 1} \\ 0_{1\times 2} & 1 \end{bmatrix};$$ (13.9)

M is an extended matrix of the form

$$M = \begin{bmatrix} M \\ 0 \end{bmatrix}.$$ (13.10)

Taking into account the introduced notation (13.5–13.10) and the results which are presented in the article [6], the algorithm for calculating the transfer function matrix that minimizes the functional (13.4) is represented in the form

$$W = (B + \Phi \cdot M)^{-1} \cdot (-A + \Phi \cdot P)$$ (13.11)

The equation (13.11) includes polynomial matrices A, B defined as a result of the representation of the block matrix H

$$H = \begin{bmatrix} O_3 & P & -M \\ P_* & -R & O_{3\times 1} \\ -M_* & O_{1\times 3} & -C \end{bmatrix}$$ (13.12)

in the following form

$$H = V_* \cdot \Sigma \cdot V;$$ (13.13)

where

$$V = \begin{bmatrix} E_3 & -S & N \\ O_3 & P & -M \\ O_{1\times 3} & A & B \end{bmatrix}; \Sigma = \begin{bmatrix} O_3 & E_3 & O_{3\times 1} \\ E_3 & O_3 & O_{3\times 1} \\ O_{1\times 3} & O_{1\times 3} & -1 \end{bmatrix};$$ (13.14)

R is a new weight matrix, which is equal to

$$R = \begin{bmatrix} E_2 \\ 0_{1\times 2} \end{bmatrix} R_0 \begin{bmatrix} E_2 & 0_{2\times 1} \end{bmatrix}.$$ (13.15)

The matrix Φ in expression (13.11) is a variation matrix of transfer functions, which is derived from the equation

$$\Phi = -(T_0 + T_+ + G_0 + G_+) \cdot D^{-1}, \tag{13.16}$$

in which $T_0 + T_+$ is a fractional-rational matrix, which is a stable part of the Weiner separation result (splitting) [20] of the matrix T

$$T = {}_{22*} \cdot (M_* \cdot P_*^{-1} \cdot S_* - N_*) \cdot D; \tag{13.17}$$

$G_0 + G_+$ is the stable part of the matrix G Weiner separation result

$$G = -{}_{22*} \cdot M_* \cdot P_*^{-1} \cdot R \cdot (S'_{\psi\varphi} + S'_{\varphi\varphi} \cdot P_*) \cdot D_*^{-1}. \tag{13.18}$$

The analysis of equations (13.17) and (13.18) proves that for successful separation it is necessary to define two fractional-rational matrices ${}_{22}$ and D. The matrix ${}_{22}$ is found from the expression

$${}_{22} = (B + A \cdot P^{-1}M)^{-1}, \tag{13.19}$$

and the matrix D is the stable result of Weiner factorization [11] of the transposed spectral densities matrix for dilated perturbation vector ζ

$$DD_* = S'_{\varsigma\varsigma} = \begin{bmatrix} E_2 & M_\psi \\ 0_{1\times2} & 1 \end{bmatrix} \begin{bmatrix} S'_{\psi\psi} & 0_{2\times1} \\ 0_{1\times2} & S_{pp} \end{bmatrix} \begin{bmatrix} E_2 & 0_{2\times1} \\ M_{\psi*} & 1 \end{bmatrix} + PS'_{\varphi\varphi}P_*.$$
$$\tag{13.20}$$

Thus, the choice of the optimal controller synthesis algorithm is substantiated. This algorithm makes somebody possible to determine the matrix of controllers transfer functions W.

13.4 Optimal Combined Stabilization System's Synthesis

Taking into account the results of identification of the CCA dynamics model presented in the article [2] and the notations (13.9), (13.10), the following form of the polynomial matrices M and P is found:

$$M = \begin{bmatrix} 3.2(s + 3.10^{-4})(s + 0.005) \\ 0.15(s + 5.10^{-4})(s + 0.001) \\ 0 \end{bmatrix}; \tag{13.21}$$

$$P = \begin{bmatrix} (s + 4.10^{-3})(s + 0.003) & 0 & 0 \\ 0 & (s + 4.10^{-3})(s + 0.03) & 0 \\ 0 & 0 & 1 \end{bmatrix}. \tag{13.22}$$

The transposed matrix of the spectral densities of the extended perturbation vector (13.20) is found by substituting the original data into this equation, namely the transposed matrices $S'_{\psi\psi}$, $S'_{\varphi\varphi}$ and the spectral density S_{pp}

$$
S'_{\psi\psi} = \frac{\sigma_{\psi 2}^2}{\pi} \left[\begin{array}{c} \dfrac{1.77 \cdot 10^{-7} |s^2 + 0.025s + 6.25 \cdot 10^{-4}|^2}{|s^2 + 0.06s + 3.6 \cdot 10^{-3}|^2} \\[3mm] \dfrac{1.63 \cdot 10^{-8}(s + 0.004)(s + 0.02)(s + 0.07)(s + 0.06)}{(s + 0.09)(s^2 - 0.06s + 0.0036)} \end{array} \right.
$$

$$
\left. \begin{array}{c} \dfrac{1.63 \cdot 10^{-8}(-s+0.004)(-s+0.02)(-s + 0.07)(-s + 0.06)}{(-s + 0.09)(s^2 + 0.06s + 0.0036)} \\[3mm] \dfrac{2.71 \cdot 10^{-8} |(s + 0.003)(s + 0.08)|^2}{|(s + 0.09)|^2} \end{array} \right],
$$

$$
\tag{13.23}
$$

$$
S'_{\varphi\varphi} = \frac{\sigma_{\psi 2}^2}{\pi} \begin{bmatrix} \gamma_1^2 & 0 & 0 \\ 0 & \gamma_1^2 & 0 \\ 0 & 0 & 30\gamma_1^2 \end{bmatrix}, \tag{13.24}
$$

$$
S_{pp} = \frac{\sigma_{\psi 2}^2}{\pi} \frac{5.21 \cdot 10^{-5} |s^2 + 4.8 \cdot 10^{-4}s + 9.10^{-8}|^2}{|(s + 5.10^{-5})(s^2 + 1.2 \cdot 10^{-4}s + 1.6 \cdot 10^{-7})|}, \tag{13.25}
$$

where σ_ψ^2 is the variance of the hydrogen pressure in the receiver, found from the experimental data $(388\ kPa)^2$. The structure of this spectral densities matrix was found out at three values of the "noise-signal" ratio γ_1^2: 0.1, 1.0, 10.

Since changes in the contents of carbon monoxide and carbon dioxide affect the quality of food hydrogen equivalently, the weight matrix R_0 is taken as the unit matrix, and the value of the weight coefficient C, which limits the control signal's change, varies within the range of 0.1–10.

A substitution for the initial data (13.21) and (13.22) into the expression (13.12) and the application of the algorithm [21] for factorization in Matlab allowed to find matrices A, B for three values of the coefficient C (0.1, 1.0, 10). For the coefficient C which is equal to 0.1, these matrices have the following form:

$$
A = [-0.22s^2 - 0.86s + 0.35 - 0.018s^2 - 0.07s - 0.007\ 0];
$$
$$
B = 0.72s^2 + 0.71s + 0.3. \tag{13.26}
$$

A substitution for the obtained data (13.21–13.26) into formulas (13.16–13.20) made it possible to determine the set of matrices Φ minimizing the functional (13.4) for different coefficients C and γ_1^2. For the case, when these coefficients are equal to 0.1, this matrix Φ is equal to

$$\Phi = [0.22s^2 + 93.8s + 213 \quad 0.0182s^2 + 7.75s + 261 \quad 2.10^{-4}s - 0.037]$$
$$\times \frac{1}{0.0001s^2 + 0.000374s + 2.46}. \qquad (13.27)$$

A substitution for the obtained values of the matrices A, B (13.26), and Φ (13.27) into the equation (13.11) made it possible to determine the structure and the parameters of the optimal transfer function matrix of the regulator W in the form

$$W = \begin{bmatrix} \dfrac{k_1(s + \omega_1)(s + \omega_2)}{s^2 + \omega_3 s + \omega_4^2} & \dfrac{k_2(s + \omega_5)(s + \omega_6)}{s^2 + \omega_3 s + \omega_4^2} & \dfrac{k_3(s + \omega_7)}{s^2 + \omega_3 s + \omega_4^2} \end{bmatrix}, \qquad (13.28)$$

where k_i, ω_j are parameters that vary depending on the values of the coefficient C and the noise-to-signal ratio γ_1^2.

The analysis of the optimal system's quality and its comparison with the existing automated system showed that the application of the optimal system would result in decreasing the carbon monoxide and the carbon dioxide content dispersions of 1.69 times and of 2.4 times compared to the system that exists.

13.5 Conclusion

In this research chapter, the authors have developed the analytic design theory of optimal stochastic stabilization systems in the frequency domain for increasing the efficiency of short-cycle adsorption of impurities in food hydrogen. As a result, they have managed to substantiate a number of vector transformations that lead the synthesis problem of an optimal control system for a multidimensional possibly unstable object with partially controlled perturbation to the design of an optimal multidimensional system of stochastic stabilization. The developed algorithm allows to take into account the perturbations and noise of the meters.

As an example, the research chapter presents the result of the synthesis of an optimal system for stabilizing the composition of food hydrogen, which has feedbacks on deviation and perturbation. The introduction of an optimal system will significantly improve the quality of hydrogen produced by the conversion method and also increase the adsorption efficiency due to the approach of full thermodynamic work to the minimum one.

References

[1] Matveykin V. G. Mathematical modeling and control of the process of short-cycle no-heat adsorption / V. G. Matveykin, V. A. Pogonin, S. B. Putin, S. L. Skvortsov. - M .: Mechanical Engineering, 140s (in Russian).

[2] Osadchy S. I. Identification of the linearized model of the dynamics of the short-cycle adsorption process / S. I. Osadchy, M. V. Yakorova, V. O. Zubenko // // "Problems of Information Technologies", Kherson: KhNTU, no. 2(018), pp. 19–24, 2015 (in Ukrainian).

[3] Blokhin L. M. Methodological bases and stages of ensuring the competitiveness of the processes of stabilization of existing mobile objects. / L. M. Blokhin, S. I. Osadchy, O. P. Krivosnenko // Bulletin of the NAU, no. 2, pp. 61–68, 2009 (in Russian).

[4] Blokhin L. N. Technology of Structural Identification and Subsequent Synthesis of Optimal Stabilization Systems for Unstable Dynamic Objects / L. N. Blokhin, S. I. Osadchiy, Yu. N. Bezkorovainyi // Journal of Automation and Information Sciences, vol. 39, no. 11, pp. 57–66, 2007.

[5] Azarskov V. N. Methodology for designing optimal stochastic stabilization systems: Monograph / VN Azarskov, L. N. Blokhin, L. S. Zhitetskii/ ed. Blokhina L. N. - K .: Book Publishing House of NAU, 2006. - 440s (in Russian).

[6] Combined method of synthesis of optimal stabilization systems for multidimensional moving objects under stationary random effects // SI Osadchy, VA Cuckoo / International Scientific and Technical Journal "Problems of Management and Informatics", no. 3 May–June, 2013, pp. 40–49.

Index

A

Adsorption 315, 316, 317, 323
Analytical solution 8, 98,
120, 135
Automatic control system 197,
200, 213, 316
Automation of control
processes 145

B

Bounded perturbations 3, 4, 14
Buckingham's theorem 98

C

Calibration procedure 254, 267
Case control 237
Closed-loop control
system 43, 54, 61, 204
Cognitive map 43, 46, 53
Computer vision 167, 169, 191
Conflict-controlled
process 17, 26, 31, 40
Control algorithms 146, 149,
273, 283
Coordinate determination
245, 254, 263, 266

D

Discrete systems 3, 67, 69, 78
Docking operations 197, 200,
206, 226

E

Efficiency 12, 145, 151, 324
Ellipsoidal estimation 181,
182, 184
Energy consumption 146, 200,
234, 279
Energy efficiency 145, 151,
235, 285
Experimental data 97, 127,
129, 140

F

Floating dock 197, 200,
222, 224
Fractional derivative 17, 26,
33, 40
Fuzzy control 158, 164, 199
Fuzzy controller 197, 199,
200, 225

G

Greenhouse 271, 275,
277, 282
Guaranteed cost control 66, 75,
80, 95
Guaranteed estimation 169, 171

H

Harvesting routes 291, 302,
307, 308

I

Identification problem 43, 47, 51, 62
Ill-posed problem 63
Impulse process 43, 53, 59, 61
Intelligent systems 233
Invariant ellipsoid 65, 80, 86, 93
Invariant sets 4, 9, 12, 14
Inventory control 65, 78, 83, 93

L

Linear matrix inequality 65, 68, 93
Logical conditions 147, 162
Logical-dynamic models 233
Lyapunov-Krasovskii functional 65, 68, 77, 93

M

Machine learning 169, 170, 175
Maximizing 271
Microclimate 282
Mittag-Leffler function 26, 28, 34, 40
Multi-assortment production 233

N

NDVI 298
Nitrogen 291, 299, 301, 302
Nonlinear dynamics problems 97
Nonlinear systems 4, 9, 171

O

Optimization 68, 86, 159, 197
Orbital data 245, 246, 267
Oscillatory regime 133

P

Point landmarks 247, 249, 257, 267
Pontryagin's condition 18, 20, 23, 34
Pose estimation 169, 170

R

Robotic complexes 272, 273, 283

S

Semidefinite programming 65, 82, 93
Set-valued mapping 18, 21, 23, 40
Simulation 46, 188, 266, 307
Situational models 145, 151, 154, 160
Smart control 145, 151, 154, 160
Space images 245, 249, 263, 267
Special method of Nondimensionization 97
Stabilization 77, 246, 315, 323
Stress indices 291, 309
Sub-satellite polygon 257, 261, 265, 267
Subspace method 45, 46
Superpositionally measurable 21, 23, 24
Supply network 65, 73, 86, 240
Synthesis 65, 78, 83, 323

U

UAV 291, 294, 304, 309
Uncooperative spacecraft 169–171

About the Editors

Vsevolod M. Kuntsevich

Ph.D. (1959), D.Sci. (1965), Professor (1967), Academician of the National Academy of Sciences of Ukraine (1992), Honorary Director of Space Research Institute of NASU, Kyiv, Ukraine.

Honored Figure of Science and Technology of Ukraine (1999), laureate of the State Prize of the Ukrainian SSR (1978, 1991) and Ukraine (2000) in the field of science and technology, S. Lebedev Award (1987), V. Glushkov Award (1995), V. Mikhalevich Award (2003).

Graduated from Kyiv Polytechnic Institute (1952). Worked at the Institute of Electrical Engineering (1958–63), Institute of Cybernetics (1963–96), Space Research Institute (from 1996).

Editor-in-chief of the journal "Problemy Upravleniya I Informatiki" published in English in the USA under the title "Journal of Automation and Information Sciences" (since 1988). The founder of the national school in the field of discrete control systems, he made a significant contribution to the development of modern theory of adaptive and robust control under uncertainty. The author of 8 books and over 250 articles. The Chairman of the National Committee of Ukrainian Association on Automatic Control (NMO of IFAC).

Vyacheslav F. Gubarev

Doctor of Science (1992), Professor, Corresponding member National Academy of Science of Ukraine (2006), Professor of Mathematic Methods of System Analysis Department, Kyiv National Technical University, Ukraine, Head of control department, Space Research Institute, National Academy of Science, Ukraine.

He has received a Ph.D. (1971) and a Dr.Sc. (1992) in System Analysis and Automatic Control from Institute of Cybernetic National Academy of Science, Ukraine.

He has taken part in several international grants with Russia Academy of Science, Moscow State University and others.

His research interests include mathematical modelling of complex systems, automatic control, estimation and identification, ill-posed mathematical problems, dynamic and control under uncertainty, spacecraft control systems.

He enters in editorial staff of several journals such as Journal of Automation and Information Sciences, Cybernetics and Systems Analysis. He is vice-chairman of Ukrainian Association of Automatic Control which is NMO of IFAC.

Yuriy P. Kondratenko

Doctor of Science (habil.), Professor, Honour Inventor of Ukraine (2008), Corr. Academician of Royal Academy of Doctors (Barcelona, Spain), Professor of Intelligent Information Systems Department at Petro Mohyla Black Sea National University, Ukraine.

He has received a Ph.D. (1983) and a Dr.Sc. (1994) in Computer and Control Systems from Odessa National Polytechnic University.

He received several international grants and scholarships for conducting research at Institute of Automation of Chongqing University, P. R. China (1988–1989), Ruhr-University Bochum, Germany (2000, 2010), Nazareth College and Cleveland State University, USA (2003). In 2015 he received Fulbright grant for conducting research during 9 months in USA (Cleveland State University, Department of Electrical Engineering and Computer Science).

He is a regional coordinator of Tempus (Cabriolet) and Erasmus+ (Aliot) projects, principal researcher of several international research projects with Spain, Germany, P.R. of China et al. and author of more than 140 patents and 12 books (including edited monographs) published by Springer, River Publishers, World Scientific, Pergamon Press, Academic Verlag, etc.

He is a member of the National Committee of Ukrainian Association on Automatic Control, as well as GAMM, DAAAM, AMSE UAPL and PBD-Honor Society of International Scholars, visiting lecture at the universities in Rochester, Cleveland, Kassel, Vladivostok and Warsaw.

His research interests include intelligent decision support systems, automation, sensors and control systems, fuzzy logic, soft computing, modelling and simulation, robotics, elements and devices of computing systems.

Dmitriy V. Lebedev

Doctor of Science (habil.), Leading Researcher at International Research and Training Center for Information Technologies and Systems of National Academy of Sciences of Ukraine and Ministry of Education and Sciences of Ukraine.

He has received a Ph.D. (1971) and a Dr.Sc. (1986) in Technical Cybernetics and Information Theory from Institute of Cybernetics named after V. M. Glushkov of National Academy of Sciences of Ukraine.

His research interests include strapdown inertial navigation systems and automatic control in aerospace.

He is a member of the National Committee of Ukrainian Association on Automatic Control, as well as a full member of the Academy of Navigation and Motion Control (Ukrainian branch) and author of more than 120 articles and 3 monographs written with co-author A. Tkachenko.

Vitaliy P. Lysenko

Doctor of Technical Sciences (habil.), Professor, Honored Worker of Education of Ukraine; Head of the Department of Automation and Robotics Systems of the National University of Life and Environmental Sciences of Ukraine (Ukraine, Kyiv), Academician of Higher School Academy of Sciences of Ukraine, Member of the National Committee of Ukrainian Association on Automatic Control, Member of the Expert Council of the section "Energy and Energy Efficiency" (scientific area "Scientific and technical research problems of electrotechnical and electromechanical complexes and systems") of Ministry of Education and Science of Ukraine.

His research interests include automated control systems of electrotechnical complex biotechnical objects in agriculture using intelligent information processing algorithms (biotechnical objects – objects containing a biological component), he is author of more than 300 publications, including more than 20 patents and 20 books (including author's chapters in monographs).

Printed and bound by CPI Group (UK) Ltd, Croydon, CR0 4YY

23/10/2024

01777696-0014